THE POETICS OF NATURAL HISTORY

The Poetics of
NATURAL HISTORY

CHRISTOPH IRMSCHER

with a foreword and photographs by
Rosamond Purcell

RUTGERS UNIVERSITY PRESS

NEW BRUNSWICK, CAMDEN, AND NEWARK, NEW JERSEY, AND LONDON

Library of Congress Cataloging-in-Publication Data

Names: Irmscher, Christoph, author.
Title: The poetics of natural history / by Christoph Irmscher ; with foreword
and photographs by Rosamond Purcell.
Description: Second edition. | New Brunswick, New Jersey : Rutgers University
Press, [2019] | Revised edition of: The poetics of natural history : from
John Bartram to William James. c1999. | Includes bibliographical
references and index.
Identifiers: LCCN 2018053822 | ISBN 9781978805866 (paperback) | ISBN
9781978805873 (cloth) | ISBN 9781978805880 (epub) | ISBN 9781978805903
(web pdf) | ISBN 9781978805897 (mobi)
Subjects: LCSH: Natural history—United States—History—18th century. |
Natural history—United States—History—19th century. |
Naturalists—United States. | Natural history—Catalogs and
collections—United States—History—18th century. | Natural
history—Catalogs and collections—United States—History—19th century. |
Natural history literature—United States. | Poetics—History—18th
century. | Poetics—History—19th century.
Classification: LCC QH21.U5 I75 2019 | DDC 508—dc23
LC record available at https://lccn.loc.gov/2018053822

A British Cataloging-in-Publication record for this book is available from the British Library.

For more information on Rosamond Purcell's photographs please see Notes on the Photographs.

 Photography and all associated expenses made possible
by a generous grant from Furthermore: a program of the J. M. Kaplan Fund

www.rutgersuniversitypress.org

Manufactured in the United States of America

For Dan and Lauren

CONTENTS

ILLUSTRATIONS

FOREWORD
Rosamond Purcell

In 1999, when I first saw the illustration of the watercolor *Snowy Owl* by John James Audubon on the cover of Christoph Irmscher's book *The Poetics of Natural History*, I felt a wave of sympathy—and relief. For the previous twelve years or so, I had been working, often struggling, to infiltrate natural history collections, mostly on my own and with a camera. I was on a personal mission to study the old specimens from an artistic rather than from an academic point of view. Here, finally, on a book about the early founders of such collections, an artist's vision of owls appears on a scholarly book and the word "poetics" is linked to natural history.

For the Rutgers University Press reprint of this mesmerizing book, I was invited to offer a small group of photographs taken in natural history museums. I have shuffled through my archives in search of images that might echo, however tangentially, the vividness of Christoph Irmscher's style as exemplified in his gallery of besotted collectors as he describes their particular passions. He holds them all in an empathetic embrace.

Past events, as he describes them, seem immediate, while the sense of time itself advances and retreats—without ellipses. The length of time it takes to shoot one hundred birds is much shorter than the length of time it takes to mount a single dead a bird on wires and paint its likeness before it rots and must be discarded. The time it takes to find a rare plant in the wilderness, to propagate that plant, the months to send it live across the ocean, the hours or days to paint a portrait of a President, the years or lifetime to make a museum, the eternity to overcome a fear of serpents.

In re-reading this book written twenty years ago, the collectors seem as three-dimensional as ever and not like wax replicas. But knowing, as I do now, the effects of long, uninterrupted tracking shots and other cinematic devices, I can sense a similar intensity of focus in each of Christoph Irmscher's biographical sketches. The first edition of *The Poetics of Natural History* appeared some years before the technique of virtual reality became a fact of modern life. Perhaps back in the late 1990s, a sophisticated pair of magic glasses would have been found in this passionate writer's arsenal: virtual reality glancing backwards.

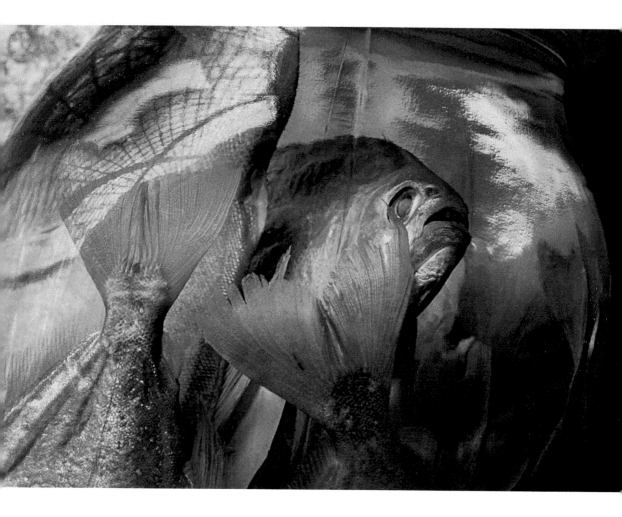

NOTES ON THE PHOTOGRAPHS

THE GREAT EGRET (page iii–iv)
In the case of the great egret (*Ardea alba*) Audubon made a drawing only—he did not ever make a full-scale print. When I consider a beautiful skin of the Great Egret from the Western Foundation of Vertebrate Zoology (WFVZ), I imagine that this is the very bird in the drawing by Audubon and that it had not been discarded after the painting was complete, but that someone, a scavenger for such relics, had spirited it away. This is, of course, just a wishful thought. According to the records at the WFVZ, this Great Egret, an adult male, was collected on March 14, 1991, from twenty-eight kilometers south of El Boliche, Ecuador. It does not matter for the sake of this story who collected it. Here it is, white bird traveling, sui generis, on a long flight. Credit: *Egg & Nest* by Linnea Hall and René Corado, photographs by Rosamond Purcell, Harvard University Press, 2008.

PIRANHA (page xii)
On the Thayer expedition to Brazil, William James dreaded each time the ship was forced to stop, because whenever this happened ". . . the Prof is sure to come around and say how very desirable it wd. be to get a large number of fishes from this place, and willy nilly you must trudge . . . If there is anything I hate it is collecting." (21 October 1865; *CWJ* 4: 127–128) Afterimages of these specimens endure—colors flashing from the scales on the fish and shining out of the gloom behind the wide-angle surface of the glass jar, in which the reflected bars from the window look like fish scales.

The Museum of Comparative Zoology, this stronghold of Agassiz's institution, holds thousands of ichthyological treasures from Brazil. The fish in this very jar were perhaps caught and pickled by William James himself, but as he writes, not with much joy. Here, on this shelf, beside the piranha in a bottle, was a catfish from the same site, attributed in handwriting to "William James."

The right angle, time of day, and all other ephemeral "rightness" for an eloquent photograph is a one-time thing—for no matter how the light comes in, it lingers for minutes, not hours. But the light is not the only thing that changes. Last week, at least twenty-five years after I took the photograph shown here, I returned to the museum's department of ichthyology hoping once more to see the head of this particular piranha, open jaws showing ominous blunt teeth, rising between and above the tails of its fellow fish in the orange-colored preservative bath. Perhaps James had also captured this specimen.

Nothing ever stays the same. About twelve years ago, the alcohol in the specimen jars from the Thayer expedition was changed and the sunset colors are now gone. While both species and subspecies of piranha remain in their original groups, in the process of decanting and refilling the glass containers, their orientations inevitably shifted. As I have no record of the collection number for that particular piranha or for its species, there will be small chance of finding it again. Unlike James searching in murky or clear running rivers to scoop up nature's bounty, I have no authority to go spelunking into the jars for museum data or even to catch a glimpse of the memorable face of this fish among so many pallid corpses. Credit: *Finders, Keepers: Eight Collectors* by Rosamond Wolff Purcell and Stephen Jay Gould, W. W. Norton, 1992

MONKEY WITH BOOK (page 2–3)
Mounted primates in early museum collections are sometimes posed holding fruit or nuts. The articulated skeleton of this squirrel monkey, seated on a painted table among a shower of crab-apples in a makeshift garden, holds a small, page-furled notebook from the museum archives. The notebook, found a long time ago on a beach, was kept in the collection as an instructive example of the action of waves and tide. Its pages, foliated like layers of ancient rock, evoke Darwin's famous text on the fragmentary nature of the fossil record: "I look at the natural geological record, as a history of the world imperfectly kept, and written in a changing dialect; of this history we possess the last volume alone, relating only to two or three countries. Of this volume, only here and there a short chapter has been preserved; and of each page, only here and there a few lines."[1] The elements in this Dutch still life combine in a symbolic language of transitory gestures: the passage of years (or waves), the brief visible manifestations of crabapples in that summer and in that place, and the attempt to reanimate a sentient primate by placing a historical document in its bony hand. Credit: Universiteitsmuseum, Utrecht.

GUILLEMOT EGGS (page 14–15)
In Labrador on a raw day in June, Audubon landed on an island "covered with Foolish Guillemot" where "eggers" were raiding the rocks for the hapless birds' eggs. They claimed that, by visiting this and other islands, to have (that day) taken eight hundred dozen "The Black-backed Gulls were here in hundreds and destroying the eggs of the Guillemots by thousands."[2]

Judging by the variety of colors, degree of pointed shapes, and collector labels, these eggs came from different sites. They are, as a friend said, "like separate worlds." The markings on each are unique to the bird that laid it, and if photographically "unrolled" into a Mercator-like projection of the circumference, each would exhibit a different landscape of scrawls and markings. These are not human signatures, no matter that we might try to see them so. Credit: *Egg & Nest* by Linnea Hall and René Corado, photographs by Rosamond Purcell, Harvard University Press, 2008.

KERERU (page 156–157)

The extinct Norfolk Island kereru, a subspecies of the New Zealand pigeon *Hemiphaga novaseelandia spadiceia*, is indigenous to a remote island in the South Pacific. A pair of this species of pigeon ended up in the vast collection of the Naturalis museum in Leiden. Captured and first described in 1801, the bird was never found in the wild again. Credit: *Swift as a Shadow: Extinct and Endangered Animals*, photographs by Rosamond Purcell, Mariner Books, 1999.

ROSAMOND PURCELL

PREFACE TO THE SECOND EDITION

When the German philosopher and wit Georg Christoph Lichtenberg mused about the possibility of "a new edition of me and my life," he said he would like the chance "to make a few not superfluous suggestions" concerning the ways in which it could be made better. What he had in mind, specifically, were improvements "concerning the design of the frontispiece and the way the work is laid out." Faced with the much smaller task of creating the enhanced edition of *Poetics of Natural History*, a book still near and dear to me, I am delighted to report that "the way the work is laid out" is indeed entirely different, thanks to the collaboration of an artist I have long revered, the incomparable Rosamond Purcell.

This book was born out of my fascination with natural history museums, places that I found, even as a child, both exhilarating and unsettling, a sentiment I share with Rosamond Purcell. In her hauntingly beautiful photographs, we enter the magical world of things—things natural or manmade, found or created, discarded or preserved, things both ordinary and weird. We leave that world transformed, with a new sense of the strangeness of the everyday. Purcell is a scientist, an artist, and a poet. The comprehensiveness of her vision reminds us of a time when disciplines had not yet separated into supposedly different modes of understanding the world. The aim of my book is to take readers back into that world, and also to show how, eventually, we managed to close the door on it. *Poetics* covers a vast range of writers, artists, scientists, and topics. For this new edition, I have, with a nod to Audubon's description of his writing, smoothed out some of the asperities of my prose, fixed mistakes, and replaced wherever possible the images used in the first edition with new photographs reflecting the vibrant color of the originals. Important new research has appeared during the last two decades. First and foremost, I want to acknowledge Michael Branch's excellent anthology of early American nature writing, *Reading the Roots: American Nature Writing before Walden*. I consider it the ideal companion volume to *Poetics*. In his selections as well as his headnotes, Branch captures the myopia of these early observers, hampered by their status as colonizers of a land that, paraphrasing Robert Frost, was theirs before they were the land. Collecting texts ranging from Columbus's log-book to the Swedish traveler Fredrika Bremer's account of reading Columbus while going up the St. Johns River in Florida, Branch also gives ample space to those luminous moments of altered perception when the boundaries between humankind and nature become porous.

When I wrote this book, gender was not a focus of my research. For readers interested in that aspect of early natural history, I recommend Tina Gianquitto's *Good Observers of Nature: American Women and the Scientific Study of the Natural World, 1820–1885* and Susan Scott Parrish's *American Curiosity: Cultures of Natural History in the Colonial British Atlantic World*, the latter of which paints a multifaceted picture of the many contributors to early nature study, including women, Native Americans, and people of African descent. Our understanding of William Bartram has been significantly enriched by the work of Thomas Hallock. His *William Bartram: The Search for Nature's Design*, coedited with Nancy E. Hoffmann, besides printing many unknown and unpublished texts by Bartram himself, offers a shimmering picture of Bartram's many passions and the best explanation we have of his pervasive influence on transatlantic culture and science. For a comprehensive overview of Bartram's art, consult Judith Magee's splendidly illustrated *The Art and Science of William Bartram*.

Since the turn of the century, there has been a steady stream of popular biographies of Audubon, but none of them measures up to the high standards of research set by Roberta Olson's magnificent *Audubon's Aviary: The Original Watercolors for The Birds of America*, which also contains the best assessment of Audubon's artistic innovations. Readers curious for more of Audubon's vigorous prose can turn to the collection I have compiled for the Library of America, *Audubon's Writings and Drawings*. Audubon's outstanding collaborator, Maria Martin, receives her due in a gorgeously illustrated biography by Debra J. Lindsay, *Maria Martin's World: Art and Science, Faith and Family in Audubon's America*. Gregory Nobles, in *John James Audubon: The Nature of the American Woodsman* recreates the scientific climate in which Audubon developed his own distinctive visual poetics.

My chapter on the fascinating power of snakes revives the contributions of John Edward Holbrook, Maria Martin's Charleston neighbor. In his excellent *Science, Race, and Religion in the American South*, Lester Stephens traces the formative influence of the circle of Charleston naturalists, to which Holbrook belonged, on American science and Louis Agassiz's misguided racial theories. My own continuing work on Agassiz—in both the full-scale biography I published in 2013 and my recent scholarly edition of Agassiz's *Introduction to the Study of Natural History*—places the insights of the final chapter of *Poetics* in a wider cultural and scientific context. Molly Rogers's *Delia's Tears: Race, Science, and Photography in Nineteenth-Century America* gives a moving and powerful account of the human cost exacted by Agassiz's use of the camera to cement the alleged differences between races. Maria Helena P. T. Machado's *Brazil through the Eyes of William James: Letters, Diaries, and Drawings, 1865–1866* fully documents young William James's participation in Agassiz's Brazilian junket. And Louis Menand's *The Metaphysical Club: A Story of Idea* outlines, among many other things, the momentous consequences for American intellectual history of James's disenchantment with the kind of science Agassiz represented. Finally, *America's Darwin; Darwinian Theory and U.S. Literary Culture*, coedited by Tina Gianquitto and Lydia Fisher, charts the myriad and clever ways in which American writers and scientists have tried to integrate Darwin's heterodox vision into the cultural fabric of a nation temperamentally averse to imagining a world without, in the words of philosopher John Dewey, "absolute origins and absolute finalities."

One of the great pleasures of my professional life has been my collaboration with the art historian Alan Braddock which was inspired by the Audubon chapter in *Poetics*. Our coedited anthology *A Keener Perception* emphasizes the environmentalist dimension of the tradition I discuss in *Poetics*, extending the debate to contemporary art. With the support of the National Endowment for the Humanities, Alan and I jointly taught seminars on Audubon at Indiana University. Our extraordinary students, who were all schoolteachers, helped me see many things differently, but they also affirmed the continuing validity of the interdisciplinary approach I had first tried out in *Poetics*.

My book saw the light of print because of the vigorous encouragement of two remarkable women, Suzanne Katz Hyman, who published excerpts from it in *Raritan*, and Leslie Mitchner, the former editor in chief of Rutgers University Press, who offered me a contract over a celebratory dinner in Harvard Square that I will remember for the rest of my life. I owe continuing thanks also to my excellent copy editor, India Cooper, as well as to the librarians who assisted me again, especially Leslie Morris and Matthew Wittmann of Houghton Library. Kudos to Zach Downey, the Lilly Library's outstanding photographer, who has replaced close to half of the original illustrations with new photographs from the unparalleled holdings of the Lilly, my academic home for more than a decade. No thanks are adequate to acknowledge Zach's contributions as well as the professionalism, warmth, and good cheer he brings to everything he tackles. Joel Silver, the director of the Lilly, has been my supporter, advisor, and friend for years now—thanks again, Joel. And, of course, my gratitude also goes to Elisabeth Maselli, my editor at Rutgers University Press. Without her advocacy, this new edition would not have happened, and she handled the complex task of assembling it with panache, grace, and enthusiasm.

My sentiments behind the dedication, which remains the same, have not changed either: my mentor Daniel Aaron helped me find my voice as a writer, and my wife, Lauren Bernofsky, has made sure that I don't lose it. I also want to acknowledge my friend Raphael Falco who always reminds me, *nel mezzo del cammin di mia vita* (or now a bit past it), to keep looking forward. My children, Nick and Julia, were not yet born when I wrote this book. Watching them grow over the years into independent, beautiful, and confident human beings has been, and will forever be, the chief joy of my life.

CHRISTOPH IRMSCHER

BLOOMINGTON, AUGUST 2018

ABBREVIATIONS

A	Charles Willson Peale, Autobiography. Transcript by Horace Wells Sellers, 1893. American Philosophical Society Library.
AJ	*Audubon and His Journals*, ed. Maria R. Audubon. 2 vols. New York: Dover, 1986 (¹1896).
AMI	*Barnum's American Museum Illustrated.* New York: William Van Norden & Frank Leslie, 1850.
AO	Alexander Wilson, *American Ornithology; or, The Natural History of the Birds of the United States, Illustrated with Plates, Engraved from Original Drawings Taken from Nature.* 9 vols. Philadelphia: Bradford & Inskeep, 1808–1814.
BA	John James Audubon, *The Birds of America, from Drawings Made in the United States and Their Territories.* 7 vols. New York: J. J. Audubon; Philadelphia: J. B. Chevalier, 1840–1844.
CJB	*The Correspondence of John Bartram, 1734–1777*, ed. Edmund Berkeley and Dorothy Smith Berkeley. Gainesville: University of Florida Press, 1992.
CWJ	*The Correspondence of William James*, ed. Ignas K. Skrupskelis, Elizabeth Berkeley, et al. 5 vols. Charlottesville: University Press of Virginia, 1992.
DI	Charles Willson Peale, *Discourse Introductory to a Course of Lectures on the Science of Nature with Original Music Composed for, and Sung on, the Occasion.* Philadelphia: Zachariah Poulson, 1800.
EV	Oliver Wendell Holmes, *Elsie Venner: A Romance of Destiny.* 2 vols. Boston: Houghton Mifflin, 1880 (¹1861).
I	Charles Willson Peale, *Introduction to a Course of Lectures on Natural History.* Philadelphia: Francis and Robert Bailey, 1800.
JB	Louis and Elizabeth Agassiz, *A Journey in Brazil.* 6th ed. Boston: Ticknor & Fields, 1869.
L	*The Life of P. T. Barnum Written by Himself.* New York: Redfield, 1855.
LC	Elizabeth Cary Agassiz, *Louis Agassiz: His Life and Correspondence.* 2 vols. Boston: Houghton Mifflin, 1885.
LJC	*Life, Journals, and Correspondence of Rev. Manasseh Cutler, LL.D.*, ed. William Parker Cutler and Julia Perkins Cutler. 2 vols. Cincinnati: Robert Clarke, 1888.
LS	Louis Agassiz, *Lake Superior: Its Physical Character, Vegetation, and Animals, Compared with Those of Other and Similar Regions. With a Narrative of the Tour by J. Elliot Cabot and Contributions by Other Scientific Gentlemen.* Boston: Gould, Kendall & Lincoln, 1850.

NAH   Holbrook, John Edwards. *North American Herpetology; or, A Description of the Reptiles Inhabiting the United States.* 2nd ed. 5 vols. Philadelphia: J. Dobson, 1842–1843.

NHC   Mark Catesby, *The Natural History of Carolina, Florida and the Bahama Islands: Containing the Figures of Birds, Beasts, Fishes, Serpents, Insects, and Plants: Particularly, the Forest-Trees, Shrubs, and Other Plants, Not Hitherto Described, or Very Incorrectly Figured by Authors, Together with their Descriptions in English and French. To Which, are My Observations on the Air, Soil, and Waters: With Remarks upon Agriculture, Grain, Pulse, Roots, & To the Whole, is Prefixed a New and Correct Map of the Countries Treated Of.* 2 vols. London: Printed at the expence of the author, 1731, 1743. Appendix, 1748.

NRA   Henry Walter Bates, *The Naturalist on the River Amazons: A Record of Adventures, Habits of Animals, Sketches of Brazilian and Indian Life, and Aspects of Nature under the Equator, during Eleven Years of Travel.* 2 vols. London: Murray, 1863.

OB   John James Audubon, *Ornithological Biography; or, An Account of the Habits of the Birds of the United States of America; Accompanied by Descriptions of the Objects Represented in the Work Entitled* The Birds of America, *and Interspersed with Delineations of American Scenery and Manners.* 5 vols. Edinburgh: Adam and Charles Black, 1831–1839.

PP   *The Selected Papers of Charles Willson Peale and His Family*, ed. Lillian B. Miller et al. Vol. 1: *Charles Willson Peale: Artist in Revolutionary America, 1735–1791.* New Haven: Yale University Press, 1983. Vol. 2, pt. 1 and 2: *Charles Willson Peale: The Artist as Museum Keeper, 1791–1810.* New Haven: Yale University Press, 1983. Vol. 3: *The Belfield Farm Years, 1810–1820.* New Haven: Yale University Press, 1991. Vol. 4: *Charles Willson Peale: His Last Years, 1821–1827.* New Haven: Yale University Press, 1996.

SDC   Charles Willson Peale and A. M. F. J. Palisot de Beauvois, *Scientific and Descriptive Catalogue of Peale's Museum.* Philadelphia: Smith, 1796.

SJ   John James Audubon, *Selected Journals and Other Writings*, ed. Ben Forkner. New York: Penguin, 1996.

ST   *Struggles and Triumphs; or, The Life of P. T. Barnum, Written by Himself*, ed. George S. Bryan. 2 vols. New York: Knopf, 1927.

T   William Bartram, *Travels and Other Writings*, ed. Thomas P. Slaughter. New York: Library of America, 1996.

TA   Alfred Russel Wallace, *A Narrative of Travels on the Amazon and Rio Negro, with an Account of the Native Tribes, and Observations on the Climate, Geology, and Natural History of the Amazon Valley*, 5th ed. London: Ward, Lock & Bowden, 1895.

V   William H. Edwards, *A Voyage up the River Amazon, Including a Residence at Pará.* London: Murray, 1861 (¹1847).

W   Charles Willson Peale, "A Walk through the Philadelphia Museum" (1805/6). Pennsylvania Historical Society.

THE POETICS OF NATURAL HISTORY

Introduction

It was on a hot Thursday in July of 1787 that the Reverend Dr. Manasseh Cutler arrived in Philadelphia, his pockets stuffed with letters of introduction to people of rank and influence. Dr. Cutler, a Yale-educated clergyman and former chaplain in the Eleventh Massachusetts Regiment, was here on important business. In May 1787, the Constitutional Convention had first met in Philadelphia, and now debates raged about the future of the American experiment. Dr. Cutler had come to rake the coals. As the newly appointed agent for the "Ohio Company," an association of Revolutionary War veterans who desired territorial compensation for the blood they had shed for America, he wanted to "secure both law and land" from the politicians in Philadelphia and New York. Specifically, he was interested in the acquisition of a tract at the junction of the Ohio and Muskingum rivers (*LJC* 1: 121).

Busy as he was, Cutler had precious little time to spare during his visit; he feared that he would not even be able to deliver his letters to all the appropriate people (*LJC* I: 254). But the good reverend was a man of many talents: besides being a cunning land speculator, he had distinguished himself as a physician, astronomer, and botanist. Wherever he went, his "entire botanizing apparatus" traveled with him. Small wonder that, on the first full day of his visit, he took time out to inspect Philadelphia's new museum, Charles Willson Peale's "collection of paintings and natural curiosities" (*LJC* I: 259).

When Cutler and his friend Dr. Clarkson arrived at Peale's museum, the collector himself was nowhere to be seen. A boy pointed them to the gallery "where the curiosities were" and promised that Peale would be with them shortly. Entering a long, narrow hallway, Cutler cautiously peered into the gallery and, through a little glass window, saw "a gentleman close to me, standing with a pencil in one hand, and a small sheet of ivory in the other, and his eyes directed to the opposite side of the room, as though he was taking some object on his ivory sheet" (*LJC* I: 259). This had to be the proud collector himself, standing amidst his prize possessions, making a sketch of a particularly interesting exhibit. Clarkson, who knew Peale personally, suggested to Cutler that they had better wait outside: "Mr. Peale is very busy. . . . We will step back into the other room and wait till he is at leisure" (*LJC* I: 259). Turning their backs to where they had just seen Peale at work, the men walked back to the other room. But there, to their amazement, they saw none other than Peale himself, hastening toward them from exactly the opposite direction. Mystified, Clarkson exclaimed:

> "Mr. Peale, how is it possible you should get out of the other room to meet us here?" Mr. Peale smiled. "I have not been in the other room," says he, "for some time." "No!" says Clarkson, "Did I not see you there this moment, with your pencil and ivory?" "Why, do you think you did?" says Peale. "Do I think I did? Yes," says the Doctor. "I saw you there if I ever saw you in my life." "Well," says Peale, "let us go and see."

Back in the gallery, they found "the man standing as before." Now Cutler was impressed, too. Looking back and forth between Peale and the figure with the ivory sheet and pencil, he "beheld two men, so perfectly alike that I could not discern the minutest difference. One of them, indeed, had no motion; but he appeared to me as *absolutely* alive as the other, and I could hardly help wondering that he did not smile or take a part in the conversation" (*LJC* I: 260).

What Cutler and Clarkson had seen in the gallery was nothing other than a full-sized wax replica of Peale himself, his self-portrait, put on display alongside his numerous other exhibits: portraits of famous people (excellent likenesses, Cutler thought); a collection of stuffed ducks, geese, cranes, and herons floating on an artificial pond ("all having the appearance of life" and "admirably preserved"); songbirds "of almost every species in America" perching on trees; enormous rattlesnakes hiding behind rocks; a gallery of wild animals ("bear, deer, leopard, tiger, wild-cat, fox, raccoon, rabbit, squirrel, etc."). To Cutler, Mr. Peale suddenly seemed like a kind of latter-day Noah, except that not "even Noah could have boasted of a better collection" (*LJC* 1: 262). But by far the most remarkable exhibit in Peale's ark was Peale himself, or rather, his copy: "So admirable a performance must have done great honor to his *genius* if it had been that of any other person," wrote Cutler, "but I think it is much more extraordinary that he should be able so perfectly to take himself" (*LJC* 1: 260).

I am interested in Peale's whimsical deception, which to me is more than a trick. In fact, it provides me with the central idea for the chapters that follow: the image of the naturalist who creates a collection and then puts himself into it, of the collector who is both *apart from* and *a part of* his collection, who is, to quote from Walt Whitman's "Song of Myself," both "in and out of the game, and watching and wondering at it."[1]

Linnaeus had hammered into the minds of aspiring naturalists the world over that the study of natural history, "simple, beautiful, and instructive," should consist mainly in the "collection, arrangement, and exhibition of the various productions of the earth."[2] Faced with a natural environment whose strange "productions" had once made the Puritan John Winthrop wince (he admitted that he had not seen "the like . . . either in *England* or *France*, or other parts"),[3] American naturalists heeded Linnaeus's advice. If anything, they even more intensely felt the "urge to collect," that is, to make comprehensible, through the selection and preservation of what appeared particularly interesting and noteworthy, a world that constantly eluded them. Often uncertain about their own place in a land that was theirs before they were the land's, they also felt the need to relate *themselves* to the collections they made. As anthropologists have maintained, it is through collections that the self that wants to "possess but cannot have it all" acquires a sense of identity.[4] Gathering plants, birds, rocks, and artifacts, American naturalists were contributing their share to the "universal census," the "general mobilization" of the natural world, and, in doing so, they defined themselves.[5]

Once safely housed in the collection, the specimens that had been so rudely extracted from their original environments often threatened to become meaningless. As the example of Peale's collection with its artificial pond and lifelike trees shows, efforts had to be made to reinvent and manipulate new contexts for them. For most of the collectors in this study, this was not an outlandish challenge; Peale and others jumped at the opportunity of transforming relatively random assemblages of natural collectibles into works of art, "aesthetic experiences."[6] In these American natural history collections, the older European tradition of the *Wunderkammer*, the cabinet of wonder, still lived on. However, as Manasseh Cutler realized, in exhibitions such as Peale's the strange and the familiar were interwoven in one complex fabric,[7] and not only the wonders of nature but also the nature of wonder were given new meaning. The collection became a version of "art as play."[8]

It is with such forms of serious fun and the memorable images they have produced in American natural history that this study is primarily concerned. To some extent, it has been my ambition to do for pre-Darwinian American natural history what Gaston Bachelard has done, in a more general way, for our notions of Rome and shelter in *Poetics of Space*. This is why, in a sense, I like to think of *The Poetics of Natural History* as a book about "daydreams"—the daydreams of American natural history collectors.[9] Since this study makes a number of assumptions about the common goals of science and literature, let me hasten to add that the embodiments of such dreams, whether material, verbal, or visual, usually took a form that was precise as well as wonderful, allowing subjective desires to meet, productively and fruitfully, with objectively perceived truths.

The waxen "Mr. Peale," in Manasseh Cutler's story, was holding an ivory sheet and a pen in his hand—an image of the collector caught in the act of producing yet another image. In collecting nature, American naturalists collected themselves, either through the physical objects they accumulated in their gardens or museums or through the words or images they created to reflect on what they and others had collected: drawings or paintings, letters, diary notes, travel accounts, formal essays, and autobiographies. Part one, "Displaying," deals with the former; part two, "Representing," addresses the latter form of collecting. The stories told in both parts share a common theme: how, in the American natural history collection, things human and things natural were, if only briefly, allowed to coincide. Such coincidences, however, became serious rather than playful after the advent of Darwinism in the United States, and the themes of the stories as well as the manner of their telling had to change.

Part one begins in John Bartram's disorderly botanical garden in Philadelphia (chapter 1), where the wild profusion of plants gathered on various collecting trips into uncharted territories became a metaphor for the insatiable curiosity of both the father and his son William.[10] Chapter 2 further explores the ways in which the collection allowed the natural history collector to play with multiple images of the self. At a time when illusionistic impressions of reality were absent from natural history collections in Europe, Peale invented the diorama, the representation of animals in lifelike poses against realistic habitat backgrounds. As Peale's autobiography and paintings show, he did not hesitate to include himself in his forever expanding collection. Part one ends in New York, with the colorful Phineas Taylor Barnum, who bought up the shabby remnants of the playful Mr. Peale's collection and founded his own "American Museum." In Barnum's tawdry "halls of humbug" (Henry James), however, the real theme was no longer nature but humanity itself. Barnum's bizarre "connecting links" set the stage on which frightening theories of human descent easily turned into harmless fun. Natural history had become completely incidental to the main purpose of the collection—the collector's self-aggrandizement (chapter 3).

Collections have sometimes been identified as narratives, as stories told about things. Part two, in a sense, deals with such collecting stories from which the immediate traces of material collectibles have disappeared, giving way, often with surprising results, to carefully crafted assemblages of words or pictures. Chapter 4 reviews a gallery of images, verbal and visual, of snakes—ranging from the sleek-bodied, aggressive rattlers that both repulsed and fascinated Cotton Mather, Paul Dudley, and William Bartram to the self-contained, gloriously aloof *Crotalidae* gracing the pages of John Edwards Holbrook's *North American Herpetology*, like beautiful objets d'art. Looking into the "large, glistening eyes" of the most dangerous North American reptile, American naturalists created texts in which they were the narrators as well as the spellbound protagonists.

John James Audubon's paintings of American birds also characteristically take the form of visual dramas, and the texts he assembled in his *Ornithological Biography* (1831–1839) offer tension-packed stories of concealment and discovery in which Audubon himself appears alternately as the killer and the savior, the destroyer and the preserver of birds (chapter 5). Audubon's *Birds of America*, one of the last evocations of America as the land of plenty, of vast unencumbered spaces inhabited by numerous species yet to be discovered, marked the pinnacle of American natural history. The final chapter of part two (chapter 6) presents its discordant swan song: it reconstructs the progress of the Thayer Expedition through the Amazon Valley, where the Harvard professor and creationist Louis Agassiz had gone at the end of the Civil War to collect exotic fish as well as himself. Evidence was needed that Agassiz was right and Darwin wrong and that the world had been created not by happenstance but according to the wonderfully complete blueprint in the Divine Mind. Like Barnum, however, Agassiz was more a hoarder than a collector. The difference between him and earlier naturalists like William Bartram can best be gauged by the fact that Agassiz rarely seemed to see anything for the first time; readily he turned from what was new to what he knew. The haphazard notes and casual travel impressions gathered in Louis and Elizabeth Agassiz's *Journey in Brazil* (1868) indicate that, in Agassiz's restless hands, collecting had ceased to be a risky search for the unknown and, instead, had turned into an affirmation of the already familiar. American natural history goes out with a whimper—with Americans amassing tons of Brazilian fish and taking snapshots of naked native women. The final image presented in *The Poetics of Natural History* is that of the young William James in Brazil, a collector *malgré lui*, watching in disbelief while locals were made to pose for shots meant to furnish Agassiz and other opponents of racial mixing with much-needed visual proof that some human beings were less equal than others.

None of the figures discussed in *The Poetics of Natural History*, with the exception of William Bartram, has so far been of great interest to the literary historian. In his *Travels through North & South Carolina, Georgia, East & West Florida* (1791), the botanist Bartram, after all, came up with some unusual metaphors (William Hedges, in *The Columbia Literary History of the United States*, disparages them as "lush hyperbole" and deplores Bartram's "grandiloquence"),[11] wrote warmly about strange flowers and Native Americans and supplied great material to better writers, such as Wordsworth, Coleridge, and Chateaubriand. Or so they say. Another naturalist featured in these pages, John James Audubon, took time off from his regular pursuits and composed a few memorable descriptions of life on the frontier, and he included bits of interesting landscape detail in his paintings for *The Birds of America*. But otherwise, mocks Edmund Wilson, Audubon's work had as "little relation to society" as the poems and stories of Poe.[12] And if Louis Agassiz, whom Van Wyck Brooks dubbed the "Johnny Appleseed" of American science, has occasionally received honorable mention in the hefty biographies of the great Transcendentalists, this is perhaps so because he once took Emerson on a camping trip to the Adirondacks.[13] (Since he gets a page or two in Ezra Pound's *ABC of Reading*, Agassiz has also put in some surprise appearances in studies of literary modernism.)[14] Finally, several critics have taken a rather censorious interest in Charles Willson Peale because they believe that he abused his museum to promote a seriously skewed view of American national identity.[15] But, as appealing as Peale, Audubon, and Agassiz might seem from the historian's point of view, who in her right mind would today turn to the hefty tomes of Peale's correspondence and Audubon's sloppy journals, with their rampant misspellings, in search of a good read"? And, in all fairness, who would expect to find, in Vladimir Nabokov's lucky phrase, "aesthetic bliss" in Peale's rambling autobiography, unfinished and unrevised as it is—a text in which the author himself apparently lost interest?

By insisting that, nevertheless, these texts are, at their best, well worth reading and rereading. I have deliberately put myself at odds with a number of previous approaches to the field. For instance, my focus on the *aesthetic* aspects of American natural history leaves me vulnerable to the charge of political naiveté and open to the complaint that I am ignoring the active involvement of American naturalists, most of them dead white males, in the conquest and exploitation of the American continent. If Mr. Peale in Manasseh Cutler's story seemed almost annoyingly happy, some critics would have pointed out to him that, in fact, he had little reason to be. In Mary Louise Pratt's eyes, for example, naturalists were the willing, if slightly befuddled, accomplices of Western imperialism. Armed with "nothing more than a collector's bag, a notebook and some specimen bottles," writes Pratt, they set out in search of "the bugs and the flowers," busily extracting every species from its native environment and callously placing it "in its appropriate spot the order-book, collection, or garden)," with its "new written, secular European name" attached to it.[16] In Christopher Looby's view, too, American naturalists were more benighted than enlightened. Thomas Jefferson, Charles Willson Peale, and William Bartram, "refusing to be distracted by the mutability of natural objects" and enamored of the rigid categories of Linnaeus's system of nature, described a fixed, homogeneous, tightly structured natural world, even where the evidence told them otherwise. Looby finds a bit of comfort only in the realization that this blessed rage for order usually foundered on the rocky roads of reality itself. Searching for harmony and peace, the Quaker William Bartram, in Looby's reading, inevitably found himself contemplating revolutionary chaos, "a nearly random set of motions, a concatenation of fortuitous processes, an intersection of unpredictable transformations."[17] Other voices, more subdued, have joined the growing critical chorus: Paul Semonin, for example, has retraced the manifold connections between American nationalism and American natural history. The stuffed birds, prehistoric bones, and botanical specimens collected by American naturalists, argues Semonin, created a disingenuous historical narrative that made white Americans happily forget the "unacceptable antiquity of Native American civilization."[18]

While I do not want to minimize (and have in fact profited from) arguments that organize themselves around the concepts of domination and appropriation, my study of American natural history charts the emergence of a Western self much less stable than such theories could comfortably accommodate. Put differently, my response to Pratt and Looby, developed more fully in the following chapters, would be that by placing themselves at the center of their collections—by letting, for example, the waxen "Mr. Peale" take over from Charles Willson Peale—collectors often successfully, and interestingly, "decentered" themselves.

My argument that these premodern collecting narratives are worth reading compels me to take umbrage with yet another camp of critics: those who would fault me for seeing literariness where it isn't and overlooking science where it is. In the first wide-ranging survey of natural history writing from Captain John Smith to John Burroughs, Philip Marshall Hicks warns us, sternly, that "great prose is not to be expected in the natural history field."[19] And even though literature, the "field" from which Hicks and others presumably expect the enjoyment of "great prose," has undergone extensive re-definition since 1924 when Hicks penned his remarks, critical opinion has apparently not swayed much in favor of "the art of natural history." Consider Pamela Regis's recent attempt to rehabilitate Bartram's *Travels*, along with Thomas Jefferson's *Notes on the State of Virginia* or J. Hector St. John de Crèvecoeur's *Letters from an American Farmer*, as "works of science" rather than "works of belles lettres."[20]

Be that as it may, I do not find the opposite view very useful either. In his excellent study of eighteenth-century naturalists, *In the Presence of Nature*, David Scofield Wilson eagerly foregrounds the *literary*—or, rather, the semiliterary—element in natural history at the expense of its scientific credibility. He carefully distinguishes between natural history and "nature reportage," designating the former as the province of true systematic science and the latter, cheerfully embraced by American naturalists such as John Bartram, as the genre in which free-lance, self-educated journeymen, only peripherally acquainted with the grand theories that justified their haphazard activities, wrote down their "simple observations" meant to aid minds greater than theirs in the further refinement of their overarching systems and abstractions. "Nature reporters" wrote essays and travelogues rather than handbooks and scientific essays. No wonder that, leaping from intuitive descriptions of the "when, where and who of occurrences" to "guesses of why," these reporters, instead of dutifully contributing to a fuller understanding of the myriad phenomena of nature, sometimes ended up perpetrating hoaxes instead of promoting the facts.[21]

Wilson's argument does not require a detailed rebuttal here; suffice it to say that while Linnaeus, the author of the magisterial *Systema naturae*, did not think it beneath him to write a vivid account of his travels through Lapland,[22] a simple "nature reporter" and traveler like John Bartram contributed notes on American plants to the 1751 edition of Thomas Short's handbook *Medicina Britannica* and felt secure enough in his knowledge of taxonomy to disagree with some of Linnaeus's authoritative classifications: "Poor Lineus . . . I always thought he crowded too many species into one genus" (*CJB* 414).

In this book I have consistently preferred the term "natural history" to such misleading alternatives as "nature reportage" or "nature writing." I have done so for a reason. While my own theoretical program could perhaps be described as a form of literary anthropology, I do not engage in any of the usual celebrations of the liberating power of "literature" that mark and, in my opinion, mar current approaches to the field.[23] In regarding the texts and visual images created by American naturalists as products of the *instructed* imagination, I refuse to treat "art" and "science" as opposites. Such a distinction would be historically inadequate, anyway. Up to the beginning of the nineteenth century, literature and science, for example, were generally understood as equal partners in a unitary endeavor.[24] In 1799, it is true, Wordsworth had sneered at the naturalist, "a fingering slave" who would coolly "peep and botanize / Upon his mother's grave."[25] But a few years later, the American scientist Benjamin Smith Barton, "the greatest botanist of his age" (Daniel Boorstin), when giving a speech before the Philadelphia Linnean Society, still saw nothing wrong in cheering natural history as the science "of just and happy arrangements; and of *beautiful and correct theories*" (my emphasis).[26] It was only by the middle of the century that the natural sciences—as the cognitive encounter with a clearly definable material world and as the proper province of specially trained professionals—had clearly separated themselves from the arts, which now had become the acceptable mode of describing one's existential encounters with a social and personal world.[27]

It would of course be wrong, as Lawrence Buell has warned us, to look at all of early nineteenth-century thought as leading up to Darwin—an assessment "more true poetically than literally."[28] But perhaps it is not inappropriate to regard the publication of Darwin's *Origin of Species* (1859) and its subsequent reception in America as marking the official end of certain accustomed ways of speaking about "Nature" (a word that Darwin, in the second edition of his book, changed to lower case). Now that it had become clear that, like the barnacles, the ducks, and the bears, human beings, too, were subject to the endless processes of struggle and change that had shaped the history of the organic world, they could no longer perceive themselves as being both part of and apart from the entangled family network of nature. Forever branded with the "indelible stamp" of their own "lowly origin," it now suddenly seemed that humans had little reason for arrogance.[29] As Darwin, a superior writer himself, knew only too well, his theory raised crucial problems for the form in which such insights could and should be rendered. Among other things, Darwin tried to cleanse his language of all traces of human will and intention, often unsuccessfully so, as Gillian Beer has demonstrated.[30] It is mainly as a symbolic, cautionary figure, therefore, that he appears at the end of chapters 4 through 6. The publication of *The Origin of Species*, a book in which humankind had disappeared from the center of inquiry, marked the point at which natural history collectors began to lose control over their happy self-projections, fearing that perhaps they themselves would soon be nothing more than exhibits gathering dust in collections that had become obsolete.

The Poetics of Natural History attempts to remove some of these layers of dust and to describe pre-Darwinian American natural history at its best—namely, in those moments when it successfully straddled, in the words of Vladimir Nabokov, an experienced butterfly collector himself, "the high ridge where the mountainside of 'scientific' knowledge joins the opposite slope of 'artistic' imagination."[31] Located at the crossroads of Linnean taxonomy and belles lettres, wavering between the demands of precise description and the seductions of narrative, American natural history, as I would summarize the central thesis of this book, only superficially avoids what it very often becomes—a form of autobiography.

A final word about my general method. I have not striven for completeness, nor have I attempted to offer a continuous history of pre-Darwinian American natural history.[32] Historians of science will miss some names; historians of literature will deplore the omission of others. In part this reflects my ambition to promote less widely known and read figures over those more frequently admired and written about, but some of these absences also result from a need to reduce and condense material so vast that it has defied my own "rage for order." Collecting entails selecting, since, as James Clifford has reminded us, one "cannot have it all." In fact, this is how my own role in the following chapters could most usefully be described—as that of a collector of curiosities, a guide through a cabinet of collectors, through a museum of naturalists that has room for more. Being a sort of collector myself, I realize that, like other collectors, I shall have to live "with a great deal of ambiguity" about my "prospects for producing closure."[33] I do not apologize for the overabundance of detail in my account; after all, the collectors I write about were convinced that the grand story of nature could be told only by the restless accumulation of little things. Readers of this book will inevitably construct their own stories from what they see, read, and remember. They should feel free to design their own paths through my collection and know that, while my chapters—the rooms of my museum, as it were—are part of one slowly unfolding argument, or a continuous tour, each of them should be worth a separate visit, too.

In sum, the purpose of the tour I offer in *The Poetics of Natural History* might most simply be described thus; to help dispel the still widespread assumption (here articulated by Mark Kipperman) that only "belles lettres" reveals to us our presence as desiring, storytelling beings within the concrete world of experience."[34] After all, what Dr. Manasseh Cutler brought home with him from his visit to Mr. Peale's museum, where the collector had duplicated himself and thus duped his visitors, was . . . *a good story.* Peale invented it, and this is how Cutler retells it: as a carefully staged tale of deception, starting with a perfect illusion and a complete withholding of information, and ending in the final amused and amusing disclosure of the truth. "To what perfection is this art capable of being carried," Cutler exclaimed, imagining that his relatives—and perhaps he himself, too—could be similarly preserved for future generations, "in perfect like-ness." Thus Manasseh Cutler's story about "Mr. Peale" even has a moral, expressing the hope, prevalent in all of antebellum American natural history, that the collector art could defeat the "ravages of time" (*LJC* 1: 260). It would have bothered Mr. Peale or Dr. Cutler only slightly that, in this view, there is little indeed that separates preserved humans from stuffed ducks.

PART ONE
DISPLAYING

CHAPTER 1

"America Transplanted"

JOHN AND WILLIAM BARTRAM

"Dog Piss Weeds" and "Lady's Slippers"

In 1733, the wealthy Peter Collinson, a textile merchant and the owner of a fine garden in Peckham on the Surrey side of the Thames, sent out an order for American seeds to a fellow Quaker, John Bartram, a farmer on the west bank of the Schuylkill River in Kingsessing, Pennsylvania, near Philadelphia. Bartram had been warmly recommended to him by two friends, Dr. Samuel Chew and the scrivener Joseph Breintnall, a poet, artist, scientist, and collector dedicated to the art of making leaf impressions.[1]

It had not been easy for Collinson to find someone who shared his passion for "fine, showey specious" flowers.[2] Collinson had become hooked on plants when, as a very young child, he was sent to Peckham to live with relatives who had "a garden remarkable for fine cut greens . . . and for curious flowers." Often he visited "the few nursery gardens round London, to buy fruit and flowers, and clipt yews in the shapes of birds, dogs, men and ships."[3] Luckily, in Bartram the zealous Collinson had finally found a kindred spirit, a fellow collector who confessed to an equally "great inclination to plants," someone who, at the age of ten, had "knowed all that I once observed by sight tho not thair proper names haveing no person or books to instruct me" (1 May 1764; *CJB* 627).

Before turning to Bartram for help, Collinson had tried to interest a number of other American correspondents in "new and rare plants," apparently with little success: "What was common with them but rare with us they did not think worth sending."[4] With Bartram, the situation was different. The Philadelphia Quaker shared Collinson's love of horticulture as well as his interest in scientific experimentation. For example, Bartram pioneered the artificial hybridization of plants: 1739, he proudly told William Byrd of "severall successful experiments of joining several species of ye same genus whereby I have obtained curious mixed colours in flowers never known before" (*CJB* 120).

Not coincidentally, during the course of the eighteenth century the proper knowledge of plants had come to be of special importance in the relations between England and its American colonies. In 1769, in a new edition of his *Account of East-Florida,* William Stork asked: "How long was England, this active, enterprizing, philosophical nation, uninformed of the uses of clover, turnips, potatoes, & c. without which its present inhabitants would be at a loss for subsistance?" And he argued that a similar waste of time could be avoided in His Majesty's colonies if naturalists realized the magnitude of the task that awaited them both there and at home: "The new introduction of but a single grain or plant, as the rice in Carolina, or the turnip in Norfolk, will sometimes totally change the face and condition of a country. Here therefore is a field in which the naturalist may make his science peculiarly useful. His knowledge extending through the vegetable world, informs him where every valuable plant, grain, or tree is to be found, and also in what country it is wanting, and may be propagated to advantage."[5]

The merchant Peter Collinson was instrumental in the promotion of such a global exchange of plants. Botanists have credited him with the introduction of approximately 180 species of trees, shrubs, and flowers into English gardens.[6] But Collinson's interests were not limited to botany; they included other areas of scientific inquiry as well. In 1745, for example, he sent books and electrical equipment to Benjamin Franklin, thus facilitating the experiments that finally led to the publication of the first scientific book written by an American to be widely discussed in Europe: Franklin's *Experiments and Observations on Electricity, Made at Philadelphia in America* (1751). With some justification, historian Raymond Phineas Stearns has claimed that "during the middle third of the eighteenth century, no one man did as much to stimulate science and to establish intercommunication among scientists as Peter Collinson," and he has argued that "scientific circles, especially those of the New World, would have been far less tightly knit together without the pen and the generous assistance" of this Quaker merchant.[7] Such lavish praise notwithstanding, a full list of Collinson's contributions is still lacking.[8] Even more surprising, his remarkable correspondence with John Bartram, called "wonderful" by at least one critic, has not received the close attention it deserves.[9] If Linnaeus believed that natural history concerned itself first with "collecting," this is what these letters were all about. But the items Bartram and Collinson incorporated into their collections presented problems of their own. Admittedly, plants—alone among organic specimens—offered the advantages of being relatively easy to preserve in a moisture-retaining medium and having the ability to replicate themselves from seeds, bulbs, and cuttings.[10] However, as Brenda Danet and Tamar Katriel argue in their essay on play and aesthetics in collecting, it is generally "easier to dominate an inanimate object . . . than a living organism such as a plant; plants grow, change shape, die."[11] Some of the excitement in the correspondence between Collinson and Bartram indeed derives from the unpredictability of the things they collect. But Bartram's and Collinson's letters are not only thematically concerned with collecting; they are in themselves a *collection*, an assemblage of botanical queries and insights, personal observations and complaints, as well as notes and reflections on the similarities and the emerging differences between England and America.

Classical rhetoricians liked to distinguish between the *epistola negotialis* and *epistola familiaris*,[12] and one of the most interesting aspects of the epistolary the exchange between Peter Collinson and John Bartram—in an age dominated by the "cult of familiar letter-writing"[13]—is that here the public and the private continually intermesh. The form of the letter had long been used in science writing (one need only think of Galileo's *Letters on Sunspots* [1613]), and the influence of this tradition on the fashionable eighteenth-century pursuit of botany became visible in publications such as Jean-Jacques Rousseau's *Lettres élémentaires sur la botanique* (1771–1773). We would be mistaken to believe that Collinson and Bartram, simply because they were scientific autodidacts and literary amateurs, used the letter form without self-consciousness. Bartram, for one, realized that all his observations were of potential interest to the Royal Society, of which Collinson was an active member, and he knew that his friend would edit and publish whatever he thought could be of interest in their correspondence.[14]

In his first surviving letter to Bartram, dating from 1734/35 (*CJB* 3–6), Collinson ecstatically acknowledged the safe arrival of one of the precious shipments that were so gratifying to his collector's instincts. "I am very much oblig'd to thee for thy Two Choice Cargos of plants which Came very Safe & in good Condition, & are very Curious & Rare & Well worth my Acceptance." He showed his appreciation for the hardships John Bartram had incurred to gather these plants, "the great pains & many Tiresome Trips to Collect so many Rare plants scattered att a distance." As a kind of compensation, he announced the gift of a "Callico gown" to Bartram's wife and "some odd Little things that may be of use amongst the Children & family." Collinson also had some practical advice to offer and, playing mentor to what he mistakenly assumed was his new friend's "untutored" mind,[15] interspersed technical notes on plant collecting and preserving with more general scientific instruction. Get your plants when in bloom, "with their Flowers on and with their Seed Vessels fully Form'd," he told Bartram, since it was only "by these two charistics" that "the genus is known that they belong too." Then preserve them carefully. For that purpose Collinson was mailing his newfound friend "Two Quires of Brown & one of whited Brown paper. "Thus, when Bartram saw "curious plant[s] in Flower," all he had to do now was dry his specimens, preferably two of a kind, between the brown paper sheets and insert them into the quire of whited paper, and they were ready for their journey to England. The duplicates were intended for the American's "Improvement in the knowledge of plants." Collinson promised that he would "gett them nam'd by our most knowing Botanists & then Returne them again—which will Improve thee more then Books for it is impossible for any one Author to give a General History of plants."

Still in the same letter, Collinson waxed rhapsodic about the beauty of the plants Bartram had sent him: "The Warmth of the ship & Want of Air had Occasion'd the Skunk Weed to putt Forth Two fine Blossoms, very beautiful it is of the Arum Genus" (*CJB* 4). But, ever the efficient merchant, he soon returned to practical patters, reminding his correspondent to number the specimens he collected and shipped to England so that "Wee have only to write to Thee for the same seed to such a number to send over again" (*CJB* 6).

Once Bartram's precious cargo arrived in England, Collinson divided and distributed the collections among his gardening friends and aristocratic subscribers such as the avid horticulturist Lord Petre, who had consented to paying Bartram an annuity of ten guineas for his work. However, Collinson kept some of the most promising seeds for his own garden, which he hoped would soon, with his American friend's help, surpass every other collection of plants in England. "I took every Thing out of the Boxes Myself," Collinson told Bartram in January 1756, "and planted every one with My own Hands for I never trust the Servants with such Curiositys procur'd with Care & Trouble & I Carefully Search every Lump of Moss, least a small root should escape." While he was planting, "Robin Red Breasts" serenaded him with their sweet song (*CJB* 392).

Most of Collinson's wishes were fairly specific. "Pray look out for a plant or two of White Cedar," he asked Bartram on 6 April 1735. When "a little dried," the leaves of the white cedar smelled like cinnamon, he wrote, anticipating the pleasure of future ownership, because if this wonderful tree survived in his garden, it would be "the only one in England" (*CJB* 86). Some of Collinson's own customers might have had pragmatic reasons for ordering through him; in 1760, for example, Collinson requested "a good Quantity of your Common Locust Seed" to be sent to Bohemia, where it was to be sown as "fodder for Cows & Horses" (*CJB* 493). Personally, however, he liked to pose not as seed peddler but as the refined, contemplative aesthete, as the knowing, insatiable connoisseur of the manifold colors and contours of Nature's beauties, which the gardener's art could only enhance, never destroy. "There is no end of the Wonders In Nature," exclaimed the sixty-eight-year-old Collinson; "the More I see the more I covet to see" (25 July 1762; *CJB* 565).

The Society of Friends, at a provincial meeting in Leicester in 1705, had seen the need to warn its members of having "too great superfluity of plants and too great nicety of gardens." Friends were encouraged to pursue their botanical interests "in a lowly mind."[16] But the Quaker Collinson was unabashed in his love of beautiful flowers and trees. For him, they were artistic compositions. In June 1761, he excitedly reported to his friend in Pennsylvania: "The Ivy, Laurel or Broad Leaved Kalmia . . . is one of the finest Ever Green shrubs that is in the World—the stamina are Elegantly disposed in the angles of the Flowers—what a pretty blush its Bullated Flower Budds—But in a few Days will the Glorious Mountain Laurel or great Chamaerhododendron appear with its Charming Clusters of flowers prethee Friend John look out sharp for some more of these Two fine plants—for one can Never have too Many" (*CJB* 520). Consider Collinson's description of the pleasures of a greenhouse, sent to Bartram in February 1768, just a few months before Collinson's death. In the past weeks he had enjoyed "Winter without but Summer within." In spite of the frost outside there had been "plenty of Primroses & Poylanthos Some Violets & Single Anemonies & Wall flowers & Stocks in abundance. . . . I was Delighted with these Beauties on My Table within Doors, whilst snow covered the Garden without & then out of my Parlour I go into my Greenhouse 42 foot long which makes a pretty walk to smell the sweets of so many odoriferous plants" (*CJB* 699). Surrounded by sweet-smelling flowers and seeing "the thick blanket of snow outside, Collinson sensed summer in his soul and soared. Feeling firm ground under his feet again, he concluded his verbal flight of fancy by remembering his overworked correspondent: "My Dear John thou Loves my Long Stories but I should be ashamed to write such insignificent stuff to anybody Else commit it to the flames" (*CJB* 699).

Sadly though he carried his passion to the heights of art, Collinson was not alone in his appreciation of rare flowers: "So great is the Itch," he sighed in a letter written to Bartram on 25 December 1767, "that a poor Raged Shoemaker a Weaver or a Baker will give half a Guinea or a whole one for a New flower such is the infatuation" (*CJB* 693). Small wonder that Collinson's garden was visited by robbers not once but twice, in 1762 and 1768. On both occasions, Collinson's favorite flowers, orchids of the genus *Cypripedium,* also known as Lady's Slippers, were among the spoils of the intruders, who apparently knew exactly what they were doing. Collinson was vexed enough to offer a considerable reward (ten guineas) to anyone who could give information about who might have "unlawfully plucked up or taken away" his beloved "exotick plants."[17]

Not surprisingly, the affluent plant fanatic Mr. Collinson and his well-heeled customers rarely realized how urgent matters were financially for Bartram. Payments were often late; mockingly Collinson excused himself that "to pay their arears our Gentry & c are but Slow & I have So many things to think on that I forget—untill thy Bills come" (22 March 1751; *CJB* 319). What clearly was a luxury for Collinson remained an economic necessity for Bartram, who time and again had to remind his correspondent of the need to maintain his large family and of the difficult circumstances under which he was compiling his collections: "I have many small children to be provided for with food & raiment so that I can hardly leave home" (*CJB* 198).[18] Nevertheless, Bartram always remained eager to please. Constantly exhorted by an impatient Collinson ("please to Go On Collecting as Seeds happens to be in thy Way"; *CJB* 10), he spent more than thirty years gathering and shipping seeds, bulbs, root pieces, cuttings, dried plants, turtles' eggs, and insects' nests to England. In return, he received not only money and the seeds of European plants for his own garden but also gifts of various kinds: clothes ("this may serve to protect thy outward Man," Collinson commented after he had sent off a suit in 1741/42; *CJB* 187), a microscope (10 March 1743/44; *CJB* 234), and a pomegranate tree ("Don't use the Pomegranate inhospitably," Collinson gently warned his friend; "a stranger that has come so farr to pay his respects to thee, don't turn him adrift in the wide world"; 10 December 1762; *CJB* 580). And while Bartram remained "Mr. Bartram" to other correspondents closer to home, the respectful forms of address used in the first extant letters to Collinson quickly became more familiar, and the two men began to sign their letters as "thy loving friend" or "thy real friend" (*CJB* 9, 72, 97).

Peter Collinson was genuinely appreciative of Bartram's collecting efforts on behalf. If his friend's shipments allowed him to turn his gardens in Peckham and, after 1749, in Mill Hill, Middlesex, into the finest of their kind in Europe,

Bartram's written observations on rattlesnakes, insects, and mussels also helped him to establish his own reputation at home as a tireless promoter of science. He soon began to take a personal interest in his correspondent's well-being and constantly reminded him not to exhaust himself in his collecting. Like a worried parent, Collinson wrote to Bartram in January 1735/36: "My Dr. frd, I only mention these plants—but I beg of thee not to neglect thy more Material affairs to oblige Mee" (*CJB* 14). He feared that Bartram's "gratified Disposition" would carry him too far "to the Neglect of thy family affairs" (20 January 1737/38; *CJB* 79) and, from the safe distance of his comfortable home, deplored his friend's want of caution and circumspectness. When Bartram and his horse nearly came to grief on one excursion, the disgruntled employer and worried father in Collinson joined hands: "For thy Disaster in passing the River pray be very Carefull for the Future and Look before thou Leaps" (3 February 1741/42; *CJB* 181). Collinson, for one, knew that Bartram's trips were not for him. And although he assured his correspondent that, were he there, he would certainly like to keep him company, he also had to admit that the strange climate ("Heats & Colds") and "above all the Danger of Rattlesnakes" would so curb his "Ardent Desires to see vegitable Curiosities that I should be afraid to venter in your woods" (3 February 1735/36; *CJB* 18), Collinson much preferred to "*Read* & Travel" (*CJB* 70; my emphasis).

Notwithstanding such genuine concerns for his dauntless seed supplier's well-being, Collinson appended yet more litanies of requests even to these cautionary letters: "Send a quantity of seed of the Birch or Black Beech. . . . Send Mee a good root of the Swallow-wort, or Apocinon, with narrow Leaves & orange Coloured flowers, & of the pretty shrub call'd Red Root, and of the Cotton-weed or Life Everlasting & some more seed of the perannual Pea, that grows by rivers. . . . pray send Mee a walking-cane of the Cane-wood" (January 1735/36; *CJB* 16–17). An exemplary mixture of barely suppressed greed and solicitude can be seen in a letter sent in January 1751, in which Collinson circuitously but nevertheless firmly expresses his dismay that a valuable part of a shipment of growing plants had apparently been pilfered by greedy sailors: "I am Sensible of the great trouble & Difficulty to procure these plants as they grow So remote So cannot Desire thee to Renew them again—but if thy Friendship for Mee should prompt thee to Do it—then add some More Rose Laurel but yett unless thy Health permits and other agreeable Circumstances Concur—pray don't, think about them" (*CJB* 312). The message wouldn't have been lost on Bartram.

In fact, Bartram seldom failed to deliver, sometimes indeed at considerable risk to his own health. In May 1761, he nearly broke all his bones, he said, while gathering a "fine parcel of holley berries" for one of Collinson's colleagues, James Gordon.[19] When the branch of the tree to which he was holding on suddenly snapped, he "fel to ye ground" and suffered "grevous" pain. He found himself in a "dark thicket"; no house was near, a "very could sharp wind" blew, and he realized he had "above 20 mile to ride home." He left it to Collinson's judgment (and imagination) "what A poor circumstance I was in & yet my arm is so weak that sometimes I can hardly pull on my cloaths." But Bartram, who by then had already toured a large part of eastern North America, was not easily deterred: "Yet I have A great mind next fall to go to Pittsburgh in hope to find some curious plants there" (22 May 1761; *CJB* 517–518). In a subsequent letter to Collinson, he even came up with a witty simile to comment on his recent mishap: "Climbing trees is over with me in this world & in ye next I rather chuse to fly like an angel to search for vegitables in realms unknown to mortals" (14 August 1761; *CJB* 534).

After their first exchanges, the correspondence between Bartram and Collinson continued unabated for thirty-five years, regularly punctuated on Collinson's part by reminders such as "pray some more white Cedar" (*CJB* 38); "pray remember the Terrapins" (*CJB* 319); "pray remember the Faba Egyptica" (*CJB* 324); "pray send seed of the Smaller flowerd Rudbecia for that is my Favourite" (*CJB* 480). On Bartram's side, letters would contain repeated assurances that he had indeed dispatched the plants requested ("I have shiped . . . to thee 26 boxes," *CJB* 413). Not surprisingly, Bartram sometimes tired of these ceaseless requests for new specimens. Irritably, he wrote to Collinson regarding yet another order he had received from his subscribers: "I have sent them seeds of allmost every tree & shrub from Nova scotia to Carolina." Did they think he could produce "new ones"? But he also professed stoical resignation unto his collector's miserable fate, a life of toil and trouble but also of many, if intangible, rewards: "If I die A martar to Botany Gods will be done" (14 August 1761; *CJB* 534).

Just the same, wasn't it annoying that Bartram's British correspondents would write to him as freely for plants "as if thay thought I could get them as easy as thay do ye plants in ye European gardens" (8 November 1861; *CJB* 538)? Yet so it must have seemed to European collectors. Surveying the boxes sent from America that were piling up in the harbor, Collinson predicted in 1756: "After this Rate England must be turned up side down & America transplanted Heither" (*CJB* 392). Collinson was not joking when he told his "Dear John" that a time would come when all the "Country productions" of America would be "Naturalized in our gardens" (10 April 1767; *CJB* 682).[20]

While the transportation, delivery, and receipt of plants remained the back-bone of the entire correspondence, Bartram's and Collinson's letters ultimately touched upon much more than just the successes and failures of a transatlantic seed exchange. In a century that considered the letter as a "short-cut to the heart," to use Ian Watt's characterization of Samuel Richardson's technique in his epistolary novels *Pamela* and *Clarissa*,[21] Collinson and Bartram still pre-ferred to use it as a detour. In their written conversation, flowers became the vehicle for the communication of more intimate sentiments. In all this, their letters seldom lose their freshness and unconventionality. Consider the letter of 4 January 1764, in which Collinson apologizes for having misplaced his friend's previous communication and explains the circumstances under which he has just recently rediscovered it: "Having Changed my Cloaths the day I received yrs which I think was the 20th Xbr or thereabout," he writes, "I could not devise where I had put yr Lett. till this very morning, I found it in the pocket of the Coat I now have on" (*CJB* 624).

This lively epistolary conversation between two remarkable men who never met was conducted against the background of two worlds drifting irrevocably apart (Collinson and Bartram, too, always distinguished between *"your* country" and *"my* country"). The steadily developing friendship between "my dear John" and "my dear worthy Peter," the increasing emotional intimacy between two passionate collectors, was invariably defined in terms of shared fears about ship-ments that failed to arrive or impatient observation of the plants that withered or, if luck would have it, thrived in their gardens. So many of them were like the mountain laurels and the azaleas, which Collinson found terribly difficult to cultivate in his garden: "how few . . . can be reconciled to Our Soil & Climate if one out of 3 or 4 Succeeds it is Well—& the Shrub Honesuckles or Azaleas are almost as Ticklish in their Natures—in some places they do Well in others not at all" (12 June 1761; *CJB* 520–521).

In Bartram and Collinson's correspondence, the precariousness of the "ticklish" plants' transatlantic fates became a metaphor for the vicissitudes of their own un-usual friendship, their form of teasing camaraderie, vacillating as it did between the extremes of deeply felt harmony and pouting dissent. "We friends that love one another sincerely," judged Bartram in December 1745, "may by an extraordi-nary spirit of sympathy not only know each others desires but may have a spiritual conversation at great distances one from one another." He was confident of the spiritual bond of love between him and his correspondent, which was manifested in the plants they sent each other: "If I love thee sincerely & thy love and friendship be so to me thee must have a spiritual feeling and sense of what particular sorts of things will give satisfaction & doth not thy actions make it manifest? for what I send to thee for thee hath chosen of just such sorts and colours as I wanted" (*CJB* 269). In his *Essays*, Francis Bacon had referred to gardening as the "purest of human pleasures," the "greatest refreshment to the spirits of man,"[22] but in the transatlantic exchange of plants and letters between John Bartram and Peter Collinson, horticulture became much more than the "Rational amusement" that Collinson once declared it was (*CJB* 324). Wherever they could, Bartram and Collinson deliberately fortified their carefully cultivated microcosm against the interventions of the outside world, such as the French and Indian War, and created their own self-enclosed garden worlds where American and European plants peacefully coexisted and the only danger came from the unpredictability of nature itself. In May 1761, after William Pitt, the king's war minister, had announced his resignation, Collinson sent a note of sympathy to his friend, a large number of whose flowers had been destroyed by a harsh North American winter. "Thy Account of the loss of so many fine plants," he wrote, "Effects Mee more then the Loss of Pitt" (22 May 1762; *CJB* 560).

About politics Bartram and Collinson sometimes disagreed. Bartram, who had lost his father in a raid by Tuscarora Indians in 1711, now believed, not at all in accordance with the peaceful spirit of his fellow Quakers, that the only way to cope with the restive "Indians" was to "bang them stoutly" (23 October 1763; *CJB* 612). The liberal Collinson's terse response was that he could fill a letter with instances of "our arbitrary proceedings, all the colonies through; with our arbitrary, illegal taking their lands from them, making them drunk, and cheating them of their property" (6 December 1763; *CJB* 615). But such differences of opinion did not alter their understanding that what mattered between them, what "pleased" them both was questions of gardening. Having strongly criticized Bartram's views on Native Americans, Collinson swiftly resumed their conversation where they had left off: "So much for a touch of Politicks Now we change the Scene to something that Pleases us both I can tell thee Gordon has raised the fine stately broad-leafed Silphium" (*CJB* 616).

Anxiously, the two collectors compared notes, and the differences in their impressions served as sad reminders that, while their botanical interests spanned the continents, the thousands of miles separating their two gardens inevitably kept them apart. Usually, experiences of a merely autobiographical sort were omitted from their communications, the private language of the self being superseded by the public language of botany. But there were exceptions. "Dear

Aflicted friend," wrote Bartram in July 1753, after hearing that Collinson's wife had died, in a letter that betrayed his personal concern even through the elevated diction and sentences that seemed more carefully crafted than usual: "her dear sweet bosom is cold her tender heart the center of mutual love is motionless her dear arms are no more extended to embrace her beloved, the partner of his cares & sharer of his pleasures must no more sit down with her husband at his table Oh! my dear friend. . . . I lost an inocent loveing wife which I lived with above 4 years I thought my loss could never have been made up" (*CJB* 347).

The bulk of Collinson and Bartram's correspondence, however, was concerned not with their own family lives but with the private lives of the plants they collected. Granted, sometimes it was hard to keep the two apart. These two men were giving each other flowers, literally as well as figuratively. "I have a sprig in flower of the Kalmia in water & it stares mee in the face all the while I am writeing—saying, or Seems to say, as you are so fond of Mee tell my Frd J Bartram who sent Mee to send some More to keep me Company for they will be sure to be well nursed & Well treated" (12 June 1761; *CJB* 521). The eighteenth century, as Ann Shteir has shown, considered botany a pursuit more amenable to the "pliant fingers" of women than the "clumsy paws of men,"[23] and it is interesting to see how in their botanical conversations Bartram and Collinson both yielded to and resisted the impulse to feminize themselves.

Although Collinson rather charmingly denigrated one of his own letters as a miscellany "of matters as they come into my Noddle" (*CJB* 134), it was obvious that Bartram usually enjoyed his friend's epistles, however rambling they were, and that in fact his only wish was that they "had been as long again" (*CJB* 88). Collinson's epistles were not just long, they were—if the pun is permitted—longing, too. The "Enchanting Spot thou describes," wrote Collinson in 1750, "Makes Mee Wish to be there . . . to see the Common Laurel & Rose Laurel together" (*CJB* 307–308). In November 1758, Bartram informed Collinson that he had recently measured the striped maple (*Acer pennsylvanicum*)[24] in his garden to find it had again "advanced one foot in height." Referring to the vertical lines inscribed in the bark of the tree's branchlets, Bartram observed that the new shoot was "delicately striped with white & red," while last year's shoot was a "silver stripe on A light green" and the previous year's growth was "silver on A dark green." He went on to describe the enchanting play of colors on the trunk of an older tree in his garden, which contained "bright silver stripes down to ye root intermixed with dark green" and "light green brown & yelow." The bark on old maple trees was "quite smooth." Abruptly Bartram apologized to Collinson, explaining that he would not have been "so perticular" in this description "to set thee A longing for it" had Collinson not given him "strong hopes" that he had such a beautiful striped maple growing in his garden, too: "I am sure I sent thee one or two some years past" (*CJB* 441). And "a-longing" the recipient of this letter certainly was. Bartram's "perticular description" had made Collinson's mouth water, and one can imagine his frustration when he finally reported, in March 1759, that "My Striped Bark Mapple has not all the Colours thou mentions phaps it may as it grows older" (*CJB* 458).

Collinson the horticulturist did not, of course, only suffer disappointments. In June 1762, Collinson's mountain magnolia (which he had "raised from Seed about 20 years agon") flowered for the first time. He was jubilant. "I presume [it] is the first of that species that ever flowerd in England," he wrote to the friend who had sent him the seed (*CJB* 562). Another high point in the correspondence between Bartram and Collinson came at the end of May 1763, when John excitedly informed "dear Peter" that "My garden now makes A glorious appearance I have A fine anonis with A large spike of blew flowers in full bloom which I gathered in Potemack 3 years ago. . . . my great Carolina saracena is in full bloom. . . . it is a glorious odd flower A goldish color & striped" (*CJB* 594). A week later, Peter reported the extraordinary coincidence that his sarracenias,[25] along with his magnolias, were in glorious bloom, too: "I am in high Delight my two Mountain Magnolias are Pyramids of Flowers—almost the Extremity of Every Branch is a flower My Short & Long Leafed Saracena Growing Close together are both in Flower & make a Charming Contrast the One Red the other a Golden Hue Well mightest thou say, how fine they looked to see a Number together" (8 June 1763; *CJB* 594).

Notice Collinson's reference to the "Charming Contrast" between two of his plants in flower—something that, by all accounts, would have been a little less interesting to Bartram. The latter's own collection of flowers and trees, as a fellow naturalist from Charleston discovered, was a "perfect portraiture of himself," and this apparently not in a conventionally aesthetic sense. In John Bartram's garden, lowly weeds grew next to choice curiosities. During his visit to Philadelphia, the aptly named Dr. Alexander Garden discovered "a row of rare plants almost covered with weeds, . . . a beautiful shrub, even luxuriant among briars, and in another corner an elegant and lofty tree lost in a common thicket."[26] It would be wrong, however, to confuse this apparent lack of interest in beautiful arrangements with indifference to "beauty and pleasure."[27] The weeds Alexander Garden observed growing next to Bartram's rare plants seemed to prove that Bartram liked contrasts, too. But they were of a different kind.

This impression is reinforced by a stylized sketch of Bartram's garden and house, which was produced for Collinson by William Bartram in 1758 (ill. 1).[28] Gardens are, as Andrew Cunningham has said, "Nature re-arranged by artifice"; they represent "spaces delimited, areas which man has artificially enclosed."[29] The drawing of Bartram's garden duly emphasizes such separation and closure: the human figure walking among the trees signifies the pride of ownership, and a note scrawled along the left border of the garden draws attention to the fence, which runs "Northwest & southeast." Yet, through the inclusion and detailed depiction of what is outside the fence—boats on the river, a man fishing in the left foreground (intended as a visual counterpoint to the owner inside the garden?)—this drawing also suggests that the inside and the outside are linked, With its ordered, strenuously rigid lines of trees on the left (separated by "Walks 150 yards long") and the much larger plants, among them a wild iris in the lower right corner of the sketch, this representation of Bartram's garden seems like a parody of European notions of how plants should be arranged in a successful "princely garden" (which, as Bacon had recommended, should be divided into three discernible, clearly separated parts).[30]

If Bartram's and Collinson's letters often appeared more like a jumble of botanical notes and businesslike receipts than like coherent narratives, some main themes do recur. Often, specific flowers begin to assume the function of leitmotifs. For instance, if "ladies"—notably Collinson's and Bartram's wives—remained, with few exceptions,[31] absent from the men's correspondence, they at least made metaphorical appearances. From the beginning of their exchange, Collinson confessed to his infatuation with the different species of lady's slippers, all of which apparently made his "mouth Water" (*CJB* 18). On 3

1. "A Draught of John Bartram's House and Garden as it appears from River 1758." © Private Collection.

February 1735/36, he pleaded with Bartram: "These fine Lady's Slippers, don't let Escape, for they are my favorite plants. I have your yellow one that thrives well in my Garden but I much want the other sorts" (*CJB* 18).[32] He desperately wanted to own, as he wrote his friend the Mine year, "some of the fine Large Slippers thou showed Dr Witt" (*CJB* 28). Finally, in June 1740, he reported with audible relief that a recently arrived red lady's slipper had at last proved to be "a fine" plant. He praised its "very Curious flower," Which was so beautiful that Mr. Catesby had already painted it (*CJB* 135).[33] But soon we hear him complaining again. Another plant he had been given by Bartram thrived at first, but then did not flower, which Collinson believed was "owing to a Corona Solis that unluckily grows out of the Midst of It, which robs it of its nourishment, which I did not know when I planted it and now I can't remove it without Danger to both." Envious, he added that he was sure that once again Bartram's "Woods & Thickets are all Flowers" (*CJB* 158).[34]

And so it went. In June 1756, although a "fine yellow Calceolus Maria" was now again "in full Beauty" in his garden, we find Collinson still impatiently awaiting the appearance of the other, white-flowered species, "the Calceolus maria sent with the last plants" (*CJB* 406). Contemplating in November 1759 another lady's slipper that had not yet flourished, Collinson wrote to Bartram, this time more sadly than angrily: "It must be a Surprising fine Sight to See the White Calceolus near 3 foot high—but your Warmth & Soil greatly promotes Vegetation—my plant flattered Mee with 2 strong Stems but no flower—phaps next year may bring It" (CJB 475). What a shame that John's son "Billy," with his "inimitable Pencil," had not been on hand to portray "the Charming Calceolas." "Such a Rare thing which I Scarcely can hope to See, I then might have always seen It" (25 February 1760; *CJB* 482).[35] Suddenly, however, Collinson's "calceolus," this time a white one, perhaps the first that ever flowered in England, lived up to its owner's high expectations. Collinson was both overwhelmingly happy and humbly grateful: "I am Charm'd, nay in Extasie to See the White Calceolus Marina Thou sent Mee in flower" (10 June 1760; *CJB* 485). And then both the yellow and the white lady's slipper rewarded him splendidly for his efforts: "the Yellow Slipper is now in glorious flower five shoots from a Root & 2 flowers on a stalk & the white one just now peeping out of ground not half an inch High What singular difference in plants of the same tribe" (7 May 1761; *CJB* 513).

While Collinson celebrated his American lady's slippers, Bartram in Pennsylvania, with a similar mixture of rapture, anxiety, and anticipation, watched the fates of the plants sent to him from Europe: "Most of them come up ye ranunculus & anemony roots grows finely & several bore fine flowers ye flag iris grows well & two of ye bulbous is ready to flower many aconites & ye polianthus by hundreds." More than thirty carnations, all from Collinson's seeds, now flourished in his garden (24 June 1760; *CJB* 486). Confronted with such splendor, Bartram reached a decision: "Dear friend I am A going to build A greenhouse" (*CJB* 486). Collinson responded: "I will send thee seeds of Geraniums to furnish It" (15 September 1760; *CJB* 492). A year later, Bartram crowed that he could now "chalenge any garden in America for variety" (19 July 1761; *CJB* 529).

Interestingly enough, Bartram seems to have been aware that his garden, though situated in the "New World," was a work of art rather than of nature, a site of *transplantation,* an "enhancement" of nature and therefore as much an invention of "America" as Collinson's garden in England. To create it, Bartram had had to sever plants from their original "birthplaces" and place them in a new, artificial environment, next to other plants in whose company they would never otherwise have been. "Although I have mentioned so often ye naturall places of growth of many trees," he wrote to Collinson on 25 September 1740, "yet I have admired to see trees & some plants which I never observed to grow naturaly any where but in moist swampy mossy [land] & many times in ponds & runs of water. These plants he had brought out of "Jersey Virginia & several places in pensilvania," and he had planted them in his garden, where now "thay grow much better than in thair place of natural growth" (*CJB* 143). For all its natural wildness, noted by visitors such as Dr. Garden and Dr. Cutler, Bartram's garden thus was also a highly artificial, artistic composition, imbued with deeply personal significance, as is demonstrated by Bartram's obvious delight in the successful cultivation and preservation of his specimens and by his memories of how he first obtained them. Treating something as a collectable, according to Danet and Katriel, means "to take it out of its natural or original context and to create a new context for it, that of the collector's own life-space and the juxtaposition with other items in the collection."[36]

In Peckham and Mill Hill, Collinson designed his own little piece of America just as one of his wealthiest customers, Lord Petre, had attempted to reconstruct an American forest on his estate. Petre had planted "about Tenn thousand" American trees so that whoever walked among them could not "well help thinking He is in North American thickets." Was it not wonderful, mused Collinson, "to see how Nature is helped & imitated by art" (1 September 1741; *CJB* 167–168)? Bartram regretted that he could not render these and similar transatlantic experiments more fruitful by his personal assistance; European collectors like Collinson really needed someone in their gardens who understood American plants and "takes perticular care of them" (21 June 1743; *CJB* 216–217). But while Collinson was willing to grant Bartram the privilege of the native's point of view (no one, he admitted in January 1743, could "make such Discoveries as a Curious Man that is a Native"; *CJB* 229), it was clear that he ultimately preferred his own sanitized version of America to the "real thing." Collinson's aesthetic interest in American plants and animals never waned, but neither did the secret contempt he felt for Bartram's colony—which, after all, was still in its "Infancy" (*CJB* 93). "None care to Leave their Native Land," he told Bartram, after his friend had expressed a desire to acquire an assistant from England, "but those that from their bad principles & Morals cannot Live In It" (*CJB* 328). Collinson dreaded the Pennsylvanian winters that Bartram described to him and relished the comforts of living in an "Ever Green Country" (*CJB* 675) where flowers could be seen throughout the year. "The Difference is very remarkable between our Country & yours," he lectured in 1753, "for I have heard Thunder but once this year & that at a Distance whilst you have had it so Terrible all over yr Continent." The summer in England had been wet, but the harvest was good and the "Autumn Long & fine." He had gathered "Such a Nosegay on Xmas day": Bartram would have been delighted to see it. In England, to be sure, "vegitation may be said never to Cease for the Spring flowers tread so on the heels of the Autumn flowers that the Ring is Carried on without Intermission" (*CJB* 341).

Collinson was especially surprised to learn that what seemed beautiful in Europe could turn into a weed in America. In a list of "Introduced Plants Troublesome in Pennsylvania Pastures and Fields," prepared in early 1759, Bartram listed the "stinking yellow linarya" (the common toadflax, also known as butter-and-eggs) as the "most mischievous" among the plants imported from Europe (*CJB* 451). In the notes Bartram contributed to the 1751 edition of Thomas Short's *Medicina Britannica,* printed by Benjamin Franklin in America, he referred to the toadflax as "a troublesome, stinking" plant, "called by our People . . . Dog Piss Weed," which was "no Native, but we can never I believe eradicate it."[37] For Bartram, the linaria was beyond doubt the "most hurtfall plant in our pastures that can grow in our Northern climate" (*CJB* 451). Neither "the spade plow nor hoe" had been able the relentless spread of the plant over all of Pennsylvania. Since it was so annoyingly persistent, some people had rolled "great heaps of logs upon it & burnt them to ashes whereby ye earth was burnt half A foot deep yet it put up again as fresh as ever covering ye ground so close as not to let any grass to grow amongst it & ye cattle cant abide it." Thus, what had first been introduced as a fine garden flower had now become a plant "heartily cursed by those that sufers by its incroachment" (*CJB* 451). Collinson's response was laconic: "See what Climate & Soil does, the Yellow Linaria is no pest with us—I keep it in my Garden & it is very orderly, for the sake of its fine Spike of orange & yellow flowers" (20 July 1759; *CJB* 469–470).

Dionaea muscipula

The difference and distance between London and Kingsessing was nowhere more evident than in the delay with which the correspondents received, and thus were able to respond to, each other's letters. In November 1759, for example, Collinson, who had been the London agent for the Library Company of Philadelphia, recommended John Josselyn's *Account of Two Voyages to New-England* (1674) to his American correspondent. At Collinson's insistence Bartram had become a full member of Franklin's Library Company, but he could not find the book on its shelves. At least the librarian, Bartram reported back to Collinson several months later, *remembered* having seen it. "Then how happened It was not placed in the Library," sighed Collinson in September 1760, almost a year after he had first mentioned the book to his friend. He added, perhaps unfairly, "It is sad Management if they loose books as soon as they Have Them" (*CJB* 492).[38]

The worst strain on the two men's friendship was their own impatience, which seems all the more remarkable since, as gardeners, they should have been accustomed to long periods of waiting. Joseph Addison had found gardening to be a soothing pursuit ("It is naturally apt to fill the Mind with Calmness and Tranquillity, and to lay all its turbulent Passions at Rest")[39]—but then he wasn't thinking of gardeners communicating over a distance of several thousand miles. When plants went missing, Collinson and Bartram did not remain calm but tended to sulk, even though they were both well aware of the many imponderabilities of transatlantic travel. Not only shipments but entire ships were lost; rats chewed their way into the boxes and "piss'd" on the carefully packed specimens (10 June 1740; *CJB* 135). Last but not least, there was always the danger that sailors or other prying knowing people," upon "seeing these plants so carefully boxed up," might steal them "to give to their friends" (Collinson to Bartram, 20 January 1751; *CJB* 312). Needless to say, wherever liquor was used as a preservative, the risk grew exponentially: "Our London common sailors," groaned Collinson, "are a most profligate Crew." Wages were low and sailors were thirsty, as Collinson well knew: "The best sailor has but 25s a Month" (*CJB* 327).

The complaints exchanged between the two correspondents frequently concerned the nondelivery of the seeds or plants requested: "Oh why," wailed Collinson on 5 March 1750/51, "had I not a few acorns of the broad leafed Willow Okes found in the Jerseys. . . . besides to Mortifie Mee the More I had none of the broad leafed Magnolia" (*CJB* 318). In December 1751, he chided Bartram for providing only descriptions, not the necessary evidence itself: "It is all a Dead Letter without a View of the Real Subjects" (*CJB* 335). "No mention of any Accacia or Locust seed," he lamented in January 1759. "If none come, I shall almost fall to pieces" (*CJB* 456). And while he didn't disintegrate, Collinson certainly always remained "in sad distress for more Seeds." In his letter of 10 March 1759, he demanded to know why on earth Bartram was "tantalizing" him "about the Dwarfe Oaks." Other things, too, had been missing from the recent shipment. Collinson and his son were "Sadly Disappointed being in hopes of Seeing Some Grafts of the True New Town pippin but there was none." Bartram should definitely "remember another year—for what comes from you are Delicious Fruit" (*CJB* 459–460). Never again, begged Collinson in October 1762, let a "Letter pass without a Specimen" (*CJB* 571).

When it was Bartram's turn to grumble, he demanded to know why his faithless friend took so much time in acknowledging receipt of the boxes that he had sent him: "We have had four ships arived from london without bringing me any letter from thee" (20 July 1760; *CJB* 488). *Pace* Stearns, Bartram never was the humble recipient of Collinson's favors.[40] His letters—which could be as petulant, demanding, and self-promoting as they were warm and witty—often sounded like a veritable lover's complaint. "When I have endeavoured to give ye greatest satisfaction," he told Collinson on a particularly bad day, "my labours hath been ye least valued." Another veiled insult followed suit: "Indeed my good friend if thee was not A widower I should be inclined to tell thee that old age advanced as fast upon thee as upon myself" (25 September 1757; *CJB* 427).

Significantly, far from being "dead" when unaccompanied by plants, the letters exchanged between England and America soon gained a status similar to that of the seeds and plants the two friends exchanged. In fact, Bartram's and Collinson's letters often shared the fate of the natural specimens. "I have much to do to read thy Letter," moaned Collinson, "for some Mischievous Insect has Eaten thy Letter in large holes" (3 February 1741/42; *CJB* 180–181). But in spite of the knowledge of such dangers, there frequently was but little forgiveness, since the nagging doubt always remained that perhaps the expected letter—instead of having been lost—had simply not been written. "I have now but little to write haveing received no letter from thee since last fall" (*CJB* 273), sulked Bartram in April 1746—an accusation he reiterated almost a year later: "I have not received one letter from thee this long time" (30 January 1747/48; *CJB* 292). On 3 January 1758, he protested: "I now trouble thee with another letter tho I hardly know what to write haveing wrote 4 or 5 times last fall & this winter" (*CJB* 432).

Collinson generally had little sympathy for such complaints, pointing out that, although he was usually very busy, he always took time away from his work to tend to his correspondent's needs and therefore didn't deserve such treatment. "If My Friend John Bartram knew better My Affairs, my Situation in Life, My publick Business . . . He would wonder I do so well as I do." To serve his impatient friend, Collinson said, "I often Neglect my Own Affairs" (19 July 1753; *CJB* 348). But he, too, complained that he wasn't getting any mail. In May 1761, he reproached Bartram: "It is a long while since I heard from my Good friend John Bartram." Even his son was saying now, " 'Father, what is the matter friend John has quite forgot you who take so much pains to Dispose of his seeds'" (*CJB* 512). A few weeks later, he repeated his criticism and, in a self-canceling statement, claimed higher moral ground for himself: "I have no Letters from my Dear frd John. . . . I don't think I am forgotten, as my frd John is often apt to Imagine; (if no Letter comes),—I always make Allowance for accidents—of Ships being taken or Castaway" (12 June 1761; *CJB* 520). John's response is lost, but apparently he didn't take too kindly to his friend's admonishments, since in his next, agitated letter, dated 1 August 1761, Collinson pointed out that his "dear John" was always "in the same strain, of grumbling & Complaining." Bartram's "frequent censorious Temper," he felt, was "not becomeing our Friendship" (*CJB* 529–530).

Collinson willingly provided his American friend (who delighted "exceedingly in reading books of natural History or botany")[41] with reading material, but clearly he also feared that too much study would keep Bartram away from his collecting. After sending him Philip Miller's *Gardener's Dictionary,* Collinson backhandedly announced to Bartram that he would not have him "puzle" with other books: "Remember Solomons advice, in Reading of Books there is no End" (*CJB* 70). Bartram's answer from a country "still poorly furnished with such books" (*CJB* 197) reveals that he was indeed able to hold his own if need be. Facetiously he said that he had taken Collinson's "advice about books very kindly." But then he added, "I love reading such dearly and I believe if solomon had loved women less & books more he would have been a wiser & happier man then he was" (May 1738; *CJB* 89).

In his letters, Collinson excelled in puns and alliterative play: "I have pleasure upon Pleasure beyond Measure with Peruesing my Dear Johns Letters" (*CJB* 591). He delighted in atmospherically charged descriptions: "I am here retir'd all alone, from the Bustle & hurry of the Town Meditateing on the Comforts I enjoy; and whilst the old Log is Burning the fire of friendship is blazeing" (6 December 1763; *CJB* 615). Strutting on the stage of his own obviously superior rhetoric, Collinson took pride in being a more accomplished writer than Bartram, criticizing the aptness of his correspondent's descriptions as well as, on a more mundane level, his spelling whenever he saw fit. Bartram, who liked to pose as a "plain countrey fellow [who] is for useing freedom & sincerity" (*CJB* 384), remained unperturbed. For him, language was a means to an end rather than an end in itself: "Good grammar & good spelling may please those that are more taken with A fine superficial flourish then real truth." His chief aim had always been, said Bartram, "to inform my readers of ye true real distinguishing characters of each genus & where & how each species differed from one another of ye same genus." If Collinson did not like what he had received and found "that my discriptions is not agreeable with ye specimens," he should send them back immediately "that I may correct them . . . for I have forgot what I wrote" (3 November 1754; *CJB* 374–375).

His own disclaimers notwithstanding, John's epistolary language on occasion soared, too. In April 1755, he was particularly pleased with Collinson's "relation of thy fine curious flowers in thy garden" and, praising his friend's "sweet disposition," rewarded him with a passage foreshadowing his son's fascination with waterfalls and cascades in *Travels* (1791). Bartram said he wished Collinson would keep him company

> in ye grand & spacious temple amongst ye lofty chains of mountains ye craggy precipices of elevated rock embellished with various shrubs pliant evergreens out of thair uneven surfaces & ye gloomy shaded vails . . . ye purling streams & glittering cascades ye level plains ye concealed humid bosoms discharging numerous pearly drops perpetualy trickling down or ye shore of ye mighty ocean ye great metropolis of ye finy tribe where we view ye rolling waves dashing against ye shore & breaking into steam rising up in vapour which is colected in humid fleeces in ye vast expance. (*CJB* 381)

Despite their irritation over missing specimens and lost letters, the intimacy between the two collectors, which the language of gardening and flowers allowed them to cultivate, was extraordinary.[42] "I can take a squib from John Bartram without the Least resentment," conceded Collinson. "Friends may . . . rally one another when it is not done in Anger" (31 July 1767; *CJB* 684). And both friends used plants not only to torment themselves but also to mock and even metaphorically tickle each other. A curious, insect-trapping plant in Bartram's garden, called the "Tipitiwitchet," became a comical metaphor for the two correspondents' own "ticklishness."[43] Significantly, the fly-trapping *Dionaea muscipula* later also greatly preoccupied John's son William, just as it would impress Charles Darwin in his work on insectivorous plants. (Darwin, who was intrigued by the "extreme sensitiveness to touch" of the *Dionaea*'s spiked leaves, called it "one of the most wonderful" plants "in the world.")[44]

In a recent essay, Daniel McKinley relishes the salacious references to the Tipitiwitchet in Collinson's and Bartram's correspondence and claims that the two men were obsessed with this rare collector's item because its leaves so strongly resembled the female sexual organ, an impression captured onomatopoetically in the strange name, which McKinley believes Bartram must have coined himself. The shared jokes about the Tipitiwitchet, as a metaphor for feminine sexuality, might indeed have helped Bartram and Collinson reassert their masculinity as collectors and to prove that—in the words from the preface to Robert Thornton's *New Illustration of the Sexual System of Carolus von Linnaeus* (1799)—botany was indeed a "manly puzzle" rather than the trifling pastime of women.[45] However, in an age that ascribed sensitivity, even souls, to plants,[46] another aspect of the Tipitiwitchet was also important to Collinson and Bartram. In December 1767, Bartram wrote to Benjamin Rush, a professor at the University of Pennsylvania Medical School, that he was "much pleased" with Rush's discovery of nervous sensation in plants, "which I hope by A more diligent search will lead you into ye knowledge of more certain truths then all ye pretended Revelations of our Mistrey mongers & thair inspirations." It seemed certain now that many animals were "endowed with most of our faculties & pashions & several perticular intelect beyond many of our species," but about the feelings of plants only little was yet known (*CJB* 690). If Crèvecoeur later wrote, in *Letters from an American Farmer*, that men were like plants, the unstated premise was, of course, that, vice versa, plants could act and react as if they were human beings.[47] Indeed, as one important instance of the effect of human-like passions, of the "appearance of A wonderfull sensation" in the vegetable world, John Bartram in his letter to Rush singled out "ye Tipiwitchit of Carolina." Years later, William Bartram would celebrate the Venus flytrap precisely because it seemed to represent a blurring of the boundaries between the vegetable and the animal world. Flowers were alive, too: "Can we after viewing this object," wrote William, who included it in one of his most impressive drawings, "hesitate a moment to confess, that vegetable beings are endued with some sensible faculties or attributes, familiar to those that dignify animal nature?" (*T* 19).[48]

The *Dionaea* was first mentioned to Collinson by Governor Dobbs of North Carolina: "we have," he wrote on 2 April 1759 from near Brunswick, North Carolina, "a Kind of Catch Fly Sensitive which closes upon anything that touches it."[49] In a letter to Bartram written in July 1762, Collinson demanded to see specimens of this "Wagish plant,—as Wagishly Described" (*CJB* 565). Months later, he again reminded Bartram how "impatient" he was to see for himself what the Tipitiwitchet was really like, this plant about which he had heard so much and which he devoutly wished dear Billy Bartram would soon sketch for him (*CJB* 571). After Bartram informed Collinson that the Tipitiwitchet had made quite an impression on a French visitor who, upon beholding it in his garden, couldn't stop laughing (29 August 1762; *CJB* 570), Collinson's testy response continued the innuendo begun in his friend's letter: "Whilst the French Man was ready to Burst with Laughing I am ready to Burst with Desire for Root, Seed, or Specimen of the Wagish Tipitiwitchet Sensitive." If Collinson did not find a specimen with Bartram's next letter, his friend should never dare write him again: "It is Cruel to tantalize Mee with relations & not to send Mee a Little Specimen, in thine of the 15 of Augst nor in thine of the 29th—it shows thou hast no sympathy or Compassion for a Virtuoso—I wish it was in my power to Mortifie thee as much" (10 December 1762; *CJB* 580).

In January 1763, the "virtuoso" Collinson, to his immense gratification, finally received his precious plant and, against his own normal impulse to hoard what was rare and hold on to what was beautiful, even passed on a sample to Linnaeus in Sweden, making him part of the circle of desirous friends: "O, Botany Delightfullest of all Sciences there is no End of thy Gratifications—All Botanists will Joyn with Mee in thanking my Dear John for his unwearied Pains to Gratifie every Inquisitive Genius—I have sent Linnaeus a Specimen & one Leafe of Tipitiwitchet-Sensitive—Only to Him, would I spare Such a Jewel—pray Send More Specimens I am afraid Wee can never Raise It—Linnaeus will be In raptures at the Sight of It" (30 June 1763; *CJB* 600). Collinson's worst fears were soon confirmed: the nurseryman James Gordon did not succeed in raising even a single one of "the Many Vegitable Wonders" from the seeds John Bartram had sent him. "With all his skill," sobbed Collinson, he "cannot bring it to light" (*CJB* 631). But Bartram himself did not fare better with what Linnaeus (or one of his pupils) had termed a *Miraculum naturae*. "Ye tipitywitchet is very difficult to raise or keep mine that I thought intirely killed last winter is revived but is so small & weak as hardly to [be] sensible of being tickled" (23 September 1764; *CJB* 638).[50] Although both collectors ultimately failed to keep the "wagish" plant alive, either in England or in America, it still served to define their relationship, to the extent of excluding others now perceived to be less sensitive—among them Arthur Dobbs, the Tipitiwitchet's discoverer, who had recently, at the age of seventy-three, married again. It would be in vain, wrote Collinson nastily, to ask Dobbs "for seeds or plants of Tipitiwitchit now He has gott one of his Own to play with" (30 June 1764; *CJB* 633).

2. "Venus's Fly-trap, *Dionaea muscipula*." Engraving by I. H. Seymour from Barton, *Elements of Botany* (1803), pl. 7. Courtesy, The Lilly Library, Indiana University, Bloomington.

The unusual friendship of John Bartram and Peter Collinson ended abruptly only a few years later, in a way that forcefully reminded Bartram of the distance between England and North America. "Wee have a very Melancholy Time," Collinson had written to Bartram on 6 July 1768; "Rain more or Less at least Every Day if it continues I don't know what will become of Us" (*CJB* 706). A few weeks later he was dead. Bartram heard of his "dear friend's" demise only months later, as he mourned in a letter to Benjamin Franklin on 5 November 1768: "Now I hear Peter is dead and I have not received a line from his son or any of my other Correspondents since" (*CJB* 708). John Bartram himself passed away nine years later, in September 1777. By that time, the "sportive vegetable" (William Bartram) had already developed a life of its own, both in England, where the naturalist John Ellis had become its champion, and in America, where William Bartram's friend Benjamin Smith Barton reproduced Ellis's illustration of the *Dionaea muscipula* in his *Elements of Botany*, the first botanical textbook published in the United States (ill. 2). He knew, said Barton, of no other plant "as limited in its range." As far as he could see, it was almost exclusively confined "to an inconsiderable tract of boggy country in the state of North-Carolina, about Wilmington."[51] In the correspondence of John Bartram and Peter Collison, the *Dionaea muscipula* had, for the first time, transcended the limitations of its boggy origin and become, briefly, quite "considerable."

William Bartram's Gallery of Plants

In the 1750s, a significant portion of John Bartram's correspondence showed a concern that, at least superficially, had nothing to do with the "Delightfullest of all Sciences," which had for so long provided a link between himself and his friend Collinson. Or so it seemed. John needed Peter's help, since he was worried about the most promising yet problematic of his eleven children—his son William, who was fond of sketching plants but little else. Billy, as his father usually called him, and his twin sister, Elizabeth, were born on 9 April 1739 at Kingsessing. As a youth, he soon showed a distinct talent for drawing but, alas, no real interest in any other activities that might gain him a "handsome livelihood." In September 1755, John Bartram wrote to Collinson, "Botany & drawing are his darling delight," adding that he was afraid that his son couldn't "settle to any business also." Surveying, the father mused, would at least give him the chance to pursue his botanical inclinations, but then, remembered John, didn't we already have "five times more surveyors . . . then can get half employ" (*CJB* 387)? Wearily John suggested several other possibilities to his reluctant son, among them an apprenticeship with a doctor, a project that was quickly abandoned. The despairing father finally complained to Alexander Garden in Charleston that, though there were "several" books on "Physic or Surgery" in his personal library, he couldn't "persuade" his son "to read A page of either" (*CJB* 403).

Collinson, ever the successful merchant, agreed with John that Billy Bartram ought to give up his "Darling amusements in some Degree, that he may Attain a Knowledge in some art, or Business by which He may with Care & Industry, Support himself in Life" (February 1756; *CJB* 394). He loved the younger Bartram's drawings ("There is a Delightfull natural freedom through the whole & no minute pticular omitted," *CJB* 393)—all the more so since many of them were excellent "surrogates for specimens" that he coveted but for various reasons could not possess.[52] Nevertheless, he had a better suggestion for Billy's professional future—printing: "It is a pretty Ingenious Imploy— never lett him reproach thee & say Father if thou had putt Mee to some Business by which I might gett my Bread I should have by my Industry Lived in Life as well as other people Lett the fault be his, not thine, if he Does nott" (20 January 1756; *CJB* 393). But the famous Benjamin Franklin advised against printing as a career path, pointing out that he himself "was ye only printer that did ever make A good livelyhood by it in this place" (*CJB* 404), and proposed engraving instead.

Billy Bartram finally tried his hand at storekeeping; from 1761 to 1765, he lived in the household of his uncle, the North Carolinian planter Colonel William Bartram. His father still felt there was little cause for celebration, however, especially when he compared Billy with his younger brother: "My John," he sighed in 1764, again in a letter addressed to Collinson, "is A worthy sober industrious son & delights in plants but I doubt Will[iam] he will be ruined in Carolina every thing goes wrong with him there" (*CJB* 622). What particularly bothered John was that William was a reluctant collector, too, as is indicated in a reproachful letter the father sent to his son in North Carolina in December 1761: "I dont want thee to hinder thy own affairs to oblige me but thee might easly gather A few seeds when thee need not . . . turn 20 yards out of thy way to pluck them" (*CJB* 543).

In 1765, thanks to Peter Collinson's efforts, George III made John Bartram his botanist to the now vastly extended realm of the colonies in North America. In February 1763, as a result of the treaty of Paris, Britain had gained Quebec, Florida, and all of North America east of the Mississippi: "See what a Complete Empire Wee have now got within ourselves," an exhilarated Collinson wrote to his American friend (*CJB* 582). Bartram, whose eyesight was failing, knew that he needed an assistant if he really wanted to explore the new Florida territory on the king's behalf. What better candidate was there than his own son, with his penchant for plant sketching? "Thee wrote to me last winter & seemed so very desirous to go there: now thee hath A fair opertunity so pray let me know as soon as possible." In fact, John even threatened that if Billy didn't answer soon he would find somebody else to go with him: "Pray let me know soon whether thee will come or not that I may provide myself with another Companion" (7 June 1765; *CJB* 652).

And Billy came, although this joint endeavor of father and son did not immediately resolve Billy's professional problems. After a brief unhappy interval in 1766 as the owner of a rice and indigo plantation on the St. Johns River (in "the least agreeable of all places that I have seen," one visitor reported to John Bartram; *CJB* 670) and the attempt to survive on what Collinson and others paid him for his drawings, William asked his father to finance another trip to Florida. John flatly refused: "I don't intend to have any more of my estate spent there or to ye southward upon any pretense whatever" (15 July 1722; *CJB* 749).

Help finally arrived from one of his father's correspondents, John Fothergill, a distinguished London physician and the "owner of a pretty large collection of valuable plants," in which North American plants flourished "exceedingly" well.[53] Fothergill himself gratefully traced his interest in plants back to Peter Collinson: "It was our Collinson who taught me to love plants," he wrote to Linnaeus. "He persuaded me to create a garden. Many items he gave me himself; others I collected as opportunity occurred." Collinson was the reason that Fothergill was now "burning with a love of plant life."[54] And it was Collinson who, shortly before his death in 1768, had recommended William Bartram's paintings to Fothergill, who admired them and thought "so fine a pencil . . . worthy of encouragement."[55] Though he confessed that his interests were less scientific than "selfish,"[56] Fothergill became to Billy what the late Peter Collinson had been to John. Through an agent, the Charleston physician Lionel Chalmers, he guaranteed Billy ("if his life be spared") "a good livelihood by sending boxes of plants and seeds to Europe" from the "less frequented parts of America" (October 1772; *CJB* 753). Fothergill also agreed to pay William Bartram the same amount his father had received from the king (fifty pounds) plus shipping expenses and a bonus for each painting he sent.

In March 1773, a few months before the Boston Tea Party, William Bartram left Philadelphia for Charleston. He spent most of the next two years traveling through coastal and northeastern Georgia (where he witnessed the treaty negotiations with the Creek Indians) and Florida, where he went up the St. Johns River. In 1775, he crossed northeastern Georgia into Cherokee country in northwestern South Carolina and the mountains of North Carolina. In July, he left Georgia for Alabama, traveling as far as the Mississippi, "the great sire of rivers" (*T* 346), which he reached in October 1775. William returned to Georgia in January 1776 and continued his exploratory trips until the fall, when he finally embarked on his overland trip back home. He arrived in Philadelphia in January 1777—only a few months before his father's death. During his southern sojourn, William Bartram had sent no fewer than 209 dried plants and fifty-nine zoological and botanical drawings, and he had also written a detailed report on the "outward furniture of Nature" in the South. This article, which extolled "the natural productions of these countries," was intended as much for Fothergill's "particular amusement" as for William Bartram's own benefit: "By doing this I shall not only have an opportunity of exercising the noble virtue of gratitude, but shall have an opportunity of knowing the merit of my labours" (*T* 480).[57] After William's return, his "Report to Dr. Fothergill" gradually grew into what has justly been called "the most astounding verbal artifact of the New Republic,"[58] *Travels through North & South Carolina, Georgia, East & West Florida, the Cherokee Country, the Extensive Territories of the Muscogulges or Creek Confederacy, and the Country of the Chactaws, Containing an Account of the Soil and Natural Productions of Those Regions; together with Observations on the Manners of Indians,* published in Philadelphia in 1791 and in London in 1792.

For Billy's father, writing had been merely a tool. In fact, John Bartram once remarked to Franklin about one of his own travel journals, perhaps a little defensively, that while there was some "satisfaction" to be gained from it, those readers looking for descriptions of the "artificial curiosities" would inevitably be disappointed (10 April 1769; *CJB* 709–710). (Think of the temples, pyramids, bridges, obelisks normally found in "our modern travailers Journals"!) Franklin's response was characteristic. If he could find "in any Italian Travels a Receipt for making Parmesan Cheese," such unexpected knowledge would give him "more Satisfaction than a Transcript of any Inscription from any old Stone whatever." Franklin emphatically recommended that his despondent friend sit down quietly "at home, digest the Knowledge you [have] acquired, compile and publish the many Observations you have made, and point out the Advantages that may be drawn from the whole" (9 July 1769; *CJB* 713–714). Sit down John Bartram never did. However, William Bartram (of whose spelling competence Peter Collinson, the vociferous critic of the elder Bartram's writing skills, had taken an equally dim view)[59] completed what his father had apparently never even contemplated: a polished, comprehensive, book-length account of his natural history explorations in the South. On August 1791, more than fifteen years after he had returned from his southern sojourn, William Bartram's *Travels* was deposited for copyright in Philadelphia. And while John Bartram, the inveterate defender of plain, serviceable prose, cheerfully unconcerned about "good grammer & good spelling," had wanted his descriptions to be exact and nothing else, his son's goals, judging by the curious result, were remarkably different.

Many scholars have speculated as to why it took William so long to finish the text of his *Travels,* especially since there is evidence that as early as 1783 he already kept some kind of a manuscript on hand that he would show to visitors.[60] Did his accident delay him? (In 1786, William fell nearly twenty feet out of a cypress tree while collecting seeds in his garden and fractured his right leg, permanently incapacitating it.) Or had he decided to wait until botanists in England had come up with more conclusive classifications of the specimens he had sent to Fothergill? Or was it, after all, his desire to produce fine writing that kept him from completing his book? Benjamin Smith Barton had assured him that there would be a popular market for his narrative ("*Natural History and Botany* are the fashionable and favorite studies of the polite as well as of the learned" in Europe), and it was to Barton, with his high expectations, that William Bartram despairingly complained in 1787 that the work was still no more than an "improper Embryo."[61]

Somewhat "improper" Bartram's *Travels* still remained even after it had matured into a form that seemed fit to print. Some contemporary reviewers expressed concern about the author's "rhapsodical effusions" and his "very incorrect" and "disgustingly pompous style," which appeared to smother the plethora of interesting new botanical facts he had to offer.[62] Modern commentators have alerted us to the unreliability of William Bartram's rather cavalier chronology, his carelessness even when he *did* supply dates.[63] They know that Bartram's facts are no longer "of much value today," but they don't care. Instead, they quote Thomas Carlyle's praise for Bartram's "wondrous kind of floundering eloquence," invite us to sample the author's "breathless run of prose," and commend Bartram for his insights into Native American cultures.[64] But it seems that even today there can be no real consensus on Bartram's only book. If John Seelye, L. Hugh Moore, and Charles H. Adams see *Travels* as "a work of art,"[65] a critic like Pamela Regis, for instance, believes that it is first and foremost a "work of science."[66]

Travels has retained its power to confuse readers.[67] And for good reasons, too. Bartram's strange book is a ragbag of different and apparently unreconcilable genres. Many sentences, in their staccato rawness, remind the reader of the diary sketches or, in more modern parlance, the collection of field notes from which they originated. "Arrived early in the evening at the Halfway pond" (*T* 156); "Crossed a delightful river, the main branch of Tugilo, when I began to ascend again" (*T* 278); "Crossed Rock-fish, a large branch of the North West" (*T* 383). Long lists of natural specimens (which John Seelye, for example, finds boring) seem to drift off into tautological irrelevance, ending with phrases such as "with many others already mentioned" (*T* 167) or "and others already noticed" (*T* 171). The easy flow of sentences is sometimes interrupted and encumbered by plodding taxonomic detail: "These clusters of flowers, at a distance, look like a large Carnation or fringed Poppy flower (Syngenesia Polyg. Æqul. Linn.), Cacalia heterophylla, foliis cuneiformibus, carnosis, papil. viscidis" (*T* 149). Straining to incorporate into English the technical terminology Linnaeus had developed, Bartram's language bulges with strange, cumbersome words. His flowers are "polypetalous" or "infundibuliform"; they have "triquetrous pericarpi"; their seed is "cuneiform"; their leaves are "lanceolate," "serrated," "sessile," or "pinnatifid" or terminate "with a subulated point."

Travels seems to be first and foremost a natural history collector's account, not a tidy narrative. It is sprinkled with references to the serious but serendipitous activity that provides the thematic basis of the book: "I made ample collections of specimens and growing roots of curious vegetables" (*T* 141); "I had an opportunity this day of collecting a variety of specimens and seeds of vegetables" (*T* 211); "I tarried two or three days, employed in augmenting my collections of specimens" (*T* 302). Collections invite proliferation, not concentration, and one might argue that Bartram's text should not be expected to deliver what it is not intended to provide.

However, there are many passages in Bartram's text that contradict such a reading and suggest that Bartram *wanted* to supply us with more than a dignified botanical shopping list. His careless catalogues are offset by many passages wrought of beautifully crafted sentences replete with evocative visual detail and alliterative sonority: "Behold . . . the pendant golden Orange dancing on the surface of the pellucid waters, the balmy air vibrating with the melody of the merry birds, tenants of the encircling aromatic grove" (*T* 150).[68] The "infundibuliform" flower of the hibiscus, upon closer inspection, sports a charmingly poetic "deep crimson eye" (*T* 104). Frequently, the various modes of discourse in *Travels*—taxonomic classification, botanical description, ethnographic field note, narrative report, or lyrical vignette—are effortlessly incorporated into the same sentence. A case in point is Bartram's intricate description of one of his favorite trees, the "Magnolia auriculata" (the mountain magnolia), in which seemingly opposed modes of representation smoothly overlap:

> This tree, or perhaps rather shrub, rises eighteen to thirty feet in height; there are usually many stems from a root or source, which lean a little, or slightly diverge from each other. . . . The crooked wreathing branches arising and subdividing from the main stem without order or uniformity, their extremities turn upwards, producing a very large rosaceous, perfectly white, double or polypetalous flower, which is of a most fragrant scent; this fine flower sits in the centre of a radius of very large leaves, which are of a singular figure, somewhat lanceolate, but broad towards their extremities, terminating with an acuminated point, and backwards they attenuate and become very narrow towards their bases, terminating that way with two long, narrow ears or lappets, one on each side of the insertion of the petiole; the leaves have only short footstalks, sitting very near each other, at the extremities of the floriferous branches, from whence they spread themselves after a regular order, like the spokes of a wheel, their margins touching or lightly lapping upon each other, form an expansive umbrella superbly crowned or crested with the fragrant flower, representing a white plume; the blossom is succeeded by

a very large crimson cone or strobile, containing a great number of scarlet berries, which, when ripe, spring from their cells and are for a time suspended by a white silky web or thread. (*T* 278–279)

Technical precision (indicated by the use of the requisite arcane vocabulary), hyperbolic imprecision ("very large"; "very narrow"; "very near"), and the gentle evocativeness of visual and olfactory impressions ("perfectly white"; "most fragrant") work together to create the final image of the leaves looking like an umbrella crowned with the flower's "plume." The many participles ("producing"; "terminating"; "touching"; "lapping"; "representing"; "containing") increase the curiously mixed sense of action and temporal suspension, of *dynamic stasis,* which we derive from Bartram's portrayal of this unique tree. Bartram's verbal fussiness ("tree, *or* perhaps *rather* shrub"; "lean a little, or slightly diverge"; "narrow ears *or* lappets"; "crowned *or* crested"; "cone *or* strobile"; "silky web *or* thread") contrasts markedly with the nimble playfulness he attributes to the plant's expansive yet "lightly lapping" leaves. The final reference to the bursting forth of the fruit comes as a surprise—and, while the necessity of the intrusion of time into the self-enclosed world of the beautiful flower isn't denied, the thread of silk from which the tree's fruit hangs seems metaphorically to halt the process, if only momentarily.

In William Bartram's rendering, his travels through the American South are not presented as a chronologically complete, unflinchingly systematic progress toward greater insight. Instead, they seem like a parade of equally surprising sights.[69] The suspension of the temporal by the visual is neatly illustrated by Bartram's description of a beautiful pyramid-shaped plant he discovers while sailing up the Tensaw River, the *Oenothera grandiflora.* Bartram's evening primrose opens its flowers in the dusk and withers when the sunshine returns. But the flowers that fade away are quickly replaced by new ones, and thus the plant "at the same instant" exhibits several hundred flowers, in what appears to be a ceaseless pageant of floral beauty:

I was struck with surprize at the appearance of a blooming plant, gilded with the richest golden yellow: stepping on shore, I discovered it to be a new species of the Oenothera. . . . The large expanded flowers, that so ornament this plant, are of a splendid perfect yellow colour; but when they contract again, before they drop off, the underside of the petals next the calyx becomes of a reddish flesh colour, inclining to vermillion; the flowers begin to open in the evening, are fully expanded during the night, and are in their beauty next morning, but close and wither before noon. Their [*sic*] is a daily profuse succession for many weeks, and one single plant at the same instant presents to view many hundred flowers. (*T* 330)

Bartram's multilayered discourse, instead of simply anthropomorphizing the plants it features (which is what Erasmus Darwin had done in *The Loves of Plants*), frequently allows them to retain their strange, spectacular, surprising beauty, their "singular pleasing wildness and freedom" (*T* 275). The temporal linearity of the journey is superseded by the immediacy of the visual impression; together with Bartram, the reader stands "struck with a kind of awe" before a tree, a cypress, for example, and admires the "stateliness of the trunk," which lifts its "cumbrous top towards the skies . . . casting a wide shade upon the ground, as a dark intervening cloud, which, for a time, excludes the rays of the sun" (*T* 93–94).

The structure of many chapters in Bartram's *Travels* is governed by such awe-inspiring, "striking" surprises, such "chance" finds that delight the heart of the diligent collector ("And now appeared in sight, a tree that claimed my whole attention"; *T* 123). While some passages of *Travels* therefore evince what some readers might regard as a considerable lack of focus, with accident instead of purposeful design dictating the choice and change of topics, other and usually smaller segments of Bartram's book are marked by the careful construction of narrative intensity. At their most effective, these passages perform a sudden change from the reporting of past actions to the immediate enactment of the present moment. They unabashedly ask for the reader's participation in the gradually unfolding drama of the traveler's experience. This is the case, for example, in Bartram's attempt to re-create the paralyzing fear he felt when confronted with a thunderstorm on his journey into Creek Territory: "Now the earth trembles under the peals of incessant distant thunder. . . . I raise my head and rub open my eyes, pained with gleams and flashes of lightning. . . . I am instantly struck dumb" (*T* 315). A more dramatic example is his account of fighting alligators ("crocodiles"), rendered particularly effective by his calculated use of parataxis, ellipsis, and syntactic inversion. The rhythm of Bartram's prose evokes the strange rhythm of the heavy animals' terrifying battle, in the course of which they sink, resurface, and sink again:

Behold him rushing forth from the flags and reeds. His enormous body swells. His plaited tail brandished high, floats upon the lake. The waters like a cataract descend from his opening jaws. Clouds of smoke issue from his dilated nostrils. The earth trembles with his thunder. When immediately from the opposite coast of the lagoon, emerges from the deep his rival champion. They suddenly dart upon each other. . . . They now sink to the bottom folded together in horrid wreaths. The water becomes thick and discoloured. Again they rise, their jaws clap together, reechoing through the deep surrounding forests. Again they sink, when the contest ends at the muddy bottom of the lake. (*T* 114)

The logic of Bartram's narrative is visual rather than chronological; its structure is that of a sequence of "scenes" or "animated pictures" (*T* 187). Drawing has indeed remained Billy Bartram's "darling delight." In *Travels,* the landscape unfolds around the human observer as if she were seated in a vast diorama watching a procession of sights, a pageant performed by Nature herself. "Now at once opens to view," Bartram muses while surveying the Alachua Savanna in Florida,

> the most extensive Cane-break that is to be seen on the face of the whole earth; right forward, about south-west, there appears no bound but the skies, the level plain, like the ocean, uniting with the firmament, and on the right and left hand, dark shaded groves, old fields, and high forests. . . .
>
> The alternate bold promontories and misty points advancing and retiring, at length, as it were, insensibly vanishing from sight, like the two points of a crescent, softly touching the horizon, represent the most magnificent amphitheatre or circus perhaps in the whole world. (*T* 200)

Together with Bartram, the reader gradually takes in this animated exhibition, this slowly moving museum of American nature's collected beauties. Not coincidentally, Bartram's text abounds with visual verbs such as "observe," "behold," "view," "have in view," "stand in view," "open to view," "present," and "present to view."

Significantly, Bartram's wonderful theater of vision unfolds on the microscopic as well as the macroscopic level. Bartram is equally adept at bringing into focus the small dewdrops on the buds of the passionflower and the dramatically ascending banks of the mighty Mississippi River in Louisiana:

> What a beautiful display of vegetation is here before me! seemingly unlimited in extent and variety: how the dew-drops twinkle and play upon the sight. . . . See the pearly tears rolling off the buds of the expanding Granadilla; behold the azure fields of the cerulean Ixea! . . . How fantastical looks the libertine Clitoria. (*T* 141)

> At evening arrived at Manchac, when I directed my steps to the banks of the Missisipi, where I stood for a time as it were fascinated by the magnificence of the great sire of rivers.

The depth of the river here, even in this season, at its lowest ebb, is astonishing, not less than forty fathoms; and the width about a mile or somewhat less; but it is not the expansion of surface alone that strikes us with ideas of magnificence; the altitude and *theatrical* ascents of its pensile banks, the steady course of the mighty flood, the trees, high forests . . . all unite or combine in exhibiting a prospect of the grand sublime. (*T* 346; my emphasis)

With special delight Bartram admires the translucency or, to use one of his favorite words, the "pellucidity" of the rivers he travels, some of which are "almost as transparent as the air we breathe" and transmit "distinctly the natural form and appearance of the objects" that are moving in the water or resting on their "silvery bed[s]" (*T* 197, 193). Once, upon seeing tribes of fish descend into and then reemerge from the crystal-clear waters of a fountain on an island in Lake George, Bartram finds himself faced with a wonderful trompe l'oeil painting. Clear sight, transparency, can itself become a deception, but it is a pleasing one: "This amazing and delightful scene, though real, appears at first but as a piece of excellent painting; there seems no medium; you imagine the picture to be within a few inches of your eyes, and that you may without the least difficulty touch any one of the fish, or put your finger upon the crocodile's eye, when it really is twenty or thirty feet under water" (*T* 151). The reader becomes a museum visitor; putting her finger on the eye of the alligator, she relishes the illusion of proximity and tangibility suggested by specimens that appear to be "within a few inches" of her eyes yet are "really" beyond her reach, "twenty or thirty feet" away.

On his journey through Alabama in 1775, Bartram unexpectedly fell ill, and it seems symbolically important that this "malady" (possibly poison-ivy infection or scarlet fever) dramatically affected his vision. The disorder, recalls Bartram, "soon settled in my eyes . . . causing a most painful defluxion of pellucid, corrosive water" (*T* 339). By the time he arrives at the Pearl River, the border of present-day Mississippi and Louisiana, the excruciating pain has rendered him incapable of making observations, deprives him of sleep, and causes his eyes to discharge painful rivers of rheum: "The corroding water, every few minutes, streaming from my eyes, had stripped the skin off my face, in the same manner as scalding water would have done" (*T* 340). His body now seems to him "but as a light shadow," and he begins to doubt "its reality" (*T* 341). Only after several weeks is Bartram able to open his eyes, finding that his left eye is especially injured as it "suffered the greatest pain and weight of the disease" (*T* 341). The severe pain wrecked his plans for a "pilgrimage South-Westward" (*T* 352), inducing him to return, prematurely, to Alabama.

Franklinia alatamaha

It is interesting that in Bartram's *Travels,* a book whose overflowing pages are populated with exquisite examples taken from nature's bountiful vegetable kingdom, only one plant has an entire (if short) chapter devoted to it (part 3, ch. 9). And this plant seems a perfect illustration of the "no two alike" principle, which Brenda Danet and Tamar Katriel have identified as central to a collector's ambition. The Franklin tree, or *Franklinia alatamaha,* has been deemed the "outstanding discovery of the Bartrams."[70] In William's inventory of new plants it certainly occupies a special place as the ultimate collector's dream, a precious curiosity that ironically, almost from the moment it was discovered and gathered, had ceased to be available where it grew naturally. As we shall see, William Bartram's scrupulous treatment of the *Franklinia* contradicts the scathing indictment of Linnean taxonomy as the handmaiden of political orthodoxy in early American republicanism that has been presented by Christopher Looby, who also claims that Bartram saw plants and animals less in their singularity and individuality than in their abstract, general importance.[71]

On 1 October 1765, John and seventeen-year-old William, in swampy territory along the Altamaha River in Georgia, not far from Fort Barrington, chanced upon "severall very curious shrubs." One of these, John scribbled in his journal, was a large upright plant they had never seen anywhere else. In Francis Harper's transcription, the passage reads:

> Fine clear cool morning. . . . this days rideing was very bad thro bay swamps, tupelos both sorts, & Cypress in deep water, some of which on ye borders was very full of brush and bryers, yet got safe through all. . . . we saw several dear, 2 or 3 together both young & ould, severall turkeys. . . . dined by A swamp on bread & A pomegranate, near which growed much cana indica. . . . we mised our way & fell 4 mile below fort barrington, where we lodged this night, this day we found severall very curious shrubs, one bearing beautiful good fruite.[72]

Alas, Bartram père and fils couldn't classify the shrub with "the beautiful good fruite," because so late in the year it was no longer in flower. Sometime between 1773 and 1776, however, William returned to the Altamaha River, hoping to rediscover the "curious shrub." The exact date of his visit remains a matter of academic dispute,[73] but whenever it was that William came back, it was then that he found "two or three acres" covered with these fascinating shrubs, fifteen to twenty feet tall, in bloom as well as bearing fruit. Closer inspection convinced him that this tree deserved to be placed in a family of its own. He secured propagating material for himself and for shipment to Fothergill and eventually proposed naming it after the man who had, with his father, founded the American Philosophical Society, "the illustrious Dr. Benjamin Franklin." The location he added as the plant's trivial name. Significantly, *altamaha* is a word of Indian or, more specifically, Timucua origin. As baptized by Bartram, the newly discovered plant thus combined in its name references to the American present and to the continent's more remote pre-European past.

Between the time of his return to Philadelphia in January 1777 and the publication of *Travels* in 1791, Bartram occasionally mentioned the shrub with the "beautiful good fruite." For instance, he included it—as one of "Three Undescript Shrubs Lately from Florida"—in a sales catalogue of the American trees growing in his garden.[74] *Franklinia* also appeared in a list of plants Bartram sent in October 1784 to "Mr. Pierepont" in Europe. But the tree made its first official appearance only in 1785, when John Bartram's cousin Humphry Marshall, proud owner of a garden in Chester County, Alabama, published the first book on American plants that was written entirely by an American-born naturalist and printed in the United States. Marshall's *Arbustum Americanum: The American Grove*[75] soon became a kind of "wish list" for European foresters and nurserymen.[76]

If William Stork's *Account of East-Florida* had still flaunted a paternalistically English perspective on the colonies' flora, Marshall's book now sounded a consciously American note. Acknowledging in his preface that the "productions of the Vegetable Kingdom" afforded the "principal necessaries, conveniences, and luxuries of life," Marshall fulsomely celebrated the "Science of Botany" as "that Branch of natural History which teaches the right knowledge of Vegetables, and their application to the most beneficial uses." Botany, wrote Marshall, was worthy of "the attention of every patriotic and liberal mind" and deserved "a place among the first of useful pursuits." He lamented the money spent on "purchasing Foreign Teas, Drugs, [and] Dye-stuffs" and suggested that patriotic minds devote their energies, first, to the cultivation of useful foreign plants in the United States and, second, to what he called "*the discovering the qualities and uses of our own native Vegetable productions.*" In this respect, the "luxuriant unexplored territory" in North America held particular promise, since it was there that, still unbeknownst to Americans, plants that could serve as substitutes for the potato or for tea and coffee were flourishing. His *Arbustum Americanum*, conceded Marshall, could prove to be of service to the "*foreigner*, curious in American collections," as it would enable him to take his pick from whatever seemed "suitable to his own particular fancy." For those interested in cultivating timber "for œconomical purposes," Marshall recommended "our valuable Forest Trees"; to those intent on adorning their gardens, he sang the praises of "our different ornamental flowering shrubs." Primarily, however, his little book was intended to whet the appetites of his "*countrymen*" with concise and accessible descriptions "of their own native Forest Trees and Shrubs, as far as hitherto discovered."[77]

It is in this context that we need to read the first published description of the *Franklinia alatamaha,* which Marshall had composed on the basis of William Bartram's field notes.[78] In good Linnean fashion, Bartram's original description (which accompanied a drawing of the plant he sent to the English collector Robert Barclay) first named the class and the order to which the *Franklinia* belongs: Monadelphia Polyandria.[79] The name indicates that the stamens are in the same flower as the pistil, that they are united with each other by their filaments, and that in one flower there are many stamens. Bartram went on to offer a more detailed list of reasons for thinking this new tree deserved the collector's special attention. The *Franklinia* was, he wrote, "circiter 20 pedes altus; rami alterni, Fl. alterna, oblonga, versus basin attenuata sessilia serrata, Flores terminales in axilis foliorum, sessiles magni nivea Staminum corona Staminarum corona fulgida." In Marshall's rendering, the passage reads:

> This beautiful, flowering tree-like shrub, rises with an erect trunk to the height of about twenty feet; dividing into branches, alternately disposed. The leaves are oblong, narrowed towards the base, sawed on their edges, placed alternately, and sitting close to the branches. The flowers are produced towards the extremity of the branches, sifting close at the bosom of the leaves; they are often five inches in diameter when fully expanded; composed of five large, roundish, spreading petals, ornamented in the center with a tuft or crown of gold coloured stamina; and possessed with the fragrance of a China Orange.

Marshall augments Bartram's description by remarking that, after the plant's first discovery by John Bartram, years passed before William revisited the location "where it had been before observed, and had the pleasing prospect of beholding it in its native soil, possessed with all its floral charms." Bartram collected the seeds and raised several plants from them in his own garden, "which in four years time flowered, and in one year after perfected ripe seeds."[80] A rare tree with the "fragrance of a China Orange," endowed with astonishing "floral charms"—one can imagine how the mouths of those foreigners mentioned in Marshall's preface, "curious in American collections," would have watered.

William Bartram's own version in *Travels,* written years later, deftly transforms Marshall's promotional entry into a touching narrative about the singularity of American nature: "After my return from the Creek nation," begins the chapter, "I employed myself during the spring and fore part of the summer, in revisiting the several districts in Georgia and the East borders of Florida, where I had noted the most curious subjects" (*T* 375). It was, he continues, in the course of these collecting excursions that he again came across the mysterious tree that he and his father had once discovered, but this time he was in luck—it was in full bloom. Bartram's rediscovery of the tree now turns into an encounter with his own personal past, a reference to which appears at the end of his botanical description: "This very curious tree was first taken notice of about ten or twelve years ago, at this place, when I attended my father (John Bartram) on a botanical excursion; but, it being then late in the autumn, we could form no opinion to what class or tribe it belonged" (*T* 376). Significantly, the actual classification of the tree on the basis of the new visual evidence is relegated to a footnote: "On first observing the fructification and habit of this tree, I was inclined to believe it a species of Gordonia; but afterwards, upon stricter examination, and comparing its flowers and fruit with those of the Gordonia lasianthus, I presently found striking characteristics abundantly sufficient to separate it from that genus, and to establish it the head of a new tribe, which we have honoured with the name of the illustrious Dr. Benjamin Franklin. Franklinia Alatamaha" (*T* 375).

There is more to this note than Bartram's studiously detached language reveals. At the time of the publication of *Travels*, the tree with the "beautiful good fruite" had already been definitively identified, by European botanists, as a species belonging to the genus *Gordonia*. This classification was strongly advocated, for example, by Sir Joseph Banks, the influential director of the Royal Botanic Gardens at Kew. "The Franklinia is," Banks had assured Humphry Marshall in 1789, "as you conjecture, a species of *Gordonia*. A drawing of this plant, sent here by MR. BARTRAM to MR. BARCLAY, has been compared with specimens; so that no doubt now can remain on the subject."[81] Seen from this perspective, Bartram's footnote stubbornly persists in replacing the name of the *Gordonia pubescens*, a plant relegated by Europeans to a genus named after a European nurseryman (Collinson's friend James Gordon),[82] with the combined names of an "illustrious" *American* scientist and a "beautiful" *American* river. It was on the "pensile" banks of the Altamaha, Bartram says elsewhere in *Travels*, that "the generous and true sons of liberty"—the Creek Indians—dwelt (*T* 63).

In Bartram's retelling of the tree's rediscovery and renaming, various temporal levels subtly intersect. The immediate narrative past of Bartram's second visit to the *Franklinia*'s place of growth (twice referred to as "this place") blends into the more remote past of his and his father's first visit (carelessly dated to a time "ten or twelve years ago"). Then Bartram projects himself back into the present. The inclusive "we" becomes "I," as the author reflects on the time that has passed between his two visits and the present, in which he is composing his chapter: "We never saw it grow in any other place, nor have I ever since seen it growing wild, in all my travels, from Pennsylvania to Point Coupe, on the banks of the Mississipi, which must be allowed a very singular and unaccountable circumstance; at this place there are two or three acres of ground where it grows plentifully" (*T* 376).

Bartram's comments include what is usually edited out of modern scientific discourse. *Accident* and *luck,* he freely admits, have enabled him to classify correctly, after a first and mistaken attempt at identification, this unique plant. But even so, since it does not grow anywhere else ("a very singular and unaccountable circumstance"), the *Franklinia* still remains somewhat mysterious. This is not a "narrative of objectivity, of human indifference" after the model of Newton's *Opticks*,[83] and those critics have it wrong who argue that in Bartram's *Travels* the supposedly separate modes of literary narrative and scientific description "typically occupy discrete passages."[84] Consider the detailed botanical discussion of the plant, which is skillfully embedded in, and framed by, Bartram's narrative. The *Franklinia* is, he writes,

> a flowering tree, of the first order for beauty and fragrance of blossoms: the tree grows fifteen or twenty feet high, branching alternately; the leaves are oblong, broadest towards their extremities, and terminate with an acute point . . . they are lightly serrated, attenuate downwards, and sessile, or have very short petioles; they are placed in alternate order. . . . The flowers are very *large, expand* themselves perfectly, are of a snow white colour, and ornamented with a crown or tassel of gold coloured refulgent staminæ: in their centre, the inferior petal or segment of the corolla is hollow, formed like a cap or helmet, and entirely includes the other four, until the *moment of expansion*; its exterior surface is covered with a short silky hair; the borders of the petals are crisped or olicated: these large, white flowers stand single and sessile in the bosom of the leaves, and being near together towards the extremities of the twigs, and usually many *expanded at the same time*, make a gay appearance: the fruit is a large, round, dry, woody apple or pericarp, opening at each end oppositely by five alternate fissures, containing ten cells, each replete with dry woody cuneiform seed. (*T* 375; my emphasis)

Again, Bartram combines Latinate terminology ("serrated"; "petioles"; "sessile"; "pericarp"; "cuneiform") with lyrical evocativeness and subtle sound effects ("a *f*lowering tree, of the *f*irst order *f*or beauty and *f*ragrance"; "these . . . flowers stand single and sessile in the bosom of the leaves"). His description culminates in an unashamedly subjective impression: with its flowers in bloom, writes Bartram, the *Franklinia* makes "a gay appearance." The flower's beauty is captured in Bartram's choice of metaphorical ornamentation. "Gold coloured" stamens combine to form a tassel; the "silky" inner petal cupping the others before the "moment" of the flower's "expansion" looks "like a helmet"; the *Franklinia*'s petals are "snow white."

Bartram's description is carefully and, at least at first sight, conventionally structured, moving from a consideration of the stem and the leaves to a discussion of the morphology of the flower. Syntactically, however, it is extremely concentrated: the entire passage consists of only two sentences, the various components of which are linked by colons or semicolons. This is not, as Pamela Regis maintains, an "atemporal" description;[85] in fact, Bartram's portrayal of the *Franklinia* is shaped by the same playful balancing of temporal succession and visual simultaneity ("until the moment of expansion"; "expanded at the same time") that distinguishes his other descriptions of plants in *Travels*. Because of the attention paid to painterly detail, Bartram's text seems as expansive as the flowers of the Franklin tree must have appeared to him when he first saw it in bloom.

After his return from the South, Bartram drew several images of the "living flowering Tree" he raised in his own garden. Among them is the sketch he sent Robert Barclay in 1788, to which Sir Joseph Banks referred in his letter to Humphry Marshall (ill. 3). In his preface to Franz A. Bauer's *Delineations of Exotick Plants Cultivated in the Royal Gardens at Kew* (1796), a collection of plates published without explanatory comment, Banks emphasized that each of Bauer's figures was "intended to answer it-

self every question a Botanist can wish to ask respecting the structure of the plant it represents."[86] Bartram's drawing of the *Franklinia* meets such official requirements for a correct botanical image insofar as it presents a frontal, clearly delineated image of the flowering plant (floating, as was the convention, rootlessly in white space, on blank vellum)."[87] The fruit of the *Franklinia*, which splits open when it has fully matured, boasts a structure that is truly unique within the "Tea family" (Theaceae) to which botanists now know it belongs. The dehisced capsule of the fruit, which is, along with two interior segments from the fruit,

3. "*Franklinia alatamaha.* A beautiful flowering Tree discovered growing near the banks of the R. Alatamaha in Georgia," by William Bartram. 1788. Natural History Museum, London.

portrayed in the lower left corner of Bartram's beautiful drawing, reveals this amazing zigzag appearance, not seen in any other fruit-bearing species in North America."[88]

This, to be sure, is first and foremost a scientific illustration, meant to satisfy the professional botanist's curiosity. As in Georg Ehret's botanical paintings, there is no attempt to link the plant to a larger environment or to place it in a particular landscape or cultural context."[89] But the drawing also demonstrates Bartram's considerable artistry and an impulse that goes beyond mere representation. As in his written description, the introduction of temporal, narrative elements complicates the scientific purpose of the image. Art and taxonomy meet in productive and exciting ways, The same picture comprehends various stages in the development of the plant's growth: a closed, a half-opened, and a completely expanded flower are shown within the same image. At the same time, the plant is presented as if softly moved by an imaginary breeze, which affords the viewer convenient glimpses of the plant's leaves both from above and below but also gently animates what would otherwise have been a static image. Linear precision is enriched by fine tonal effects, achieved through variation in color, the use of light and shade, and subtle stippling. The viewer admires the fine pubescence coating the five scallop-shaped, crenulate petals and recognizes the difference between the lustrous green of the leaves' upper surface and the pale green of their lower surface. As frequently in Bartram's drawings, the plant is depicted as rudely plucked rather than smoothly cut; suddenly the viewer also realizes that Bartram's *Franklinia* is not merely an illustration of different stages of fructification (which, it is true, often occur simultaneously in nature) but a meditation on the passage of time: the leaves below the flower have already assumed the yellow or reddish-orange tints of autumn and provide a startling visual contrast to the white, camelia-like petals. The gray globular fruit on the left, unlike its green counterpart on the right, seems swollen, as if heavy with seed.

In the compositional center of the drawing sits what from the Linnean taxonomist's point of view makes up the most important part of the plant: the boss of yellow stamens clustered around the plant's truncate ovary, crowned with a slender style. The "refulgent" golden shine of the stamens' compact cluster is beautifully emphasized by the shadow it traces on the "perfectly white" petals. As in the verbal description, the attention given to detail is remarkable. Bartram's drawing accurately captures the silverish fissures on the tree's slate-gray bark, the serrated margins of the oblong leaves, and, in the upper left corner, the stout pedicules from which the scallop-shaped petals, so conspicuously expanded in the center of the picture, have emerged. As Peter Collinson had said about Bartram's earlier drawings: here, indeed, there is no "particular omitted."

The refinement of Bartram's picture becomes even more evident when we compare it with the illustration furnished by the Belgian-born botanical artist Pierre-Joseph

Pl. 59

Franklinia.
Gordonia pubescens.

P. J. Redouté del.

Gabriel sculp.

4. "Franklinia, *Gordonia pubescens.*" Hand-colored engraving by Gabriel after a drawing by Pierre-Joseph Redouté. From F. A. Michaux, *North American Sylva* (1810), vol. 1, plate 59. Courtesy, The Lilly Library, Indiana University, Bloomington.

Redouté (1759–1840), the famous draftsman at the Jardin des Plantes in Paris. A plate made after Redouté's drawing was included in François André Michaux's three-volume work on North American trees (1810; ill. 4). Michaux was the son of a French botanist who had come to the new republic in 1785 and had established a botanical garden in New Jersey and a large nursery near Charleston, South Carolina. Always in search of plants they could ship to their denuded country, father and son Michaux (a kind of French version of John and William Bartram) became the first naturalists to locate and document many rare American plants.[90] The younger Michaux's magisterial *Histoire des arbres forestiers de l'Amérique Septentrionale,* or *North American Sylva,* as it became known in its English translation,[91] long remained the most comprehensive study of American trees in the Northeast. Given his interest in America's "botanical geography," Michaux naturally also visited William Bartram in his garden in Kingsessing, where he paid special attention to the *Franklinia*: "I have seen," writes Michaux, "several trees of this species in the garden of J. and W. Bartram . . . whose growth was luxuriant."[92] Michaux's plate of the *Franklinia* fully exploits the possibilities of stipple engraving, with its emphasis on dots rather than lines, which makes it well suited to the representation of fine tonal gradations and lends an illusion of tangibility and plasticity especially to the leaves of the plant.[93] Compositionally, however, Redouté's drawing is considerably less interesting than Bartram's. The artist tidily distributes the leaves over the entire length of the plant's stem (which, unlike the one in Bartram's drawing, is neatly cutoff at the bottom). Redouté, who fails to include a section of the fruit and thus obscures the genus of the plant he depicts, does not succeed artistically either. His flavorless, stiffly beautiful representation is a far cry from the startling contrast between the slim stem and the expansive crown of leaves and petals so effectively evoked in Bartram's drawing.

In Bartram's own garden in Philadelphia, the transplanted tree adapted quickly. It "is so hardy," Bartram wrote to Barclay, "as to stand in an open exposd situation . . . without suffering the least injury from our most severe frosts, when very few Plants from that country will do in our Green houses."[94] Bartram's friend Benjamin Smith Barton even singled out the *Franklinia* as an example of how American plants, too, habituate themselves to new climates: this "truely beautiful vegetable," since the period of its first introduction into Pennsylvania, had "altered the time of its flowering, by nearly two months."[95] Didn't it seem all the more remarkable, then, that such a "hardy" plant should have been found only in one small location and nowhere else in America—namely within "4 miles downstream from Fort Barrington, on the northeast side of the river," or, more specifically, on a "sand-hill bog" on the northeastern side of the road leading from Savannah to Pinch's Hill?[96] Bartram himself had said he was amazed by the *Franklinia*'s singularity, an impression he tried to re-create in his verbal as well as his visual representations, both of which combine scientific precision with artistic individuality: "I have travled by land from Pensylvania to the banks of the Missisippi, over almost all the Teritory in that distance between the Sea shore & the first mountains, cross'd all the Rivers, and assended them from their capes a many miles; & search'd their various branches Yet never saw This beautiful Tree growing wild but in one spot on the Alatamaha about 30 miles from the Sea Coast neither has any other person that I Know of ever seen or heard of it."[97]

Not surprisingly, the enchanting *Franklinia* with the "beautiful good fruite," after it was first discovered in 1765, soon acquired an "American" history of its own. Its literal singularity was enhanced by its metaphorical uniqueness. This was not lost on Michaux, who in *North American Sylva* pointed out that this tree, which "appears to be restricted by nature within very narrow bounds," had been baptized "Franklinia" after a very singular man, namely "in honour of one of the most illustrious founders of American independence: a philosopher equally distinguished by his scientific acquirements and by his patriotic virtues."[98] The fame of the *Franklinia* spread quickly, and with it came requests for specimens, especially from European collectors—a factor that might have contributed significantly to the complete removal of the plant from what was regarded as its native place.[99] In 1787, a nephew of Humphry Marshall, Moses Marshall, was asked by his English customers to supply them with "as many as you can of Franklinia." Marshall's report to Sir Joseph Banks that he had discovered some of these specimens near the Altamaha River is one of the few indications that we have of the tree's continued presence in Georgia. The Bartrams' tree with the "beautiful good fruite" was last seen in its original location in 1803, when the Scottish collector John Lyon visited the area and noted, with disappointment, that "here there is not more then 6 or 8 full grown trees of it."[100]

After Lyon's visit, the *Franklinia* remained shrouded in mystery. Was it really limited to the specimens the Bartrams discovered? Or were the few trees languishing in a Georgia swamp only the last survivors of a genus that was once widespread? In its very elusiveness, endowed as it was with a questionable place of origin that, to adapt a phrase by Wallace Stevens, could have its own origin,[101] the *Franklinia* became a metaphor for the ambiguity of American nature itself. More recently, its improbable rarity has led botanists to speculate whether or not this tree, transplanted from one part of America to another, had perhaps first been itself transplanted to Georgia. In other words, they have wondered if Humphry Marshall was right when he asserted that William Bartram had enjoyed "the pleasing prospect of beholding it in its *native* soil." Botanist Gayther L. Plummer, for example, judging from the accounts supplied by Bartram and Lyon and also from his own failed attempts in Georgia at vernalizing seeds, which he exposed "to all important difficulties," concludes that the original colony of *Franklinia* discovered by Bartram had apparently not been reproducing very well: "If it had been, survivors should have remained somewhere throughout the vast area where similar habitats abound." Put differently, the *Franklinia* "just does not behave as most native plants do."[102] Plummer suggests and then rejects various theories, all of them equally vague, which could help account for the plant's presence in the only place where William Bartram found it. Birds of passage might have imported it, as John Lyon first suspected,[103] or maybe the Spanish, or the French, or—why not?—West African slaves:

Perhaps it was brought by migratory birds . . . in which case Central and South America are likely places of origin. Perhaps the Spanish brought it; again Central and South America would be suspected. But representatives of the Theaceae are not common in Central and South America, nor in the West Indies. Perhaps the French imported it; they had tea and camellias in the seventeenth century that originated in the East Indies and were shipped to Charles Town. West Africa could be the origin of it, if it came with a group of slaves.[104]

But come to think of it, ponders Plummer, the English were the great importers of plants into America; they were the ones who had enjoyed direct trade relations with China since the 1680s, and they had looked at Georgia as a kind of utopian agricultural enterprise. However, since they were avid botanizers as well as zealous importers of tea, they probably would have recognized similarities between the *Franklinia* and other plants belonging to the family of Theaceae. And all of the above hypotheses finally fail to explain one circumstance, which William Bartram had already found "very singular." "Of all the places available to *Franklinia*," Plummer summarizes his own and his fellow botanists' quandary, "it is indeed strange that only one site was ever found."[105]

Since it had ceased to exist in its natural habitat (whether this was its place of origin or not), the *Franklinia* eventually came to signify the inevitable damage done to the integrity of nature by the interference of the human observer. Bartram's tree became a product of culture rather than nature, a metaphor for the glory as well as the pathos of the collector's art. Having a *Franklinia* growing in one's garden, especially during the Victorian era, was "to mark one as a horticultural aristocrat."[106] The exceptional status of the Franklin tree as a cultural icon is emphasized by the fact that, amazingly enough, every *Franklinia* growing in North America is somehow derived from the plant in Bartram's garden.

It was as a natural "curiosity" that the *Franklinia* appeared in various nineteenth-century publications, almost always in conjunction with references to its utmost rarity. In 1832, a woodcut of the *Franklinia*, reminiscent of Bartram's own drawings (ill. 5), graced the cluttered pages of the *Atlantic Journal and Friend of Knowledge*, a "cyclopedic review" edited and written by the eccentric Turkish-born Constantine Samuel Rafinesque, a self-declared "Botanist, Naturalist, Geologist, Geographer, Historian, Poet, Philosopher, Philologist, Economist, Philanthropist" and proud inventor of 2,700 "new" generic names

FRANKLIN TREE.
FRANKLINIA ALATAMA.

5. "Franklin Tree, *Franklinia alatama.*" Woodcut from C S. Rafinesque's *Atlantic Journal and Friend of Knowledge* 1.2. (1832). Courtesy of the Gray Herbarium, Harvard University.

Bachmans Warbler. SYLVIA BACHMANII. *Male 1 Female 2. Gordonia pubescens.*

6. *Bachman's Warbler*, by John James Audubon and Maria Martin (background). Flora: "Gordonia pubescens" (*Franklinia alatamaha*). 1833. Watercolor, gouache, and graphite on paper, 54.6 x 35.7 cm. Purchased for the Society by public subscription from Mrs. John J. Audubon. New-York Historical Society, 1863.17.185.

to be added to the Linnean system.[107] Rafinesque reported that the botanical garden founded "over 100 years ago by the elder Bartram" was still flourishing, and he thanked the present owner, Colonel Starr, for having allowed him to "examine, describe and draw" some of the rare southern plants growing there, among them the "beautiful Franklin tree, which grows with the utmost perfection in this garden. The original tree brought by Bartram nearly 60 years ago is now nearly 40 feet high. All those in other gardens come from this tree. Their sweet white blossoms and orange-like leaves make them highly ornamental and prized."[108]

Bartram's tree also appeared in plate 185 of the Double Elephant Folio Edition of John James Audubon's *Birds of America,* a work that even with the birds left out would still hold its place as a great American flower book.[109] In Audubon's original composition dating from 1833, the drawing of the *Franklinia* was provided by Maria Martin to serve as a decorative background for the representation of a pair of Bachman's warblers (ill. 6).[110] The beautiful birds and the beautiful tree have in common that Audubon had not seen them himself. "My friend BACHMAN," writes Audubon in his *Ornithological Biography,* "has the merit of discovering this pretty little species of Warbler, and to him I have the pleasure of acknowledging my obligations for the pair which you will find represented in the plate, accompanied with a figure of one of the most beautiful of our southern flowers, originally drawn by my friend's sister, Miss MARTIN." What Audubon says about Bachman's warblers could with equal justification be applied to the *Franklinia* in his composition: "I myself have never had the good fortune to meet with any individuals of this interesting Sylvia, respecting which little is as yet known" (*BA* 2: 93). Both the plant and the birds were already extremely rare at the time Audubon's painting was made; unlike the *Franklinia,* however, which today at least survives in people's gardens, Bachman's warblers are now virtually extinct, the helpless victims of the millinery trade and the depletion of the southeastern forests.

Since John Lyon last saw the *Franklinia* in Georgia, no botanist has again found it growing in its natural state, neither along the banks of the Altamaha nor elsewhere in Georgia. In *Travels* (1993), a collection of poems by W. S. Merwin whose title resonates with Bartram's, the *Franklinia* movingly appears as a metaphor for the lost beauty of America itself.[111] Merwin's "The Lost Camelia of the Bartrams" begins with a near-verbatim transcription of John Bartram's wistful recollection of the day the *Franklinia* was found, then shifts its attention from the father to the son, becoming closely based on William's description of the tree. While the modern poet replaces Bartram's technical terms with more colloquial equivalents, his lines taken as a whole do little more than bring out the poetry already present in the source. Merwin's poem ends with the chilling insight that, though the *Franklinia,* "a cultivated / foreigner," has survived "here and there" in gardens, cultivation cannot make up for the loss of its native place, of its original wildness. The final line break between the adjective "cultivated" (with its positive connotations of "refinement" and "improvement") and the poem's last word, "foreigner," throws into relief the insight that, inevitably, collections destroy what they attempt to mend; they deprive what they intend to restore.

There is no reliable portrait of the elder Bartram,[112] and we have only one likeness of the son, painted by Charles Willson Peale in 1808 (Independence National Historical Park Collection). William Bartram was almost seventy then, a revered if odd and reclusive figure, an old man whose advice was sought by other naturalists of the time but who unfailingly declined all invitations to become involved in expeditions or any other enterprises that would have required, to use Thoreau's later dictum, new clothes, preferring instead to putter around in his garden in leather breeches, tending the plants, watching the weather, and taking notes on beetles and herbal cures. Charles Willson Peale, only two years younger than William Bartram, was still greatly preoccupied with the museum that he had started in 1786, the subject of the next chapter. He was arranging and rearranging his prize exhibits, preserving specimens, and painting backgrounds for his display cases. Yet, busy as he was, he took his time with William Bartram's portrait. Peale expertly captures the gentle features of his friend's face, showing him in a characteristically contemplative mood, with his eyes turned away from rather than toward the viewer. Bartram's buttonhole sports a sprig of flowers—perhaps fictitious, perhaps a blossom of bowman's root (*Protherantherus trifoliatus*),[113] a beautiful white flower growing on slender, graceful branches, to be found mostly in the Appalachians and, we may assume, in that microcosm of the American plant world, Bartram's garden. Botanists might have argued over the identity of the flowers in Peale's painting, but what is more important here is that Peale should have included them in the first place. His portrait of Bartram is a moving tribute by one extraordinary collector to another: the recognition that, in a collectors world, the appreciation of the whole always depends on an excessive love of detail. Or, as Peale put it, much better than I could: "It is the detail alone that enchants us" (*A* 132).

CHAPTER 2

Collection and Recollection

CHARLES WILLSON PEALE

The Collector in His Habitat

In November 1818, the seventy-eight-year-old Charles Willson Peale undertook the long and arduous journey to Washington to offer his life's work, the Philadelphia Museum, for sale to the Congress of the United States of America. It would prove to be yet another failed attempt to turn a collection "of Valuable import," which had for so long been the effort of one dedicated individual, into an enterprise of relevance for the nation, funded by the nation: "I believe," Peale reported home, "that I shall not find the Congress disposed to purchase it. one Gentleman told me that *if congress was to Purchace it, not one man in many hundred thousand would see it*" (*PP* 3: 634).[1]

A brief overview of the development of Peale's Philadelphia Museum will help us understand why, in 1818, Peale so badly wanted the power-wielders in Washington to take an interest in his work. "Mr. Peale's Museum" had started in 1786 as a modest assortment of "natural curiosities" in Peale's own picture gallery on Lombard Street in Philadelphia. Peale's inspiration to collect not only specimens of humanity, in the form of portraits, but also specimens taken from nonhuman nature came from a few bones of a "nondescript" huge animal—the "mammoth"—which he had been asked to paint for a client. At the end of his first year of collecting, Peale described his new project in a letter to George Washington:

> I have lately undertaken to form a Museum and have acquired the means of preserving in the natural forms, Birds, Beasts, and Fish, my Intention is to collect every thing that is curious of this Country, and to arrange them in the best man-

ner I am able, to make the Collection amusing and In[s]tructive, thereby hoping to retain with us many things realy curious which would otherwise be sent to Europe. (*PP* 1: 464)

The scope of Peale's project soon expanded. In May 1792, Peale announced to Adrian Valck that the museum had now become his "principal object" (*PP* 2: 33). Eventually, in April 1794, Peale felt compelled to notify potential clients in *Dunlap and Claypoole's American Daily Advertiser* that, alas, he was now "so engrossed by His Museum, that he finds it necessary that he should bid adieu to Portrait painting. . . . It is his fixed determination to encrease the subjects of the museum with all his powers, whilst life and health will admit of it" (*PP* 2: 91). A move to more spacious quarters was inevitable, and in 1796 the museum was relocated to Philosophical Hall.

Peale began to think of his "well organized" collection as a kind of secular temple, the "Epitome of the World" (*PP* 2: 177). If his museum gave pride of place to specimens from the New World, this reflected only his quiet confidence that the eyes of God rested perhaps a little more favorably on the American portion of His creation.[2] As we have seen, collectors in the American colonies such as John Bartram still sent the overwhelming majority of their gatherings of plants and other specimens to wealthy patrons in England. By contrast, Peale retained his carefully preserved specimens for exhibition in his own constantly expanding collection, although he would willingly enter into exchanges with European collectors, as he told "the Citizens of the United States of America" early in 1792 in *Dunlap's American Daily Advertiser:*

America has . . . a conspicuous advantage over all other countries, *from the novelty of its vast territories.* But a small number is yet known of the amazing variety of animal, vegetable and mineral productions, in our forests of 1000 miles, our inland seas, our many rivers, that roll through several states, and mingle with the ocean.

A Museum stored with these treasures must indeed become one of the first in the world; the more so, as the principal naturalists in Europe, will be anxious to acquire our productions, by an exchange of whatever is most valuable in their respective countries and foreign colonies. (*PP* 2: 10)

It was in this spirit that Peale concluded a speech given at the University of Pennsylvania on 16 November 1799: "Natural history is not only interesting to the individual, it ought to become a NATIONAL CONCERN, since it is a NATIONAL GOOD" (*I* 12).

In 1802, Peale again moved his museum, this time to the second floor of the Philadelphia State House. Within three years' time, Peale's new museum sported a large collection of minerals, four thousand species of insects, ninety species of stuffed quadrupeds, and over seven hundred mounted birds, displayed in glass-fronted cases (to prevent "fingering").[3] "Where shall I find room to place the numerous specimens of the Animal Kingdom I expect to collect," he lamented in a letter to Nathaniel Ramsay on 3 April 1805, "when already I cannot find space . . . for what I now possess? To imbrace every sort of Articles to complete a Museum, a whole square of buildings might be occupied to advantage" (*PP* 2: 821). It is obvious why James Clifford's explanation of why people become collectors, cited in most of the recent literature on museums, can be applied to Peale's peculiar museum only with great caution. The self "that must possess but cannot have it all," writes Clifford, "learns to select, order, classify in hierarchies—to make 'good' collections."[4] Indeed, Peale's "good" collection relied heavily on taxonomic order, but this didn't keep him—who classified but never adequately learned to select—from thinking that he could and should "have it all."

Peale's 1818 trip to Washington needs to be understood in the context of the ceaseless expansion of his museum. It also proves how serious he was when he argued that natural history should more properly be regarded as *national* history and that a natural history museum should be "a National Work" (*PP* 2: 821). Christopher Looby, on the basis of a fleeting acquaintance with Peale's theory and practice of collecting, has accused him of intellectual complicity with the new political order that was solidifying itself in Washington. In this reading, Peale, like Jefferson and Bartram, helped put a forceful end to revolutionary change and "trouble."[5] Let us ignore, for the moment, that it was in fact Jeffersonianism, an ideology carefully proscribing the areas of society into which the federal government could safely venture without impunity, which hindered Peale from realizing his main goal of turning the Philadelphia Museum into a national establishment.[6] Even when taken on its own terms, Peale's museum was by no means a "display of certainties," "a world free of ambiguities, obscurities, and difficulties," marked by "perfect consensus."[7] Throughout the sixty years of its sometimes precarious existence, Peale's collection remained a contested, troublesome institution, but not because of the "elitism" it represented.[8] If Looby does not consider the extent to which Peale himself was regarded as an oddity by some of his contemporaries, William H. Goetzmann's ungenerous reading of Peale as "pure showman" also misses the point of a project that in fact did not, as Goetzmann writes, include everything from "a six-legged cow" to Egyptian mummies.[9] (The cow in question only had five legs anyway [*PP* 2: 8; *W* 40], and though Peale accepted *lusus naturae* for his collection, he claimed he did so partly out of courtesy to those who donated them and not because he really felt that they aided him in his goal of creating an institution that would be "useful . . . to the citizens of the new world" [*PP* 2: 10]).[10]

Given the ambitious definition of its owner's aims and the sheer geographical size of the country to be represented in Peale's museum, it is not surprising that, at the time Peale turned to Congress for help, his collection in the State House was already bursting at the seams. "I had not room to put the subjects I had all ready prepaired, nor even room for my Pictures" (*PP* 3: 621). It seems ironical but is only characteristic of the collector Peale that his trip to Washington, which was intended to remedy a hopeless situation, promised, rather, to aggravate it. Peale had come to Washington not only in order to sell but also to *collect*—this time *human*, not natural, "curiosities," namely the faithful "likenesses" of those national political leaders not yet represented on the walls of his museum. Of course, Peale had not given up portrait-painting entirely; he had just made it part of his natural history collecting. "A Museum containing the Portraits of those men who at the risk of their lives achieved the independance of America besides a vast number of Articles that must be interresting to all Scientific men"—could there be any doubt, asked Peale rhetorically, mixing lofty idealism with his own brand of entrepreneurial pragmatism, "that such an Establishment would not be of great importance at the seat of Government to amuse as well as to instruck those who come to spend the winter season in Washington?" (*PP* 3: 634).

Accompanied by his third wife and his niece, an aspiring painter of miniature portraits, Peale took the steamboat to Baltimore (a seventeen-hour voyage). Here he inspected an offshoot of his own museum, the collection established by his son Rembrandt just a few years earlier, and found everything to be to his satisfaction: "The neatness of his display throughout the Museum does him infinite credit," Peale wrote in his diary (*PP* 3: 612). In Baltimore he boarded a stagecoach to Washington. Upon his arrival, he promptly discovered that "many of the Gentlemen I saught was not in the City" (*PP* 3: 616). Peale was even more dismayed to learn that those who *were* there took little interest in him. President Monroe, for example, although he was supposed to sit for his portrait, was not always as eager to see Peale as Peale was eager to see him: "The President had much company last night & was kept up to a late hour, consequently he could not rise so early as usial, especially as he had a pain in his head" (*PP* 3: 619). John Quincy Adams, Monroe's secretary of state, vowed to meet the painter regularly every day at eleven o'clock for uninterrupted sittings. Notes Peale, dryly: "Of course he is engaged for tomorrow" (*PP* 3: 627–628). In fact, the entries in Peale's diary for the next weeks are punctuated by references to the busy Mr. Adams's apologies for not being able to keep his portrait-sitting appointments.

After a few weeks in the capital, a miffed Peale offered the following summary of his activities to his son Rembrandt: "The difficulty of getting Gentlemen to sett who are in public offices is too perplexing to those painters who wish to be industrous, and in the time I have lost waiting for those even who have made appointments, I could have painted twice as much as I done" (*PP* 3: 684). On 14 January 1819, when the streets of Washington were covered with snow, Peale felt ready to be "on my return home" (*PP* 3: 633). Sadly, his hopes concerning his museum had not received much encouragement either: "Respecting your inquiry about my prospects of disposing the Museum to the United States," he told Rembrandt, "I answer that I do not think the majority of the Legeslature are sufficiently illuminated to appreciate it to its intrinsic Value. . . . they will in some future day be sorry that they let it slip through their fingers" (*PP* 3: 682–683).

But not all had been in vain. Although he was unsure where he would hang them, Peale left the "muddy wilderness" of Washington (*PP* 3: 678) with a series of new portraits of distinguished men. And, at the Washington house of Colonel Bomford, yet another possible addition to his museum had been pointed out to him, a "Picture of the Animals going into the Ark Painted by Mr. Catton" (*PP* 3: 684).[11] Writing to Bomford after returning to his farm near Germantown, Peale asked if he could borrow the painting for a while. On 1 May, 1819 he excitedly acknowledged receipt of Catton's work: "I cannot do justice to the merit of the Picture by attempting a description of it, therefore I shall for bear further particulars, and only say, that it is *a Museum in itself,* and a subject . . . most appropriate to a Museum, so much so, that although I have never loved the Copying of Pictures, yet I would wish to make a Copy of it, if such would meet with your approbation" (*PP* 3: 716; my emphasis).

When Bomford's permission arrived, Peale began furiously to copy Catton's painting. Rising by daybreak, he worked ceaselessly, interrupted only by brief excursions "in the Salubrious air" of his garden on the new Velocipede he had constructed (*PP* 3: 780). Peale found the copying more "difficult & tegious" (*PP* 3: 789) than he had imagined, since Catton's painting of "Noah and his ark, with a multitude of animals collected togather" (*PP* 3: 720) was so crowded with detail.[12] But then so was his own museum. What especially intrigued Peale about the picture was its strong emphasis on Noah, who appears devoutly kneeling in the foreground of the picture—the archetypal pater familias, his face illuminated by a strong ray of light that seems to be coming directly from heaven. The ark itself is painted in rather dark colors to indicate, as Peale rightly supposed, its magnitude (ill. 7): "Noah is a principal figure in the assemblage . . . and a ray of light from above encircles him in the center of the picture, a Lamb just by his side, an emblim of Inocence, Domestic fowls on the edge of the circle of light. . . . The Ark is of a duskey lead colour, of several Stories, seen through trees, and the mind is led

7. *Noah and His Ark,* by Charles Willson Peale after Charles Catton Jr., 1819. Oil on canvas, 112.3 x 127.6 cm. Acc. no. 1951.22. Courtesy of the Pennsylvania Academy of the Fine Arts, Philadelphia, Collections Fund.

to suppose the other end extends beyond the picture. . . . All in all its a wonderful Picture, and shows that the author possessed a great deal of Natural history" (*PP* 3: 720).

Noah's central position in the painting becomes even more evident when we compare Catton's "assemblage" with Edward Hicks's well-known representation of the same theme in which Noah himself does not even appear. Instead, we see an orderly train of animals, in tidy pairs, waiting for admission into a rather quaint ark that looks suspiciously like a Pennsylvania barn (*Noah's Ark,* 1846, Philadelphia Museum of Art). Not so in Catton's painting: Noah dominates the natural as well as the human world.[13] As Roland Barthes puts it, the "possession of the world began not with Genesis but at the Flood," because this was when "man" "housed" the animals.[14]

After the lukewarm reception Peale had received in Washington, this painting promised to offer the awe-inspiring glorification of the collector, the quasi-mythical apotheosis of "man" touched by the divine light amidst the animal world—in short, an image that could warm the heart of the normally modest but deeply disappointed Peale. To him, the obvious attention Catton had paid to detail ("even the smallest of the Animals are true in character," *PP* 3: 790) must have seemed like a reflection of his own concerns as a collector. In the best Plinian fashion, Peale was convinced that an understanding of the grand scheme of nature could rest only on an infinitely precise acquaintance with the tiniest details of God's creation: "There is nothing wherein Nature and her whole power is more seene . . . than in the least creatures of all."[15]

Peale would not have been Peale had he simply copied Catton's picture: for example, he replaced an elk in the left background of Catton's painting with a small American buffalo and inserted a homely donkey between the horse and zebra in the right foreground. But the finished product, he insisted, was "in every thing else, as faithful a copy as I could execute with the aid of high magnifying powers of Spectacles" (*PP* 3: 789). The most striking aspect of Catton's otherwise mediocre work is that it does not actually give us a "Picture of Animals going into the Ark," as Peale had originally assumed. With the exception of a minuscule cow that looks as if it were being led into the shed rather than into the biblical ark, most of the animals assembled around Noah do not seem to be in a particular hurry—and this in spite of the darkening clouds in the background, the trees bending under the strong winds, and the perilous closeness of Leviathan, whose spout we see rising from the sea. The assembled animals' heads are turned toward Noah and not toward the ark. Even more significant, the other members of Noah's family have faded into the background; they are much less clearly delineated or recognizable than the little chipmunk or the tiny armadillo in the foreground of the painting.

Peale's interest in Noah was certainly not merely iconographic. In a sense, Noah was, as John Eisner and Roger Cardinal have observed, the "first collector," someone who had created his collection as "a unique bastion against the deluge of time." And, perhaps unique among collectors, he "achieved the complete set."[16] In 1638, John Tradescant created a museum called the Ark in his home in Lambeth, England, which he described "as a world of wonders in one closet shut," including "Any thing that is strang[e]."[17] Tradescant's museum or, as it should more appropriately be called, his *Wunderkammer*, was explicitly designed to inspire wonder in its gawking visitors. The metaphor of the museum as ark is an established one; in the American context, however, it now assumes additional significance. What better source of wonder could there be than a "New World" that, supposedly pristine and untouched, seemed itself to have just emerged from the deluge? The English traveler and founding father of American natural history, Mark Catesby, who first came to America in 1712, believed that nowhere else in the world did "the Signs of a Deluge more evidently" appear "than in many Parts of the Northern

8. *The Artist in His Museum,* by Charles Willson Peale, 1822. Oil on canvas, 263.5 x 202.9 cm. Courtesy of the Pennsylvania Academy of the Fine Arts, Philadelphia. Gift of Mrs. Sarah Harrison (The Joseph Harrison Jr. Collection).

Continent of *America*" (*NHC* 1: vii). As we have seen, when Dr. Manasseh Cutler, on his visit to Philadelphia in 1787, inspected Peale's collection of American animals and fossils, he thought that the collector was Noah reincarnated. Full of admiration, Cutler added: "I can hardly conceive that even Noah could have boasted of a better collection" (*LJC* 1: 262). Indeed, Peale never tired of searching for remnants of an antediluvian (but still fairly recent) America. In 1801, Peale's quest for the American "mammoth," the bones of which had originally induced him to start his natural history collecting, culminated in the excavation of an almost complete skeleton on John Masten's farm near Newburgh, New York.

Given the scope of his collecting, which spanned centuries as well as vast territories, Peale felt justified in recognizing himself in the proud yet devout pose adopted by Catton's Noah. Not surprisingly, he eventually would try his hand at the theme again, this time creating his own design (ill. 8). The differences between Catton's painting and Peale's more complex composition provide important clues to Peale's conception of his own role as a collector.

The circumstances under which *The Artist in His Museum* was painted can be summarized quickly. In 1821, the Philadelphia Museum was finally incorporated, giving stockholders the right to appoint a board of five trustees who were to run the institution. At last, Peale's collection had achieved a status that made it more than just the product of the labors of one dedicated individual. On 11 April 1821, the new trustees elected a faculty of professors to be associated with the museum. Benefits of the new system, however, were slow to come. The city of Philadelphia could not be persuaded to reduce the rent of the State Hall by more than half, and attendance at the lectures given by the new museum faculty was disappointing.[18] In 1822, Peale once again took over the directorship of the museum from his son Rubens. At one of their meetings, the trustees, in celebration of the old-new museum director, commissioned him to paint a portrait of himself that could be prominently displayed in the museum. As he explains in his unpublished autobiography, Peale resisted the idea at first—there was, of course, no room in the museum. But Peale's objections were overruled: "The picture must be painted, and room will be found for it" (*A* 446). Peale then decided that he would not "make a picture such as are usially done in common Portraits" (*A* 446). And unusual it was indeed, Peale's life-sized portrait *The Artist in His Museum.*

One major difference between Catton's portrayal of Noah and Peale's representation of himself is, of course, that the imposing figure of the life-sized collector facing us here obviously does not kneel. And the light, coming discreetly from behind Peale's towering figure as well as from the front, where it is reflected by an unseen mirror onto Peale's head, is less a sanctifying message from God than a dramatic stage device. Proudly erect and with a vigor belying his eighty-one years, Peale lifts the curtain on his ark—a gesture, as he writes in his autobiography, intended to be "emblematical that he had given to his country a sight of natural history in his labours to form a Museum" (*A* 446). This time we see the "ark" from the inside, in an impressive perspective view of the Long Hall in the Philadelphia State House, the hundred-foot-long centerpiece of Peale's collection. The walls on the left and in the background are lined with neat cases displaying dead, expertly preserved animals; whereas Catton's collector had a newborn lamb cowering by his side, this modern Noah is surrounded by images of death. To his right, for example, we see the limp corpse of a wild turkey slumped over a case containing Peale's dissection tools.[19] On Peale's left, just below his brushes and palette, the enormous bones of the extinct "mammoth" (the jaw and tibia) create a suitable parallel. The imposing reconstruction of the complete mammoth skeleton, Peale's "Great Incognitum," can be glimpsed just behind the half-raised curtain. Although the French naturalist Cuvier had, as early as 1806, rather disappointingly identified the "mammoth" as the peaceful, herbivorous "mastodonte," the concealing curtain in Peale's painting still playfully alludes to what once had been perceived and celebrated as a mysterious "nondescript animal" (*A* 107).

In *The Artist in His Museum,* the members of Noah's family have been replaced by the small figures of the museum visitors in the background of the painting. A gentleman examines some of the birds in their display cases on the left wall; a father instructs his son (who is holding a book, probably the museum's catalogue, in his hands); a Quaker lady stares at the mastodon "in astonisshment" (Peale to Jefferson, 29 October 1822; *PP* 4: 194). It is perhaps no coincidence that the lady is the most clearly rendered of the museum visitors: Peale firmly believed that natural history was for women, too, and argued that "our daughters" had "an equal right with our sons to our instruction" (*DI* 13). Besides, her gesture of wonder demonstrates the excitement we would have felt ourselves, had the enormous skeleton (testimony and, literally, monument to Peale's "scientific zeal, and indefatigable perseverance")[20] not been hidden from our view. In fact, the skeleton of Peale's mastodon appears like an exaggerated allusion to the moralizing human skeletons that appeared in the *Wunderkammern* of the late Renaissance.

Peale's collection deliberately exploited some of the effects aimed for in institutions such as Tradescant's cabinet of curiosities, in which "wonder" was a quality of the items displayed as well as the reaction to be elicited from the museum visitors. Is *The Artist in His Museum*, then, an image of the collector as "showman"? In a sense it is, since the essence of the picture is contained in Peale's gesture of showing, and the conflation of the museum with a theater has a tradition going back at least as far as Ulisse Aldrovandi's *teatro di natura* in sixteenth-century Bologna, an institution Peale had adopted as his model.[21] Post-Puritan America had deep-seated reservations about theatrical presentations; thus, other areas of public interaction seemed predestined to take over the functions of the stage.[22]

And yet there is more to *The Artist in His Museum* than meets the casual observer's eye. As other critics have noticed, the central "trick" of Peale's painting consists in the separation of foreground from background, or, to continue the theatrical metaphor, in the division between the action on the forestage and the more subdued play enacted on the stage behind the curtain. According to Roger Stein, the curtain represents the crucial line separating the "raw" material from its "cooked" transformation into museum exhibits.[23] In a similar vein, Alan Trachtenberg has argued that there is a marked contrast between the trompe l'oeil effects of the picture's foreground, which he thinks are of a more genuinely painterly interest, and the geometrical constructedness of the background, which appeals more to our intellectual curiosity about the museum and less to our desire to actually "enter" and become part of the world of the painting.[24]

Such readings create a number of problems, because—with the exception of the lifelike, life-sized figure of the museum owner himself that serves as the painting's central axis—the *entire* picture appears somewhat artificially composed, deliberately overconstructed, as if Peale wanted to insist on its *fabricated* quality. Even the perspectival construction of the museum hall is skewed: in the left part of the painting Peale has inserted a smaller triangle into the larger horizontal triangular plane formed by the floor of the hall. Its left side is defined by the rather curiously positioned small bench behind Peale's body, while its right side is delineated by the pedestal on which the skeleton of the mammoth rests. The foreground of the painting is dominated by the large vertical triangle consisting of the mastodon's massive jawbone on the right, the wild turkey's limp body on the left, and Peale's illuminated head, which marks the triangle's apex. In a clever compositional maneuver, this central triangle and consequently also the viewer's attention are displaced through the gesture of Peale's dramatically raised hand. It carves out the shape of another triangle, projecting the viewers back into the painting and thus encouraging them to enter into the actual display room. It is no coincidence that, while Peale's head is illuminated, it is his *hand* that literally *makes* us see the museum and that reveals Peale himself as, literally, the *manipulator* of the privileged vision given to us.

Peale had a keen interest in the history of museums, as his *Discourse Introductory to a Course of Lectures* with its references to Aldrovandi and other collectors shows. There is, in fact, a curious similarity between the first known representation of a museum—the frontispiece of a book written by one of Aldrovandi's contemporaries, Ferrante Imperato's *Dell' historia naturale, libri XXVIII* (1599)—and Peale's 1822 self-portrait.[25] This beautifully crafted woodcut (ill. 9), an interior view of the Museum at Naples, opens up to the amazed viewer's eyes a kind of enchanted chamber filled with interesting items from the vast treasure troves of nature. Some of the exhibits (birds, a tiny armadillo, a seal) have been placed on the shelves that grace the left wall of the room. They provide a visual counterpart to the books piled up on the shelves along the opposite wall. The herons on the left, with their undulant necks, strike the viewer as an eerie visual anticipation of the ducks, penguins, and pelicans displayed on the left wall of the Long Room in Peale's museum. But—and here Imperato and Peale would part company—by far the majority of the Naples Museum's specimens (the fish, crustaceans, shells, and two-headed monsters) are distributed in bewildering profusion on the walls of the room. Some of them fill, like the stars in the sky, every available inch of the ceiling, which also sports, in sinister defiance of the laws of gravity, the upside-down carcass of a huge alligator. (From a modern perspective, it is ironical that the fish are rigged up not below but *above* the water birds, their natural enemies.) However, as in Peale's later painting, the compositional energy of this woodcut derives from a gesture of showing that dominates the foreground: a group of elegantly dressed courtiers, one of whom extends a large pointer, is captivated by the display of a stuffed pelican atop the shelf on the left, which appears to be stabbing its own chest.[26]

In Peale's self-portrait, the absolute centrality of this traditional gesture of showing is shrewdly displaced though a variety of compositional tricks. As if to offset the tri-

9. Ferrante Imperato, *Dell' historia naturale, libri XXVIII* (1599), frontispiece. By permission of Houghton Library, Harvard University.

angulations in the left part of the picture, Peale makes use of another, smaller inverted triangle in the right foreground, which creates a telling correspondence between the immense bones of the mastodon and the well-formed leg of the collector Peale. The bones synecdochically represent the finished skeleton, which, of course, we cannot yet see: partially "clothed" by the curtain, it becomes the elegantly clad collector's larger, ghostly brother. Seen in this context, the mastodon's giant hind foot seems to be a playful comment on Peale's neatly shod left foot.

The emblematic significance of the wild turkey on Peale's right hardly needs pointing out. Thomas Morton had already praised the "sweetness" of the American species of that bird, which he alleged by far surpassed the English turkeys, "feede them how you can."[27] Benjamin Franklin went so far as suggest that America should adopt the turkey rather than the eagle as its national symbol. A life-sized representation of the wild turkey was the first plate produced by John James Audubon in 1827 to herald the publication of his *Birds of America,* and Doughty's *Cabinet of Natural History and American Rural Sports* called it "the most beautiful and interesting bird" of the United States, a true "Native of North America": "Many attempts have been made to introduce the Wild Turkey, in its native state, on several preserves of game in Europe, but . . . they have not succeeded."[28] Framed by an American turkey on his right and by the bones of the American "mammoth" on his left—what better way could there have been for the proud proprietor of a national American museum to present himself to the public?

Russell W. Belk and Melanie Wallendorf have argued that the collection allows "the collector to play with multiple images of the self and multiple images of others," and they have called for an examination of collections "not just statically in terms of the objects contained in the collection, but also dynamically in terms of the collector's interaction with these objects."[29] Such an insight would not have come as news to Peale; it is in fact the central idea of this self-portrait, deliberately not titled "The Naturalist in His Museum" but *The Artist in His Museum*. To Peale, a museum was more than a room filled with natural curiosities. As he once asked, rhetorically, "Can the imagination conceive any thing more interesting than such a Museum?" (*DI* 35).

Peale's self-portrait is characterized by other playful correspondences. Taken together, these visual puns question the neat separation of foreground and background emphasized in standard interpretations of the painting. Look, for example, at the sparse tuft of white hair on Peale's head, which alludes not only to the portraits of white-haired gentlemen above the first row of cases in the upper left corner of the painting but also to a "portrait" of a bird in the top row of display cases on Peale's right: the white-feathered head of the bald eagle (Peale preferred the term "white-headed eagle").[30] Significantly, Peale's eagle is itself an allusion to a drawing by Alexander Wilson that was in turn partially inspired by a bird Peale had kept in "a large wire cage on the top of the Hall" (*A* 221). The proprietor of the Philadelphia Museum apparently had enjoyed an unusually close relationship with this unusual pet: "The Eagle had been so long domesticated, that Peale could without fear stroke him with his hand, nay it knew him so well that when he was walking in the State House Garden, it would utter cries expressive of his pleasure on seeing him."[31] Like Peale, Wilson ridiculed the epithet "bald" when applied to a bird whose head was in fact "thickly covered with feathers." As absurd as the names bestowed on the "Goatsucker" and the "Kingfisher," it had been occasioned merely "by the white appearance of the head when contrasted with the dark colour of the rest of the plumage" (*AO* 4: 89). The white appearance of the head, as opposed to the dark color of the body—a fair description nor only of the bird but also of the figure of the man in the painting! The fact that Peale, unlike the misleadingly named bird, is partially bald adds a distinctly comical note to his self-representation.

Another important visual correspondence in *The Artist in His Museum* is established by Peale's raised arm: the movements of the figures of the visitors in the background are all smaller, subdued variations on Peale's own prominent gesture. The Quaker lady is raising her hands; the lecturing father is using his left arm for emphasis, while his right hand rests on his son's shoulder; the unidentified visitor in the background folds his arms across his chest. Finally, when we compare the position of Peale's legs with the legs of the man intently studying the bird cases, the latter suddenly seems like a younger image of Peale projected into his own museum—not as its owner but as one of its visitors. More amusingly, Peale's convex figure and his black coat allude playfully to the penguin exhibited in the foremost case on the bottom row to the left.

The care Peale took in arranging and rearranging the various compositional elements of his portrait reflects the care that had gone into the making of his museum. It is important to remember that Peale had no real models to follow when he started planning his exhibition of "natural curiosities" in the 1780s. As Joel Orosz points out, the collection established by the Library Company of Charleston in 1773 under the tutelage of John Bartram's friend Alexander Garden was the first American exhibition to advertise itself as a "museum." However, this exhibition was destroyed by a fire in 1778. Contemporary models seemed less than impressive: Peale certainly wanted to distance himself from comparatively haphazard accumulations of treasures large and small such as Pierre Eugène Du Simitière's "American Museum," on which he commented witheringly in his autobiography: "Mr. DuSimetere . . . had a fondness for natural history, and he possessed a small collection of Butterflies, and had some snakes in Voials with spirits. He also collated some few articles of antiquity . . . but he made no attempts to preserve either birds or quadrupeds. He had a few coins, perhaps they were the most valuable of his collection" (*A* 99).[32]

Peale wanted his museum to be different, a well-planned, well-organized institution for the gentle "diffusing of knowledge," as he declared in his unfinished manuscript for a guidebook, "A Walk through the Philadelphia Museum" (1805/6; W 119). Peale flaunted his own professionalism by adopting the Linnean system as a basis for his collection—a decision that seems even more daring given the sketchiness of his own education. Peale was a saddle and harness maker and therefore had but "small Latin and no Greek": "Charles was now put to school; but the Latin Language, which he had begun in his fathers lifetime, was to be no part of his Education; and after he had learnt Arithmetick, writing, &c., before he was 13 Years of age he was bound an apprentice to a sadler, and was kept diligently to his trade for several years" (*A* 2). Predictably, Peale soon came to resent the arcane language of Linnaeus, calling it "Folly . . . to still retain the Latin terms so difficult to remember," which he felt were only meant to intimidate students.[33] His own knowledge of classical taxonomy came mainly from a popular book by Richard Pulteney—a concise and not uncritical overview of the writings of Linnaeus. Pulteney stressed the paradoxical intention behind the imposing systematic edifice erected by Linnaeus and his disciples, which he said was intended not only to give people a better idea of the *whole of nature* but also to lead them to a clearer understanding of *specific differences*—that is, of the concrete, minute details of God's creation.[34] He frankly acknowledged that the Linnean system, in its attempt to represent the living world in its entirety, was a cultural achievement more than an absolutely accurate description of nature-as-it-is. But it was a *useful* construction, one that, above all, was to be praised for its great mnemonic value: "It is pleasing to observe," exclaimed Pulteney, "how well many of the natural classes are kept together in the Linnean system, the characters of which enjoy the advantage of being very easy and simple to retain in the memory."[35]

Among Pulteney's summaries of the writings of Linnaeus was also a reference to the *Bibliotheca Botanica* (1736), a treatise classifying the authors of botanical books. If books could be classified like plants or animals, perhaps the analogy went even deeper? One of the models for Linnean taxonomy was indeed the library—an image that, according to philosopher Hans Blumenberg, depends on, and sharply conflicts with, the old metaphor of the Book of Nature.[36] In 1813, admission tickets to the Philadelphia Museum,

10. Admissions Ticket to Peale's Museum, dated 10 April 1813, designed by Charles Willson Peale. Courtesy of the Massachusetts Historical Society, Boston.

designed by Peale himself, still depicted the resplendent rays of light emanating from the open pages of the Book of Nature (ill. 10); the same image had also graced the title page of the *Scientific and Descriptive Catalogue of Peale's Museum*, prepared in 1796 by Peale and the French naturalist Ambroise Marie François Joseph Palisot de Beauvois (1752–1820). If Linnaeus had opened the book of nature to the world, Peale, the sole owner and proprietor of the Philadelphia Museum, was now issuing reprints of that volume to his visitors. The boy in the background of Peale's portrait, museum catalogue in hand, personifies the essential premise of Peale's museology and, even more important, Peale's belief that no special training would be required to "read" his museum. In his *Discourse Introductory*, Peale included the brief, but significant, etymological reflection that "museum" had originally been a term for a library, the part of the palace in Alexandria where Ptolemy Philadelphus stored "things that have an immediate relation to the arts" (*DI* 16–17) Peale's ark is, properly considered, such a library: looking at nature—like looking at Peale's emblematic portrait—became a reading experience. Incidentally, a brief account of a visit to Peale's museum by a Moravian sister further reveals just how much Peale wanted his visitors to see *and* read. "I went about the rooms with my spectacles on under my bonnet, so that I could read the finely written labels," wrote Catherine Fritsch in 1810. "I took much pleasure in reading whence all these curiosities came, and who had presented them."[37] Is it more than a coincidence that at least some of the display cases in the Philadelphia Museum had originally held books?[38]

Several decades later, Herman Melville would offer the definitive criticism of the notion of sorting living beings as if they were books, when he made Ishmael propose his wacky taxonomy of whales. Ishmael finds his work a "ponderous task" ("no ordinary letter-sorter in the Post-office," he claims, would have been up to it), but he prevails. Having "swum through libraries" in the course of his cetological research, it seems logical to him to read whales "as books," too: "According to magnitude I divide the whales into three primary BOOKS (subdivisible into CHAPTERS), and these shall comprehend them all, both small and large." The "Bibliographical system," moving from folio to duodecimo size, is the only one that is "practical," since, as Ishmael puns, it is the only one that "boldly" seizes the whales "bodily," that grasps them in their "entire liberal volume."[39] As late as 1859, in an essay excerpted from his *Contributions to the Natural History of the United States of America*, the Harvard naturalist Louis Agassiz would still seriously assert that in the order of nature, animals were "but the headings of the chapters of the great book" that the student of natural history was reading.[40] Melville relentlessly exposes the irony inherent in the concept of a Book of Nature that had become so subdivided as to have lost its former power as an image. His insidious trick is, of course, that he stresses the figurative quality of the Book of Nature by insistently literalizing the metaphor: Ishmael does not actually "read" the whales; like Samuel Pepys, who took pleasure in ordering the books in his library according to their different sizes, he looks at the books in his whale library in terms of their physical properties, not their content.

Peale was still innocent of such ironies, and yet this complication of the metaphor of the Book of Nature—which effectively turns the museum of natural history into a library in which its various chapters can be conveniently consulted—can also be seen in his 1822 self-portrait, a painting that he explicitly said was intended to tell a "Story" (*PP* 4: 166). We have already noticed how Peale's compositional strategies encourage us to "read" meaning into visual parallels, to contextualize them, as it were. Verbal displays, whether in the form of framed explanations, labels, or written catalogues to be purchased by the visitor, belonged as much to Peale's museum as did its visual exhibits. As Paula Findlen has pointed out, verbal descriptions of museums became popular in the late sixteenth century because they extolled the collector's own status: "Written by the naturalist himself," the museum catalogue "displayed his erudition. Written by another scholar, it conveyed the status of a collector who had earned the right to commission a description of his work."[41] Werner Hüllen has even argued that we should begin to see the catalogue as a literary genre in its own right, as a reproduction in another form of the "spatial arrangement of exhibits in the sequence of entries" that reflects the general principle of natural history. Like the collector, the museum catalogue abstracts "an item of nature . . . from the ecological context of its natural existence and defines it *per se.*"[42]

Peale's 1796 catalogue, coauthored with Beauvois, exemplifies Hüllen's definition and yet also resists some of its most important implications. Many of the entries are acutely (and, it seems, superfluously) conscious of their own status as *texts* ("We shall dwell no longer on this animal," *SDC* 5; "These animals, being already so well known, it is thought unnecessary to add any thing in this place," SDC 41) They strain beyond the conventions of the descriptive list toward the establishment of narrative *contexts*, creating icons of life where the "implied visitor"—a term that best reflects an understanding of museum catalogues as narrative texts—finds herself confronted with motionless indices of death. "This monkey died since the Catalogue was begun, and is preserved in the Museum, in a posture that was familiar to it" (*SDC* 7); "this animal while living was sprightly, pettish and passionate" (*SDC* 8); "The black colour of this membrane"—Peale is talking about a bat's wings—"forms a beautiful contrast with the colour of the body . . . which, when the animal was alive, was of a reddish or flesh colour" (*SDC* 15); "One would scarcely believe that so small an animal . . . could have produced cries so strong and sharp as those with which it frequently alarmed the whole neighbourhood" (*SDC* 29); "Although a small animal, it would bravely defend itself when attacked. It continued alive several months at the Museum" (*SDC* 36).

11. *The Long Room, Interior of Front Room in Peale's Museum,* by Titian R. Peale. 1822. Watercolor over graphite pencil on paper, 35.5 x 52.71 cm. Detroit Institute of Arts. Founders Society Purchase, Director's Discretionary Fund Accession Number 57.261. © The Detroit Institute of Arts / Bridgeman Images.

Most entries in the catalogue mention the origins of the specimens and sometimes even try to evoke, with a few, quick verbal brushstrokes, their native environment: "Opossums are natives of the new continent only, are commonly found in the woods, and in holes which they dig in the earth, or in hollow trees" (*SDC* 43). In some passages, the clash between the static descriptive and contextualizing narrative modes is particularly evident: "This monkey, which from its form and gentleness, is one of the prettiest we know, is in height about two feet" (*SDC* 8–9). At least in principle, Peale wanted, as he told Thomas Jefferson, to reach out to "all Classes of the People" (*PP* 2: 431). Thus personal comments appealing to the visitors' "common sense" are frequent. For example, Peale offers the following folksy criticism of the Linnean system, which leads him to compare the hyena with the spaniel: "Indeed one finds it difficult to believe, that these pretty little spaniels, the inseparable companions of our fine ladies, who experience in this little animal so sincere and so lively an attachment, should have the same origin with so hideous and ferocious an animal as the savage, or wild dog of the Cape of Good Hope" (*SDC* 26).

After Peale purchased his own printing press in October 1804, he was able to produce and reprint his catalogues for immediate distribution among his visitors: "I shall now give some thousands of *Guide's* to the Museum in a year" (*PP* 2: 771). Peale's new "Guide to the Philadelphia Museum" proudly emphasized that the birds on display, unlike those exhibited in other collections elsewhere, uniformly appeared before backgrounds "painted to represent appropriate scenery; Mountains, Plains, or Waters, the Birds being placed on branches or artificial rocks . . . producing an uncommonly elegant display" (*PP* 2: 761–762). Upon closer inspection, the cases on the left wall of the Long Room reveal Peale's insistence on contextualizing, "telling" detail: some of the birds are indeed sitting on branches or rocks. Titian Peale's watercolor of the Long Room (1822), which his father used when preparing the background for *The Artist in His Museum*, shows that some of the birds were displayed in the process of spreading their wings (ill. 11). In "A Walk through the Philadelphia Museum," Peale draws the visitor's attention to his llama, represented "in the act of spitting through the fissure of his upper lip, which he used to do when he was alive in the Museum," and to the porcupine, portrayed "in its natural attitude when eating" (*W* 6, 27). The careful arrangement of the birds' bodies in lifelike poses, the painted backdrops in the cases, the mounting of quadrupeds "in their natural attitudes" (*PP* 2: 761)—these were Peale's crucial curatorial innovations.

Peale usually separated the precious skins of his birds from their bodies immediately after they had been killed, "hooking out," as he elaborates with relentless explicitness in his autobiography, "the brains, eyes and tongue, and as much of the flesh as possible" (*A* 112). In his fight against the "dermests and moths" that were "the enemies of most museums" (*A* 109), he used a combination of antiseptics and arsenic to preserve the skin and feathers, often nearly poisoning himself and other members of the family in the process.[43] He sculpted the body of smaller birds out of cork and cotton and put artificial eyes in their heads ("made of glass and painted of the proper colour . . . or black wax droped on a piece of card"; *A* 112). To support the animals' necks and legs, he developed artful wire constructions, which he then wrapped in oakum or tow. The bodies of the larger birds he usually formed entirely out of annealed wire. As he explained in a letter to the botanist Stephen Elliott, "practice will make all things easey" (*PP* 2: 1179). With his quadrupeds, Peale used an essentially similar technique, "twisting the wires of the fore & hind legs to a wire which passes through the Scull" (*PP* 2: 1179). But he was especially proud that for the "very large Animals" he had perfected a method of preservation "not done in any other Museum." He carved their limbs in wood to re-create the muscles as they had looked "in the living Animals . . . which cannot be imitated by the most expert artist by stuffing" (*PP* 2: 1179). Peale could think of "little more that human invention can devise to perfect such a Work" (*W* 7).

Though he did not yet use the requisite terms, Peale's conservatorial procedures effectively anticipated the artful naturalistic animal groups designed in the 1880s for the Milwaukee Public Museum by Carl Akeley (who is often and wrongly supposed to have pioneered the technique).[44] One of the first known instances of the use of the term "habitat," signifying "the natural place of growth of a plant in its wild state," occurs in the third edition of the English naturalist William Withering's *A Botanical Arrangement of British Plants* (1796). The phrase "habitat group" (later replaced by the term "diorama") was introduced only much later, apparently by Frank Chapman, curator at the American Museum of Natural History in New York, for a conservatorial practice that was becoming increasingly familiar to museum visitors. When F. A. Lucas, the director of the American Museum of Natural History, stressed in 1914 that "habitat groups" represented "actual scenes," he merely repeated (unbeknownst to him) the theatrical concept already encapsulated in Peale's dramatic raising of the curtain in *The Artist in His Museum*.[45]

In her study of the "habitat diorama," Karen Wonders notes the absence of these and similar contextualizing exhibition techniques in all of eighteenth- and nineteenth-century Europe (with the exception of Scandinavia) and pays tribute to the American collector Peale for his inventions. But even the most comprehensive definition of "diorama" listed by Wonders ("a display of free-standing objects arranged against a perspectival background so as to create an illusionistic impression of reality")[46] does not capture the full complexity of Peale's animal groups. In his gallery, Peale had been experimenting with panoramic pictures of landscapes, illuminated by shifting light effects, even before he started to collect natural history specimens, and the same fusion of artistic illusion and constructive engineering shaped his animal habitat groups with their evocative backgrounds.[47] Peale was aware that his method was probably unique: "It is not the practice, it is said, in Europe to paint skies and landscapes in their cases of birds and other animals, and it may have a neat and clean appearance to line them only with white paper, but on the other hand it is not only pleasing to see a sketch of a landscape, in some instances the habits of the animal may be also given; by shewing the nest, hollow, cave, or the particular view of the country from whence they came" (*A* 318).

The logic of the museum collection, if we believe Susan Stewart, demands that we forget the origin of the items collected.[48] However, in the design of his habitat groups as well as in the composition of his catalogue texts, we see Peale in conscious opposition to a practice that sacrifices the detail to the idea of the whole. He is intent on conjuring up exactly the kind of scenario that would constantly remind museum visitors of the context out of which each single one of his museum exhibits had come. Peale presents animals frozen in characteristic poses and simultaneously attempts to capture the moment before the freeze, thus creating narratives out of descriptive stasis. From Peale's extensive correspondence, his writings, and eyewitness accounts such as Manasseh Cutler's, we derive a fairly accurate idea of what Peale's "world in miniature" must have looked like. Cutler was impressed by Peale's skillful landscaping: the mound of earth on which Peale had placed his quadrupeds; the trees whose boughs were loaded with birds; the artificial pond full of fish and waterfowl and surrounded by a beach, complete with shells, lizards, frogs, water snakes, and rocks. Peale's animals all had, marveled Cutler, "the appearance of life" (*LJC* 1: 262).

Some of Peale's preservation techniques are vividly illustrated by his son Titian's drawing of the "Missouri Bear Group" from the mounting in the Philadephia Museum (1819–1822; ill. 12) and plate 11 ("Grisly Bears") from the first volume of John Doughty's *Cabinet of Natural History and American Rural Sports* (1830; ill. 13), both obviously based on the same display. There is, first of all, the deliberate attempt to provide varying perspectives on the animals by arranging their bodies in different attitudes. One of the bears has apparently spotted the human "voyeur," whereas the other still appears as if arrested in a pose of contem-

12. *Missouri Bear*, by Titian R. Peale. Watercolor from mounting in the Philadelphia Museum. Ca. 1819–1822. By permission of the American Philosophical Society, Philadelphia.

13. *Grisly Bears*, by F. J. and C. A. Watson, Philadelphia. Chromolithograph from Doughty's *Cabinet of Natural History and American Rural Sports* 1 (1830). Photograph courtesy of Leo Boudreau.

plation, oblivious to the presence of outsiders. Peale's mounted animals are shown in interaction with their natural-artificial environment, which thus becomes more than just a background. The bear on the left in Titian's drawing (which appears on the right in Doughty's engraving) is grabbing a branch of a tree, which, in pictorial terms, serves as the central axis of the composition and, in a more practical sense, must have helped Peale to stabilize the bodies of his bears. Through the simple device of representing two of a kind, Peale succeeds in convincing his visitors that even "the most dreadful and dangerous of North American quadrupeds" have something resembling a social life.[49] Thus, his composition helps to humanize even the ferocious grizzly bear.

The "personal" histories of these two bears are well known. Thomas Jefferson had received them in 1807 from Captain Zebulon Pike, who had bought them from an Indian during his failed southwestern expedition in search of the Red River. Pike and his men transported the bears, sometimes on their laps, for several thousand miles without significant problems. Apparently, the animals became extremely docile and playful and followed Pike's men around the camp like dogs. When Jefferson donated them to Peale, Pike recommended that they be kept together, without chains, and not be "confined assunder."[50] The sad sequel to the story of the playful cubs is narrated in a chapter of *American Natural History*, written by the zoologist and physician John Godman, Rembrandt Peale's son-in-law. As adults, reports Godman, the bears proved so dangerous that they were a threat to other animals:

> In one instance an unfortunate monkey was walking over the top of their cage, when the end of the chain which hung from his waist dropped through within reach of the bears; they immediately seized it, dragged the screaming animal through the narrow aperture, tore him limb from limb, and devoured his mangled carcass instantaneously. At another time a small monkey thrust his arm through an opening in the bear cage to reach after some object; one of them immediately seized him, and, with a sudden jerk, tore the whole arm and shoulder-blade from the body, and devoured it before any could interfere.[51]

After such violence, the peace reigning in Peale's artful arrangement of the "Missouri Bear Group" returns the animals to an environment in which they (and the human observer) can feel "at home" again.

The aim of Peale's "Stupendous labour" ($W7$) was not simply to naturalize again what human interference had destroyed. Ivan Karp errs when he claims that natural history exhibits differ from other displays in that they "are not produced by human agents who have goals and intentions."[52] Peale would have been the first to admit, with delight, that his stuffed, posed, and carefully placed animals were as "made up," as manufactured, as were his portraits. His "habitat groups" triumphantly fused both taxidermy and painting, science and art. By creating lifelike, three-dimensional sculptures made of wood and wire—what he dubbed "statues of the Animals with real skin to cover them" (*W* 7)—and by presenting them against painted two-dimensional perspectival backdrops, Peale successfully simulated life while reminding his audiences of his own superior skills of construction. His displays imitated reality by incorporating bits of reality itself (real furs, real claws, real stones). As Manasseh Cutler noted, all the exhibits "were real, either their substance or their skins finely preserved" (*LJC* 1: 261). Thus, Peale's compositions were both mediated "realistic" representations—in more technical terms, icons of reality (based on "resemblance")—and immediate presentations (based on "identity," the use of "the thing itself").

Susan Stewart insists that the collection deprives things of their individuality by subjecting everything to the rule of the collector's totalitarian self. Again, this generalization does not seem to apply to Peale's museum, in which the collector apparently did not hesitate to make himself part of his collection. An amazed Cutler could not "discern the minutest difference" between the original and the waxen copy (*LJC* 1: 260). Peale's statue has not survived, but his 1822 self-portrait could be seen as yet another attempt to include himself in his museum: rather than asserting his exclusive right to define his collection, Peale seemed to be telling his visitors that the collection also defined *him*.

It seems only natural that a portrait-painter should have decided to make the background an integral element of his museum displays, too. Peale had been painting men and women in their homes for a long time. In his autobiography he pokes fun at an incompetent fellow painter who, in order to be able to accept more commissions, would just rapidly and roughly sketch out people's faces and then attempt to finish his portraits at home. Not surprisingly, his unfortunate colleague afterwards discovered that he couldn't remember which background went with what painting: "It is advisable for artists," mocks Peale, "always to finish the drapery of their portraits directly after painting the likeness, otherwise there may not always be produced so much harmony in the picture" (*A* 105). Portraits do not only respond to questions such as "Who am I?" or "What am I like?"[53] They also often invite us to think about individuals in relation to the world they inhabit ("Where do I belong?"). It was Peale's innovative idea to conceive of his exhibits in the natural history museum as if they were portraits. In attempting to re-create, to *portray* nature as it originally was, Peale's museum inevitably humanized it.

If we now return to *The Artist in His Museum* and bear in mind Peale's conception of natural history as portraiture, it seems possible to argue that Peale applied some of the same ideas to his own representation in the museum—but with the opposite result. His own lifelike portrait of the human collector in the natural history collection "naturalizes" man by presenting him in his habitat. Just as the animals appear in natural attitudes, against backgrounds painstakingly delineating their habitats, the human collector Peale appears, as large as in life, in his "natural" habitat. It is ironical, though, that this natural milieu now is a deeply cultural *humanized* one.[54] Peale is apart from his collection, yet a part of it. But rather than deploring the hybridity of his displays, Peale took delight in their unreal realism, their realistic unreality. His emblematic painting alludes to, "quotes," the diorama concept, just as his own dioramas, represented in the background of the painting, allude to and quote nature. Such a precarious interweaving of nature and culture Peale himself would have regarded as the ultimate proof of his abilities as an artist.

Stewart assumes that the collection, which is there solely for the pleasure of the acquirer, foregrounds lucky capture rather than strenuous production and eschews references to the labor that has produced it.[55] But in *The Artist in His Museum*, Peale's taxidermic tools as well as his palette and paintbrushes (the instruments needed for the creation of a successful "habitat group"), indices of Peale's labors, appear at compositionally important points of the picture. The occasion of the painting—the wish of the museum trustees that the collector display himself in his own museum—called for self-aggrandizement, but there is evidence enough in the available correspondence that Peale himself would not have wanted to separate the *activity* of collecting from its (always preliminary) result, the *collection*. Recently, natural history museums have been criticized for inviting their visitors all too often just to "look at what we have," instead of exhorting them: "Look at what we do." Peale's museum still harbored the hope that both of these perspectives could be reconciled.[56]

Now that ecologists have realized that the relationship between a species and its habitat is much more complex than can be expressed by the demeaning term "background," some theorists argue that the diorama has become obsolete.[57] In his "self-diorama" showing the artist in his museum, Peale still took his work on the background seriously, which is why he had Titian draw his watercolor representation of the Long Room (ill. 11) as a model: "The minutia of objects makes it a laborio[u]s work; it looks beautiful through the magnifiers. Coleman seeing it yesterday, says it deceived him. He thought he was viewing the Museum in the looking Glasses at the end of the Museum" (*PP* 4: 170–171). As if it were one of his specimens, Peale exhibited his "unusual" portrait in the Museum, where his minutely detailed copy of Catton's *Noah and His Ark* was already on display. As happened to Coleman Sellers, visitors would enter the museum and, confronted with Peale's painting, would see themselves enter the museum. Peale's painting transports them beyond the collection and yet places them right back into it. As Georges Bataille once said, the real content of the "vessel" museum is its visitors.[58] This strangely circular experience, in which the inside becomes the outside and the outside the inside, is addressed in the painting itself, which presents itself as *a self-portrait of the portraitist surrounded by portraits*—by portraits, that is, of men *and* animals. It is fitting and in line with the inclusiveness of Peale's ideas about collecting that the museum visitors could also "collect" themselves by having their own portraits made—in the form of profiles cut out by the physiognotrace, a silhouette machine developed by the inventor John Isaac Hawkins and operated by Peale's ex-slave, Moses Williams.[59]

Preservation: Washington's Pheasants, Franklin's Body, and Peale's Teeth

In 1772, Peale painted the first of several portraits of George Washington. Posing in the uniform of a colonel of the Virginia Regiment, Washington evidently did not enjoy the experience. He wrote to Jonathan Boucher that during the long sessions he was "now and then under the influence of Morpheus, when some critical strokes are making," and added that he felt "the skill of this Gentleman's Pencil, will be put to it, in describing to the World what manner of Man I am" (*PP* 1: 120). Nevertheless, this was not the last time Washington would have his portrait painted by Peale. In 1779, for example, the state council of Pennsylvania requested that Washington, now the commander in chief of the Revolutionary Army and worried about desertions, the depreciation of American currency, rampant corruption, and declining enlistments, would sit for a "full-length picture," which, according to the *Pennsylvania Packet* (28 January 1779), he "politely complied with . . . and a striking likeness was taken by Mr. Peale of this city" (*PP* 1: 303).[60] At Washington's explicit request, Peale also produced a miniature copy of this portrait, expressing the hope that "the likeness will be Striking to the Lady your Sister, for I have taken much pains to have it a just Resemblance" (*PP* 1: 314).

Vowing never to hold public office, in December 1783 Washington returned to Mount Vernon as a private citizen. The conversation about faithful "likenesses" between him and Charles Willson Peale continued, but soon references to portraits of humans would interestingly intermesh with discussions of the preservation of animals—notably, several golden and silver pheasants, which Washington accepted in 1786 as a personal gift from his protégé, the marquis de Lafayette.[61] Washington liked to experiment with nature and, in his semiretirement at Mount Vernon, spent a considerable amount of time trying to breed mules using a rather uncooperative jackass he had received from the king of Spain.[62] Nevertheless, Peale's letter of 31 December 1786 must have seemed unusual to him, and not only because it suggested that the pheasants he had just received might soon pass away. In fact, Peale was quite specific about how, in case of such an unfortunate-fortunate event, the bodies of the birds should be packed up in order to be rushed to the only man in North America able to preserve them adequately: "Having heard that you have been presented with some beautiful Birds of China I take the liberty of requesting in Case of the death of any of them, to have them packed in wool and put in any sort of packing Case, and sent by the Stages to me, and I will preserve them in the best manner I am able, and either send them back to you, or place them in my Museum, as your Excellency may please to direct." Peale went on to explain that his reason for such an unusual request was that "such beautiful and rare things should not be wholly lost," which would inevitably happen were they to fall into the hands of "persons not sufficiently skilled in the manner of preserving"—of persons other than himself, that is (*PP* 1: 464). That these birds originally hailed from China (even though they had been bred and raised in the king's aviary in France) apparently did not bother Peale. Although he wanted his museum to be an American institution, his collector's desires were cosmopolitan; it was his hope that "in some future day, this Museum will . . . contain specimens of the intire Animals, belonging to every part of the Globe" (*W* 41).

Washington's response to Peale, written on 9 January 1787, was not without a shade of irony. He admitted that Peale might have been right to assume that his pheasants probably would not live long and happy lives in America and added (which must have made Peale cringe) that one of them, in fact, had already died and, alas, had been swiftly trashed: "I cannot say that I shall be happy to have it in my power to comply with your request by sending you the bodies of my Pheasants, but expect that it will not be long before they will compose a part of your Museum, as they all appear to be drooping. One of the Silver Pheasants died sometime before the rect of your letter, & its body was thrown away, but when ever any of the others make their exit, they shall be sent to you, agreeably to your request" (*PP* 1: 465–466).

On 15 February 1787, one of the golden pheasants obligingly made his "exit" and was quickly packed up in wool and sent to Philadelphia. "I am afraid," worried Washington, "the others will follow him but too soon" (*PP* 1: 466). Peale's response came at the end of February. Complaining about the "vexatious" delay in receiving the bird's body, he nevertheless confessed his excitement about being "able to preserve so much beauty. Before this time I had thought those Birds which I have seen in the Chinese paintings were only works of fancy, but now I find them to be only aukered . . . Portraits" (*PP* 1: 473).

The implication is, of course, that Peale's conservatorial skills, working with the material itself rather than abstracting from it, would produce much less "awkward" portraits. Peale bewailed the fact that the lives of such beautiful animals "cannot be preserved" and added speculations of his own on the natural history of pheasants: "I did not find the body very lean, the musles of the Thighs were strong, which with smallness of the Wings, makes me think that they run fast and fly but little." In the manner of the true collector, he had his mind set on the acquisition of yet further specimens: "When you have the misfortune of loosing the others . . . be pleased to order the Bowels to be taken out and some Pepper put into the Body, but no Salt which would spoil the Feathers." Since Peale was a dreamer as well as a pragmatist, he gracefully moved, within a single, smoothly anticipatory sentence, from mere wish to immediate fulfillment and encouraged Washington "to have some directions put on the box which would prevent delay on the Passage of them." The change of topic in Peale's next sentence was surprising even to the writer himself, which is perhaps why he prefaced it with an explanatory "I mean": "Another labour which I have lately undertaken I hope will give you Pleasure, I mean the making of Prints in Mezzotinto from my Portraits of Illustrous Personages."

On 13 March 1787, Washington sent yet another pheasant—this time, ignoring Peale's directions, by ship ("it will meet with as quick & safe a conveyance," he explained; *PP* 1: 474). Peale was appalled. His disappointed response treads the thin border between gratitude for the provider's wish to indulge the desires of the collector and more than slight irritation at the fact that Washington had not obeyed his explicit instructions: "Your obliging favor of the 13th I . . . received on the 28th. The Pepper I believe preserved the body from being thrown overboard." But Peale thought he could still repair at least some of the damage: "My Anticeptic Powders I hope will preserve the remains, yet not so perfect as I could wish as many of the feathers fell off." Peale's summary of the lesson to be learned from this experience was stern: "I believe the conveyance by the stage waggon with a particular direction will be the most certain" (*PP* 1: 475). And Peale did not lose his hope for yet further shipments of pheasants from Mount Vernon. Two years later, in January 1788, he wrote from Baltimore reminding Washington that, should more of these rare birds die, "you will oblige me much by sending their remains" (*PP* 1: 477).

Peale's pheasants, their feathers only slightly ruffled, are still on display in the Museum of Comparative Zoology at Harvard.[63] But I have not summarized Peale's and Washington's epistolary exchange on Chinese pheasants because of its anecdotal value. The modest point I want to make is that when discussing his preservation of "natural curiosities" Peale employs some of the very same language he uses when referring to his portrait-painting. Not only do allusions to his painting and his preservation work continually overlap in Peale's diaries, sometimes amusingly,[64] but we also find the same terminology ("preservation"; "faithful likeness") applied to both activities.

In 1792, Peale thought he had "exerted himself to his utmost abilities" in his ceaseless attempts "to collect and preserve articles of natural history." He resolved to enlist the "aid of gentlemen of influence" as "Visitors" to the museum, i.e., as members of a kind of advisory board that would help send the museum on its way to becoming a "great national Institution" (*A* 191). In a programmatic statement written for distribution among these "gentlemen of influence" ("My Design in Forming this Museum"; *PP* 2: 12–19), Peale praised his own superior preservation skills and in rapturous tones anticipated what, given the right support, his collection might soon look like: "The sight, alone, of our large animals, collected together in good preservation, surely would be pleasing if not also instructive." To his audience, which included the naturalist Dr. Benjamin Smith Barton and two members of the U.S. Congress, James Madison and John Page, as well as the anatomist Dr. Caspar Wistar, Peale made it clear that he had his heart set on the preservation of one particular species: "The portraits of the Presidents of Congress, the Presidents of the state of Pennsylvania, the ministers of high departments . . . I have taken and preserved, and have done as much as I have been able to afford in this line of the design." Reflecting on "man's" position in the system of nature, Peale quoted Richard Pulteney's abridgment of Linnaeus, suggesting that humans were exempt from, as well as subject to, the demands of taxonomy: "However the pride of man may be offended at the idea of being ranked with the beasts that perish, he nevertheless stands as *an animal,* in the system of nature, at the head of this order"—the "Primates"—"and as such is here described" (*PP* 2: 14).

Pulteney's statement reflects the extraordinary accomplishment and radicalism of eighteenth-century natural history, which proposed that the zoology of the human species presented in fact few differences from that of other natural species. In 1735, Linnaeus had lumped the "Simia" (and, for that matter, the sloths) together with the humans in a new order called "Anthropomorpha." Over the next decades, as if horrified by his own daring, he experimented with various names for the varietal distinctions that new anthropological "discoveries" seemed to necessitate. The 1735 edition of the *Systema naturae* distinguished between Europeans (*Homo sapiens europaeus albus*), American Indians (*Homo sapiens americanus rubescens*), Asians (*Homo sapiens asiaticus fuscus*), and Africans (*Homo sapiens africanus niger*). In 1758, the "Anthropomorpha" were replaced by the "Primates," and Linnaeus added other varieties, the "wild man" (*Homo sapiens ferus*) and the "monstrous man," *Homo sapiens monstrosus* (which included the Hottentots and the Patagonians). In the same edition of *Systema naturae,* Linnaeus also suggested that the genus *Homo* might contain at least one other species, namely the "nocturnal" man or *Homo troglodytes,* under which he synonymized the *Homo sylvestris* or "orang-outang" described in 1658 by Jakob de Bondt.[65] However, this mistake was corrected in the thirteenth edition, published posthumously in 1788. "Linnaeus seems to have been misled," wrote Pulteney, "by the accounts of credulous travellers, otherwise he would not have placed what is properly a *Simia* in the same genus with *Man.*"[66]

Such errors notwithstanding, Linnean systematics had contributed, as Phillip Sloan suggests, "to a profound blurring of the human-animal distinction." Sloan exaggerates, however, when he argues that, for Linnaeus as well as for Buffon, the one property remaining that prevented the full naturalization of human beings was *reason.* First of all, Linnaeus had a fairly exalted conception of just what it meant to be a rational creature and not an animal: "There is something within us, which cannot be seen, whence our knowledge of ourselves depends—that is *reason,* the most noble thing of all, in which man excels . . . all other animals."[67] Second, ever since Edward Tyson cut up the body of the first anthropoid ape to have made it all the way to England, other factors had entered into the debate on what exactly made human beings human. Tyson had concluded that the "great Agreement . . . between the *Orang-Outang,* and a *Man*" made his ape (which was not an orangutan but in reality a chimpanzee) a good candidate to be the "*Nexus of the Animal and the Rational.*" But Tyson also argued that, though this highest of subhuman creatures in the chain of being was equipped with the organs of speech, it still lacked the ability to use them.[68]

This idea was eagerly adopted by Buffon ("la langue et tous les organes de la voix sont les mêmes que dans l'homme, et cependant l'orang-outang ne parle pas"), because it supported his claim that the gap separating the ape from the man was in reality "immense."[69] However, if overcoming the divisions between apes and humans depended on whether or not apes actually made use of what they, too, had been equipped with, Buffon's "immense interval," at least in the eyes of some eighteenth-century thinkers, no longer appeared so wide. "Parle, et je baptise," the cardinal de Polignac exclaimed when faced with an orangutan in his cage in the Jardin du Roi, according to Diderot's *Suite de l'entretien,* a sequel to *Le Rêve de d'Alembert.*[70] The fact remained that in the system of nature, nonhuman creatures had never come so uncomfortably close to what Buffon still persisted in calling "l'être le plus noble de la création."[71]

As if frightened by this prospect, classical taxonomists, when talking about *Homo sapiens,* soon began to preach *difference-in-sameness* more than *sameness-in-difference.* Perhaps some of the varieties of *Homo sapiens* designated by Linnaeus were closer to the animals than others. For example, in the 1758 edition of the *Systema naturae,* Linnaeus himself already suggests that, while *Homo sapiens europaeus* is "ruled by customs" ("*Regitur* ritibus"), *Homo sapiens afer* lets himself be dominated by "caprice" ("*Regitur* arbitrio"). He further asserts that African women are "sinus pudoris," without shame, and have "*Mammae lactantes prolixae*," that is, breasts that lactate profusely, while their men are lazy and smear grease on their bodies.[72]

Peale, for one, was exhilarated by the possibilities of Linnean taxonomy and the perspective it offered on man as a "natural curiosity." A fairly established way of representing man in a natural history museum was, of course, through portraits: "By good and faithful paintings the likeness of man is perhaps with the greatest precision handed down to posterity" (*PP* 2: 14). Now Peale proposed the next logical step:

> There are other means to preserve, and hand dawn to succeeding generations, the relicks of such great men. . . . The mode I mean, is the preserving their bodies from corruption and being the food of worms: this is by the use of powerful anticepticks. Altho' perhaps it is not in the power of art, to preserve these bodies in that high perfection of form, which the well executed painting in portrait, and sculpture can produce; yet the *actual remains* of such men as I have just described, must be highly regarded by those, who reverence the memory of such luminaries as but seldom appear. (*PP* 2: 25)

The first specimen Peale had tried to preserve for his museum had been Dr. Franklin's cat. "Very soon," Peale recalls in his autobiography, "Peale found it necessary to preserve many things which were perishable that was presented to him, and Doctr. Franklin sent to him an angora cat which he had brought from France, an attempt was made to preserve it, but for want of knowledge . . . it was lost" (*A* 109). Now that his knowledge of the procedure had increased, Peale no longer regretted the loss of Franklin's cat; it was a shame, Peale suggested to the Visitors of the Philadelphia Museum, that he had not proposed "the means of preservation to that distinguished patriot," i.e., Franklin *himself*, who, Peale mused, could have perhaps been prevailed on to "suffer the remains of his body to be now in our view."[73]

Whether Peale knew it or not, Franklin would have been amenable to the idea that Peale so emphatically proposed to his distinguished audience. In a letter written in 1773 to Jacques Barbeu-Dubourg, Franklin reflected on the possibility of reserving plants in quicksilver so that they might keep their smell and color, and he claimed that he had heard of American flies drowning in Madeira wine and regaining life in Old England upon being hit by the rays of the European sun. Franklin went on to suggest that he himself, possessed with the ardent desire to see America "a hundred years hence," should be pleased to be preserved in liquor, a method he said he found infinitely preferable to an ordinary death, "un mort ordinaire."[74]

Washington's pheasants, Franklin's body—as collector's items they both point to the view Peale had of his museum as something that would be more than merely a metaphor for the world outside but would, in actuality, bring together, within its walls, the world itself. Frederik Ruysch's sixteenth-century *Theatrum anatomicum* in Leiden had included a parade of fetal skeletons, shocking to the modern viewer in their deliberate incongruousness. Festooned with beads and clad in lace, these babies rested in their glass jars, contentedly asleep, next to another form of display invented by Ruysch: gruesome tableaus composed of body parts.[75] Ruysch's "theater" turned death into art, a chillingly beautiful memento mori. Peale's eighteenth-century sensibility aimed for the opposite effect: his museum was an evocation not of death but of life. His displays celebrated not the flight of time but the permanence of human craftsmanship.

The editors of Peale's correspondence claim that "only one incident of the exhibition of an embalmed body is known."[76] It is certainly true that philosopher Jeremy Bentham, at his explicit request, was turned into an "auto-icon" after his death and then exhibited in University College, London, where he is still on display today. But other incidents of human bodies on display are known, and Peale was not easily deterred.[77] In 1806, he wrote to the members of a church committee in New York, expressing a lively interest in the "preserved" body of a child, which he had heard was in the possession of the church and which he wished to make part of his museum ("which I flatter myself will be considered a National Advantage"; *PP* 2: 946). He appealed to the gentlemen's "desire to aid the acquirement of knowledge in the science of Natural history" by providing him with this "relick of art and nature" (*PP* 2: 945–946, 947). Peale was unambiguous about his main qualification for making the request, which even in 1806 must have seemed a little out of the ordinary: "The preservation of human bodies has for many years engaged the thoughts of some of my leisure hours, & I have devised various means to effect it, some more perfectly than others" (*PP* 2: 946). Peale's brother-in-law Philip DePeyster on whose assistance in the matter he depended, soon had disappointing (and gruesomely precise) news for Peale. Not only had the gentlemen declined to help, it had also turned out that the body had simply been "embalmed" or rather kept in a tin-lined coffin with spirits, an amateurish procedure; the "Spirits had Considerably wasted and a part of the forehead which was dry had began to decay" (*PP* 2: 948). No body for Peale.[78]

Although he was never successful in obtaining the embalmed bodies of humans, Peale's natural history collection at various times included references to the human species. In 1796, the museum harbored the skeletons of a man and a woman from the Wabash nation. The catalogue admits that there was "nothing remarkable" about these bones and goes on to relate, in lieu of more pertinent scientific information, a story "of a singular fact, relative to these Indians." The two Wabash were a man and his wife, allies to the Americans in the Revolutionary War. They produced a child, "the fruit of their mutual love." After the sudden death of the mother, the child was fed for a while by American soldiers; when the father also died, it suddenly disappeared. Army surgeons, "wishing for anatomical subjects, dug up the dead Indian, and to their astonishment found the child they had before sought in vain, placed between the knees of its deceased father" (*SDC* 3). Reproached for their "barbaric" behavior, the other members of the tribe defended themselves by explaining that "the parents of the child being dead and none left to teach it to hunt, or make use of the bow and arrow, they had sent it to its parents" (*SDC* 3). The story is a horrifying reminder of the gratuitous hunt for the bodies, skeletons, and crania of Native Americans that was soon to develop into a favorite pastime of naturalists, doctors, and anthropologists and continue unabated throughout the nineteenth century.[79] However, the point of Peale's story, which stresses the "affection" and the "mutual love" that had united the child's parents, is ironically not the peculiar "barbarism" of the Native American couple but the sensible coherence and consequence of their "family-centered" worldview—as strange as it might have seemed to Peale's visitors that such an attitude would have required the newly orphaned child to die and be buried alongside the father.

The Indian couple vanished from the museum's records; the "Guide to the Philadelphia Museum" from 1804 only refers to the wax figures of two Shawnee chieftains, which were displayed in the museum's Model Room together with replicas of a Sandwich Islander, a Tahitian, and a "South American" (*PP* 2: 765). Peale's 155-page "Walk through the Philadelphia Museum" mentions no such displays at all. Nevertheless, David Brigham attaches great significance to Peale's use of "figures and artifacts of human difference." It is true and deplorable that Peale did not shy away from presenting some live specimens of humanity in his museum that were meant to be perceived as "odd"—among them an albino boy (billed as a "a white Negro") and an armless painter (an "instructive and consolatory example to the world generally").[80] But we must not forget that Peale's interest in the display of humans also extended to "normally" shaped bodies. Peale's view of humanity, however much it was shaped by prejudice and the ready acceptance of social difference, remained remarkably inclusive.

In a sense, Peale did succeed in his intention to contribute to the preservation of human bodies—at least in part, or, to be more precise, in *body parts*. The collector who replaced the eyes of his birds with glass on which he painted the "Colour proper to each bird" (*PP* 2: 43) also worked diligently on the perfection of prosthetic devices for humans, notably their teeth.[81] Peale did not perfect his skills merely to make up for his own deficiencies (although he himself, by the age of thirty-four, had lost many of his own teeth). Rather, he saw his production of various dental substitutes as part of his overall project of coaxing nature into the preservation of its beauty. For Peale, the presence of man-made implants in the human body did not dehumanize a creature whose nature consisted in "making," in artifice. "How pleasant it is," he jubilated in his autobiography, "to be able to masticate any kind of viands, to assist pronunciation, and to enjoy a sweet breath." Benefiting from these comforts himself, Peale felt it incumbent on him to serve the needs of others, too (*A* 477).

Peale's own dental problems were not unusual at a time when doctors thought that the regular brushing of one's teeth could be detrimental and instead recommended rinsing out one's mouth on a more or less regular basis. George Washington, for example, was known to possess ill-fitting, clacking dentures, "terrifying-looking contraptions," made from wood and the bones of a hippopotamus, which distorted his face, an impression supposedly reflected in the famous Gilbert Stuart portrait. As embarrassing as his dentures were, Washington was, according to his biographer James Thomas Flexner, probably "no more disfigured than was then common among the elderly and prosperous."[82] Colonial travelers had already pointed out that New England's population was "pitifully tooth-shaken," a fact they attributed to bad diet or to the unpropitious climate.[83] Over the years, the standards for dental care did not improve, and neither did American prosthodontics. In 1768, the soon-to-be-famous Boston silversmith Paul Revere advertised his ability to replace lost front teeth "with artificial ones, that looks as well as Natural & answeres the End of Speaking by all Intents," but wisely did not promise that his dental replacements would answer the end of chewing.[84] Writing from Paris in 1808, Rembrandt Peale urgently encouraged his wife to see to it that she and "the Children" cleaned their teeth regularly: "The french have good teeth only because they clean them often & rince them always after eating." Rembrandt felt all the more serious about this advice as he knew that his own teeth were "bad only from the want of using the Brush & powder—and that one of the front ones has broken off in the middle & I must lose of all my upper teeth for the same foolish reason" (*PP* 2: 1107).

Peale's diaries indicate that on his collecting trips he was not only busy mounting moorhens and kingfishers but, if needed, would also "bush up" a tooth that had given a friend's daughter some trouble (*PP* 2: 61). He himself once suffered the embarrassing experience of having his own dental plate, which connected artificial maxillary teeth to "some real teeth in the front," break while he and the famous German naturalist Alexander von Humboldt dined with President Madison in 1804 (*PP* 2: 693–694). Fortunately for Peale, a discreet local gunsmith was able to help out with the necessary tools. "I riveted in a new piece and returned to our company in about ½ an hour." From then on, Peale would always carry a spare set in his pocket: "Often I have found the advantage of them and changed my Teeth while at Table without the Company Observing what I was about by holding a handkerchief before my mouth" (*PP* 2: 694).

However, what induced Peale to search ceaselessly for sturdier dental implants during the last decade of his life was not simply the desire to avoid socially compromising situations for himself and similarly afflicted family and friends. If Peale's hunt for a complete skeleton of the American mammoth had for some time centered on the recovery of a complete set of teeth, the reason for this was, of course, that teeth would have provided important clues to the dietary habits of the "Incognitum." But teeth were also one of the most important ordering devices for the class of Mammalia in the Linnean system. According to Peale's primary source, Richard Pulteney, taxonomic distinctions in Linnaeus's "*artificial* arrangement" rested on difference in the number, situation, and the forms of the three kinds of teeth," incisors, canine teeth, and molars.[85] Peale's own catalogue from 1796 begins with a table showing the organization of quadrupeds mainly according to the structure of their teeth, the various designations of which are carefully explained in a footnote. In full accordance with Linnaeus, Peale and Beauvois define the first order of Mammalia, the Primates, as having "four front teeth incisive, the upper parallel." Following the example of Linnaeus, Peale admonishes "Man" to "know thyself," but he also makes it clear that man's self-knowledge should tell him that, "merely from the organization of his physical frame," he actually *does* belong with the other animals listed in the catalogue. And a footnote reminds critics of the Linnean system whereby the bat is, in fact, "similar to man in the number and form of its teeth" (*SDC* 2).

Peale devoted considerable effort to the perfection of dental replacements for the privileged species of mammals to which he belonged. He read widely on the subject, consulting French treatises, such as Claude Jacquier de Geraudly's *L'art de conserver les dents* (Paris 1737), as well as traveling practitioners. Unfortunately, teeth made out of animal substances such as ivory, bones, the teeth of sea cows (*PP* 3: 267), and shells were all "subject to decay" in the mouth (*PP* 3: 324). Peale proposed that the best and most durable teeth could probably be made out of porcelain ("Chinea Teeth"), glazed over in a furnace—no easy achievement, as Peale discovered. In October 1822, his oven overheated, ruining the materials with which he was experimenting (*PP* 4: 187). But he did not give up easily: "I am now rebuilding my furnace," he wrote to his son Rubens, "and shall make it on an improved plan . . . so much for Porselain teeth" (*PP* 4: 189). A letter written two weeks before his death, at age eighty-six, still shows him doing "some little work at Porcelain teeth, for to be confined to the House without employment will not agree with my Ideas" (*PP* 4: 578).

It is important to see Peale's creativity in this area not just as an expression of his belief that, as Sidney Hart has said, "man's reason could be used as a tool to better his condition."[86] Rather, Peale's interest in dentistry reveals an attitude toward the human body (and his *own* body, for that matter)[87] that, at least in principle, did not differ fundamentally from his treatment of the animals he exhibited—namely the conviction that nature, even human nature, was, where lacking in life, subject to skilled improvement, "mechanical" preservation. In August 1776, after Peale had joined the militia, we find him working on Mrs. Washington's portrait as well as on replacements for the teeth Jenny Ramsey had lost; using thread, he carefully tied them to the remaining teeth in her mouth (*PP* 1: 194). In Peale's extensive correspondence with his daughter Angelica after 1802, large sections are devoted to the discussion of her dental needs and the replacements (made of "Sea-Cow's tooth") that he was perfecting for her, a task which Peale took very seriously.[88] From the context in which these debates occur, it is obvious that Peale thought of his dentistry as being on a par with his portrait-painting: "Besides doing some . . . Portraits, . . . I set to and have made a Silver bottom to those Teeth which you said fitted your mouth" (*PP* 3: 708). He promised Angelica that he would soon improve his implants: "I want you much to be with me that I might try to make your teeth better than I can possibly do at present. . . . I need not say this to urge your exertions to visit us" (*PP* 3: 261–262). When Peale finally developed a technique that allowed him to carve both teeth and base out of a solid block of porcelain, he went on the market, offering his state-of-the art replacements for $150 apiece.[89]

Peale had hoped his museum would eventually become "the best . . . in the world," a place where "the articles" were kept "in such good preservation *that time cannot alter them*" (*PP* 2: 1015; my emphasis). In a revealing letter, the sixty-nine-year-old William Thornton, head of the Patent Office in Washington, appropriately identified Peale's interest in dentistry as yet another one of his ceaseless attempts—like the Philadelphia Museum—to defy "the effects of time" and "grin the old Fellow" Death "out of countenance":

> If he knock out a Set of your Teeth, you laugh at his ill-nature, & clap in a better set.—If a man loses a *single* Tooth, he may be truly called a *Tooth less* Fellow—but you cannot be so called, though you have not a tooth in your head: yet you are no mumbler, for you can crack a nut with any young man of 20.—Though man is doomed to suffer great evils in life, Science has done much for our regeneration. A good pair of Spectacles, & a new Set of Teeth, not only enable a man to eat well, but to see what he eats. (*PP* 3: 602)

Perseverance: A Collector's Autobiography

In his famous autobiography, a contemporary of Peale's, the Italian poet and dramatist Vittorio Alfieri (1749–1804), formulated what might very well count as one of the central anthropological paradigms, and one of the primary justifications, of classical autobiography: "The principal aim of this work is the study . . . of man in general. And about which man could one speak better than about oneself?"[90] He whose proper study is mankind should best study himself. Peale's autobiography, written late in his life, subscribes to this premise—and yet, in one crucial sense, does not: Peale writes not about himself but about "Peale."

If we believe his own words, Peale did not particularly enjoy writing his autobiography. On 26 March 1826, in a letter to his son Rubens, Peale complained that writing his "life" was "a heavy job—my almost whole imployment of the winter" (*PP* 4: 523). However, he had been given to understand that there was some public interest in his "Biographical Sketch." Besides, there was the prospect of remuneration: if money could be made by the publication, Peale wrote Elizabeth Bend on 6 April 1826, he saw "no reason why I should not be benefited by the work, but more of this hereafter" (*PP* 4: 524). Alas, Peale did not have much time left for reflections "hereafter." In January 1827, the eighty-six-year-old widower went a-courting again. All by himself, he traveled to New York to woo Mary Stansbury, a teacher at the New York Institution of the Deaf and Dumb, the daughter of a former acquaintance from Philadelphia. Even though Peale managed to convince Miss Stanbury of his undiminished "corporeal vigour" (*PP* 4: 564, n. 9) and impressed her by offering her some great artifical teeth, the much younger woman announced that she would rather not become his fourth wife. On the arduous trip back to Philadelphia, the octogenarian strained his heart while attempting to lift his heavy trunk. After a brief and violent illness, which he coolly analyzed in letters to his children, Charles Willson Peale died on the evening of 22 February 1827.

The "Biographical Sketch" that kept him so busy during the last full winter of his life has survived only in draft form, but what we have is a true collector's autobiography, in more than one sense. First, it is a book *about* collecting. One particularly memorable episode (*A* 229–232) describes Peale's pursuit of insects, a new ambition that had been fueled by one of Peale's Baltimore acquaintances, the Reverend Mr. Kurtz, who possessed a "pritty collection of Beatles": "Before this time he had never taken much trouble, except to catch Papillions. This gentleman taught him to seek for subjects under *stones, raggs, dung* and *carrion,* and for the smaller insects in the *flowers* and on the *small bushes along the meadows,* and under the bark of old Trees &c., and the best time to find them, that is from ten o'clock untill about 2 o'clock, however they made sometimes their excursions until late in the afternoon" (*A* 229–230). Peale stresses the sacrifices brought upon him by this new obsession with small specimens:

To persons unacquainted with this part of Natural History, his labours must appear extraordinary, that he should day after day continue to walk for many miles in the sun, in the hotest weather, and in those places most exposed to parching heats of the sun, between the hills, and through the wet grounds amidst the briars and thick bushes, which most people would avoid for fear of snakes, ticks &c. but to gain an acquaintance with a new set of beings, for new they certainly were to him, as they arc to most men. (*A* 230)

Collectors are not like "most people"; the lure of novel insights overrules their desire for personal comfort. Peale reflects on the moral lesson taught him by his collecting of insects and, in his excitement, inadvertently switches tenses: "We commonly pass on regardless of an infinite number of beings who constantly avoid our sight. Peale frequently since he has began this work has stopped to reflect on our modes of passing through life making comparisons on the persuits of men, some have no pleasure but in their prospects of gaining wealth, some gratifying their senses rather of the brutal kind, their passions, and some few in persuits of arts and science." Gathering insects, Peale experiences his own private epiphany and vows to mend his ways as a collector: "Some collectors like Peale, have only looked for subjects large and striking to the sight, but now he declares that he finds an equal pleasure in seeking for an acquaintance with these little animals whose life is spent perhaps on a *single leaf*, or at most on a single bush" (*A* 230). Peale's insects are not merely collectibles but are endowed with a life of their own. What these wonderful passages from Peale's autobiography illustrate is perhaps the single most distinguishing characteristic of Peale the naturalist, the collector, the inventor, and finally, the writer: *a constant capacity to surprise even himself.* As long as the collector perseveres, he will be successful, even if he finds what he hasn't been looking for:

It is diverting to watch a flower as you approach a bush, and see the little being watch your approach, turning round a *twig*, or part of a flower to avoid your sight, and in an instant draw in its legs and roll off, sometimes falling from leaf to leaf to get a passage to the ground—some others as quickly take to flight, some others depending on their shelly coats remain where you first see them regardless of danger, others depending on the velosity of their flight stay untill you give them the alarm, after which it becomes almost impossible to come near them. (*A* 231)

Successful capture matters less than new insights, and those may only be gained through sheer stick-to-itiveness:

Mr. Kirtz [*sic*] had taken a fly with transparent wings, a slender Insect with a tuft of feathers on its tail, it appeared to them a rare species, they could not find any more of them at this time, however Peale went frequently to the same meadow, and there only he could find them, but although he ardently wished to possess them, and although he had at different times seen about 5 or 6 of them, yet he could not take one of them, sometimes he has had his forceps within a few inches of them, but their quick motion aided with quick sight, they always avoided the trap.

For the true collector, there are no disappointments. The pleasure of the chase is an end in itself, and unforeseen rewards are in store for those who did not covet them, "for although we do not always find the Insect we go in search for, yet our labours are oftain overpaid by the obtaining new subjects . . . perhaps before quite unknown or un-described" (*A* 231). Peale became so engrossed in his collecting that he would be totally oblivious to the passage of time: "The last morning Peale went out he intended to walk several miles before dinner time, and to extend his walk to the country seat of Mr. Valk, but in the first meadow he found himself disposed to examine the bushes attentively, and there he found so much amusement that several hours passed away before he could think of leaving those bewitching animals, and looking at his watch he found it dinner time, when he scarcely thought he had began his pursuit" (*A* 231–232).

This episode encapsulates some of the characteristic features of Peale's autobiographical project—notably the curious vacillation between, on the one hand, the narrator's studious self-detachment (effected by the consistent use of the third-person point of view) and, on the other, the appeal to the reader's empathy with the actual experiences of "Poor Peale" (*A* 29). Peale not only insists on the importance of detail ("it is the detail alone that enchants us," *A* 132), but he also dramatizes his position by careful attention to detail wherever this seems appropriate. A striking effect is achieved, for example, through the focus on what actually *escapes* the collection, the small insect sitting on a leaf, watching and, where possible, avoiding the human predator's sight. Here, Peale successfully adopts what Thoreau, claiming that "entomology extends the limits of being in a new direction," would later call the "insect view": "Nature will bear the closest inspection; she invites us to lay our eye level with the smallest leaf, and take an insect view of its plain."[91] Peale individualizes the encounter between the collector and the object of his collecting, which, since it stares back, immediately ceases to be simply that—an *object*, a mere collector's item.

If in the episode quoted here Peale's attention necessarily has to move from leaf to leaf, this is in a sense also how we need to read Peale's entire autobiography: leaf by leaf, page by page. This is the second reason why this text is a collector's autobiography. Put in more fashionable terms, in Peale's autobiography we are confronted not with a "Work" or a "grand narrative" (in Jean-François Lyotard's sense) but with a series of stories, "local narratives." Like his museum, Peale's autobiography is the sum of its parts. It is doubtful that this overall impression would have changed significantly even if Peale were to have had the opportunity for revisions and retouching.

Peale called his autobiography a "Novel" (*PP* 4: 513), and the careful reader will indeed notice parallels with earlier eighteenth-century fictional texts, particularly Henry Fielding's *The History of Tom Jones* (1749). If the recommendation of "prudence" has been seen as the overriding ethical purpose of Fielding's novel, the guiding principle behind Peale's autobiography is a more modest one and could perhaps best be described as "perseverance." The "picaresque" novel of Charles Willson Peale's life has poor, honest Peale himself as its mostly happy but somewhat disorganized protagonist, a reluctant hero, who has to go through a series of trials of his "fortitude." Like Fielding, Peale is not unafraid of "Chasms" in his story, cheerfully acknowledging the "Blanks in the grand Lottery of Time."[92] But unlike the "tyrannical" narrator of *Tom Jones*, the presiding consciousness in Peale's third-person autobiography has no general purpose, no "Matters of Consequence" in mind, which would easily justify his many omissions. The true model for Peale's narrative must be found elsewhere. We may assume that at some point, having retraced his life up to 1824, Peale lost interest in his project, but it is no coincidence that his memoirs survive only in draft form, a little incoherent in places and often incomplete (as first drafts tend to be), but always fresh in the immediacy of the telling, which is palpable in almost every line. Accident has taken the place of providence, which has been displaced to the function of a distant, absent-minded, none-too-helpful stage manager who is a bit confused about the proper order of events. Or, to use imagery that appears to be more appropriate to the present context: Peale has left many "holes" in the text of his autobiography—the way a collector would attempt to leave some free space on his shelves for objects yet to be discovered or to be acquired, if time and luck will allow. A case in point is the gap in Peale's narrative where he proposes "some more particulars" concerning one of his friends and sponsors, John Beale Bordley, yet fails to deliver on his promise. Titian Peale later explained that his father had left the pages "blank until he felt he could do full justice to the benevolent . . . friend" (*A* 139).

If the German Enlightenment wit Georg Christoph Lichtenberg found himself wishing for a chance to prepare a "second edition" of his life,[93] Peale's biographical sketch from the outset seems to have been conceived for future revision. And this is not an incidental but an inherent characteristic of Peale's autobiographical procedure, in which everything seems more provisional than final: Peale is not recollecting but in fact *collecting* his life. Even in its written form, the life of Charles Willson Peale does not really develop; rather, it *continues*, as a series of "collecting adventures."

If progress in Peale's life can be measured at all,[94] it becomes evident in Peale's cease-less accumulation of useful knowledge—knowledge, for example, about the mechanics of a watch he owned as a boy: "The Watch getting out of order, he was obliged to pay 5 s. to have it put in repair; not long after the watch was again wrong and the boy think-ing it too heavy a Tax to pay out of his hard earnings, determined to try if he could not save the expence by repairing the Watch himself, and in this attempt it may easily be imagined he did the Watch but little service, however by this essay, he acquired knowledge of the principles of such Machines" (*A* 3). This episode acquires further significance when applied to Peale's unusual autobiographical narrative as a whole: in a peculiar way, for Peale the experience of time is not ontological but resides in the wheels and springs that make a watch tick.

In an attempt to substantiate the "close relation" between narrativity and tempo-rality, Paul Ricoeur has emphasized that, far from reflecting a chronological series of instances in time (indicated, for example, by the ticking of a watch), narrative always *reckons with time.* It elicits patterns or configurations out of mechanical succession and thus, as in Saint Augustine's autobiography, eventually transcends episodic time by means of *repetition.* More specifically, it makes the end present in the beginning, the beginning in the end: "The hero *is* what he was."[95] This is true even where the opposite seems to be the case. Benjamin Franklin, for one, suggested in his autobiography that he had been "the more particular in this Description of my Journey . . . that you may in your Mind compare such unlikely Beginning with the Figure I have since made there."[96] Peale was, of course, familiar with Franklin's writings,[97] and he had also read and enjoyed Rousseau's autobiography: "If any man will say, that on looking back on his past life, he knows nothing to repent of, that man I will not believe. Rousseau, that great Philosopher . . . says in his Confessions, that he suffered upwards of (60) years for the stealing of a Riband and suffering an Innocent Girl to bear the blame of the theft, he laments most bitterly this base action, and paints the consequences in lively colours" (*A* 12).[98] In his *Confessions,* Rousseau is at pains to establish not only the consequences of his misdeeds but also the lasting impact on his life of other people's actions, which makes his text more than just the "dirty maze" that he claims it is. Mlle. Lambercier chastises the boy, by nature "extremely ardent, extremely lascivious and forward," for a reason not even mentioned by Rousseau. Thus, she instills in him a lifelong taste for masochistic pleasures, indeed for the *repetition* of that first delightfully perverse experience: "Who would believe it, that this childish chastisement, received at eight years old from the hand of a girl of thirty, should decide my desires, my passions, for the rest of my days?"[99] Put more accurately, young Rousseau's accidental punishment only reinforces his natural disposition, his "violent constitution." This is different in

Peale's text; not only are Rousseau's dark desires, his illicit passions and "vivacity of feeling" noticeably absent from Peale's narrative, but what is also lacking is the experience of repetition that haunts the pages of *Les Confessions*.[100] Although he would willingly trade specimens he had preserved for that purpose with other museums, Peale's collection itself contained "no duplicates" (*DI* 34), and Peale's description of his own life is likewise free of experiences that too closely duplicate earlier ones.

For example, Peale shares with Rousseau the experience of a joyless apprenticeship, but whereas Rousseau takes to stealing, Peale soon finds himself running his master's saddlery. Hoping to set up shop himself, he eventually buys his master's equipment, thus incurring debts that bring about his ruin and make him temporarily dependent on the help of others. But Peale does not comment on the importance of these experiences for his later life; from the beginning, Peale *is who he is*, fully present in whatever situation he finds himself in; he does not become, he *perseveres*. Depending on his moods, Rousseau is "a rogue" or "a worm";[101] Peale, however, is always Peale. The sections of his autobiography dealing with "poor Peale's" boyhood and childhood have little to offer in the way of formative influences, experiences that would have instilled in him a sense of his mission as a painter—in the way Rousseau, by reading novels with his father all night long, had developed his literary sensibility early on. In his autobiography, Peale does not reckon with time, time reckons with him, and Peale persists by remaining, inexplicably but triumphantly, "Peale."

Peale's first brush with portrait-painting takes place when "at an early period of his life" an ill-advised grandmother decides to put the boy's "fondness for Pictures" (*A* 16) to good use. For him, the experience is more terrifying than encouraging: "An Uncle was dead, and Charles's Grand Mother, a very aged woman, begged him to draw her a picture from the corpse, the boy told his Grand-Mother that he did not know how to do it, she persisted that he could, if he would try. All her entreaties were in vain, the task appeared too difficult" (*A* 16). There is no attempt here on the part of the autobiographer to connect this early failure to preserve the likeness of a dead person with the wish of the mature artist to display preserved bodies in his museum. If we believe Peale's own account, it was only after he had tried his hand at several other trades without much financial success that he considered painting as a way out. Having seen a few landscapes and a portrait in an amateur painter's room in Norfolk, Peale decided that he "possibly might do better by painting, than with his other Trades" (*A* 17). Had these "miserably done" paintings been any better, "perhaps they would not have lead Peale to the idea of attempting anything in that way, but rather have smothered this faint spark of Genius" (*A* 15). And this is about as close as Peale gets to commenting on the general course of his life. What determines his autobiographical account is not, as in Rousseau's case, the happy or unhappy coincidence of chance and natural predisposition; rather, it is the accidental avoidance of potentially worse situations.

Accident is also what determines Peale's own writing process. Awareness of the erratic structure of his text once leads him to confirm the relevance of his many "descriptions." While superficially they halt the flow of the narrative, they in fact help move it along. It is the cumulative principle of his text that keeps Peale riveted to his desk hour after hour, inducing him to write "more" than he had originally planned. "These minute descriptions is more than the writer intended when he began this history but as these tasks (transactions) are considered as trials of Peale's fortitude, and when he conceived it was a duty incumbent on him, he endeavored to act with a steady perseverance, in as strict a line as he imagined would promote the common cause of America" (*A* 53).

The consistent use of the third person singular throughout Peale's autobiography (except on those pages where he quotes from his diary and obviously did not have the time to revise) serves to heighten the objectivity of the narrative, reminding the reader of other eighteenth-century autobiographies such as Johann Heinrich Jung-Stilling's *Lebensgeschichte* or Goethe's early "Selbstschilderung" (1797).[102] It is likely that the third-person autobiography written by Linnaeus in 1770 and included in the second edition of Pulteney's *General View of the Writings of Linnaeus* was a more immediate influence, even though the self-congratulatory tone of the diary-like *Vita Caroli Linnaei* ("All the child's powers, both of mind and body, conspired to make him an excellent natural historian"; "No person has ever had a more solid knowledge of all the three kingdoms of nature") is a far cry from Peale's subdued look at his own trials and tribulations.[103] However, both in Linnaeus's and in Peale's memoirs, the main advantage of third-person narration is that it conveniently dispenses with the reader's expectations of probing reflections on the part of the author. Explication, introspection, and hierarchy are replaced by mere "perseverance" and pertinacity: "Perseverances [sic] was his motto and that was the name he gave to his farm" (*A* 389).[104] Or, as Peale defines perseverance in an earlier, self-promoting passage: "That man who can go through various sceanes of distress, and bear with fortitude the rude shocks of adversity . . . certainly deserves commendation" (*A* 36). The quality of the man was that of his museum, too: "persevering industry," not good fortune and wealth, had made the Philadelphia Museum what it was (*PP* 2: 759; *W* 4).

My account of Peale's autobiography is not meant to suggest that he was a naive writer; his text, even in the unrevised state in which he left it, is in fact a fairly sophisticated performance, notwithstanding the rather low opinion Peale had of his own writerly skills ("Writing when the Person has any talents, may be a lasting source of amusement, and the writer of this sketch laments his want of abilities in that way"; *A* 36). The basic unit of Peale's narrative is not the "*whole* of a life," which Benjamin Vaughan, in a letter included in Franklin's autobiography, proposed should influence a man's conduct.[105] Rather, it is the "situation" in which the autobiographical subject happens to find himself, and the challenge is to prepare himself for life's many contingencies as best he can. We can see this principle at work, for example, in the sections of the autobiography that deal with the role Peale played in the American Revolution. A self-described "advocate . . . of the liberties of America" (*A* 70), Peale took pride in the fact that while living in London he had always refused to take his hat off when the king passed by (*A* 40). The passages detailing his subsequent involvement in the militia might have been colored by the pacifism of his later years, but they still offer a remarkably lively description of the conditions under which the American revolutionaries fought. Consider Peale's studiously precise explanation, complete with temporal qualification, of exactly why the ground on which he was forced to rest his fatigued body during his military campaigns would "sometimes" have been particularly unhealthy: "Peale was a thin, spare, pale faced man, in appearance totally unfit to endure the fatigues of long marches, and lying on the cold wet ground, sometimes covered with snow" (*A* 45). However, "temperance" and, more important, "a forethought of providing for the worst that might happen" helped him endure "this campaign better than many others whose appearance was more robust." He "always carried a piece of dryed Beef and Bisquits in his Pocket, and Water in his canteen which he found, was much better than Rum" (*A* 45).

Peale's autobiography is pervaded by a sense of the sometimes incredible luck that has usually allowed him to escape from difficult situations in one piece—combined with the confidence that, even if he had not escaped, the basically benign order of the universe would have allowed him to make the best out of a potential misfortune. One cold night during the campaign, Peale nearly fell victim to hypothermia: "The cold awakened him, and he found his right hand in an almost senseless state of feeling, which alarmed him greatly, and he emediately sett to work with rubing it in cold water, and continued this labour the remainder of the night, and by morning had recovered the feeling except in two of his fingers, many applications were afterwards used, with frequent rubings, and in something more than two months they perfectly recovered their proper feeling" (*A* 62). The auto-biographer coolly concludes: "The loss of a hand to a Person who had his living to get by his labours, would undoubtedly have been a great affliction, yet bountiful nature often gives a wonderful use of Limbs, before but little made use off" (*A* 63).

Readers wanting to put together their own portrait of "poor pale-faced Peale" and his brushes with disaster might consider a later episode from the years in which Peale kept a number of live animals in the yard of the Pennsylvania State House, among them a large and mostly friendly female elk, which, however, sometimes proved to be a bit difficult to control. Peale's report about an incident connected with this fact is truly comical yet also contains, in a nutshell, the auto-biographer's basic philosophy. The elk, bothered by boys who had taunted her (Peale's comment: "such is fun to boys"), started to bolt, and the concerned owner of the museum was forced to run out himself in order to restrain the animal: "She would suffer Peale to stroke her, he put a halter in his pocket, and getting along side of the Elk, slipped the halter over her head, she darted off in an instant, and the long rope in his pocket would not deliver, the consequence was that the animal by her quick motion threw him down, and she dragged him about 30 or 40 feet, when fortunately the skirt of his coat gave way, and thus saved his life, for had it held fast, she would have dragged him against the Trees, and would innivitably have killed him" (*A* 222). Saved by a fragile piece of cloth, Peale knows there is no lesson to be learned from the experience.

Reflections—even those concerning the need for more reflection (*A* 226)—arise in Peale's narrative because they are triggered by particular situations. In fact, the organizing idea behind his narrative is an awareness of life's many unpredictably difficult (but never fatal) situations rather than an underlying awareness of the complexity of "Life" itself. Consequently, Peale does not explicitly attempt to distinguish between what other autobiographers in the tradition of the Augustinian confessional narrative with its characteristic climactic turning point would call crucial and less crucial events. Peale did not believe that a sense of hierarchy was what we should derive from our study of natural history:

> Whether we begin with the first or last link of the chain is of little consequence; whether we first view the simple naked animals destitute of limbs, such as worms, and ascend through all the various classes of different organized matter until we reach the most intelligent creature, man;
>
> Or whether we commence our view from those standing erect, and then descend through all the gradation to such as creep in the dust,—is of little consequence, provided we proceed step by step, to trace the beauties which we shall find that each possesses, in its relative situation to other beings;—its force, its intelligence, and its activity to supply its wants, and protect its young. (*DI* 33)

Applying such a "step-by-step" principle to his own life, Peale accords as much space to descriptions of his domestic life, his museum-keeping, and his farming as he does to, say, an episode from his time as an itinerant portrait-painter when one of his clients accidentally mistreats his old, beloved dog: "This poor lame dog was resting himself on a chair, when a principal of the family cruelly threw him on the floor with violence; this inhumane act so offended Peale that he determined to make all the dispatch he possibly could to finish his portraits, in order to leave a house where his dog was not welcome to rest; and he does not remember to have painted with more expidition since he practiced the art" (*A* 182).

Throughout his autobiography, Peale is always busy doing things: "Where a person is willing to be industrious, and thereby amused . . . even on that restless Element, the Ocean, ways may be found to be so" (*A* 36).[106] Traveling to England in the winter of 1766–1767, Peale is "sick on every turn of rough water," but he nevertheless manufactures a violin for himself, "for he had some fondness for Musick" (*A* 30). Coming back two years later, he reads the captain's set of the complete works of Alexander Pope ("with other books to the number of 31 Voll.") and paints him and another passenger in oils (*A* 35). Once, embarked on a voyage to Baltimore, Peale, annoyed by the laziness of his fellow passengers, whiles away the time by developing and perfecting brushes that can be lengthened or shortened on demand:

> He brought with him a number of small tubes of brass, these he cut into pieces and screwed large wire to fit the calebere of them, and thereby he made his Pensils sticks of more than twice the usial length of such tools, for having of late accustomed himself to paint portraits with very long brushes, altho on some occasions he had found it necessary to use shorter, where small touches were necessary to the finishing the work, so that by having his pensil stick made with screws in the middle, he might use them long or short as fancy might direct. (*A* 224)

If Rousseau claims that he *felt* before he *thought*, it seems that Peale *did* before he either felt or thought. Sometimes, muses Peale in one of his rare retrospective summaries, he might have been too busy, and a feeling of time badly spent and chances for profit wasted evidently haunts him, at least occasionally, in his old age. But it is also evident that Peale's interest in newfangled contraptions did not cease even after he had established himself on his farm, where, among other things, he tried to construct a fountain for a basin below his greenhouse only to discover that his logpipes would inevitably get blocked by frogs. "These amusements cost some money and much time," Peale admits, but "the labour gave health," and "happiness is the result of constant employment." In short, Peale's "inventions pleased himself, and they gave pleasure to others, and offended none—being perfectly innocent." Peale's defense is simple but not disingenuous: "The economist will say, time, money and labour was mispent. He answers that happiness is worth millions" (*A* 394).

In other words, "misspending" is just a less generous term for what Peale himself would call the fundamental "happiness" that keeps his narrative alive. And "happiness," as Peale thought, depends not only on "good" or "bad" luck; within limits, it can be engineered. In *Discourse Introductory*, he declares that "if we are not the most cheerful . . . and happiest creatures that inhabit this globe the fault arises wholly from our want of a proper education" (*DI* 43).[107] Peale's most important contribution to happiness, his own as well as that of his fellow human beings, was, of course, the museum, but just as Peale's collection can never be complete, happiness for Peale is an endless pursuit rather than an achievement. However, lest his readers confuse happiness with exuberance, "poor pale-faced Peale" never allows it to become the presiding theme of his narrative. Consider the description of Peale's carriage ride to New York with his new wife-to-be, Elizabeth DePeyster. After the death of his first wife, Peale, now a despairing father of seven, has unsuccessfully reviewed several eligible widows in Maryland,[108] only to discover Elizabeth, a visitor in his own museum back home, "view[ing] the articles" (*A* 150). With considerable elation, he invites her to accompany him to New York, where they plan to ask her father's blessing for the marriage. It is characteristic that Peale should mention the "dust" on the no doubt rocky road to New York as a trivial but real impediment to his experiencing full contentment and that he should use this reference as a means of including and at the same time censoring those readers who probably would not have shared his own sense of modest exuberance:

> The weather was fine and the season delightful and the business Peale was going on the most pleasing that man can enjoy; the company, the lady still heightening his pleasure by her agreeable conversation, he had nothing to interrupt happiness except a little dust, and for other description he leaves it to the feelings of those who may choose to read this journal.
>
> The affectionate heart will readily enter into the true spirit of his happy situation, but the cold, the insensible soul, will perhaps paint to themselves a different scene from what it was to Peale. (*A* 152–53)

This and similar passages are indicative of the anthropology that also informs Peale's conception of himself as a collector. In a wonderful verbal sleight-of-hand, Peale once declared that he wanted to be "not unconspicuously useful" to his country (*DI* 3). Being human is a privilege, but also an obligation: "The injunction given to Adam to name the works of creation, implied a necessity to become acquainted with those works," insisted

Peale (*I* 7). Everything is "beautiful in its kind," and antipathies to any part of God's creation are a "stigma on man's reason." In fact, the worst enemy humans can have is "ennui"—the want of employment that Peale himself dreaded. According to Peale, the entire world is a museum of natural history, "an unceasing fund of entertainment, from which no situation whatever in life, could wholly deprive us" (*DI* 12). "Not unconspicuously" exhibiting himself in and through this museum, in his self-portrait as well as in his autobiography, Peale seemed like a modern Noah for whom the Deluge was less a reason for worry than a source of delight since it allowed him to build his wonderful collection. What Peale tells us is that he lived his life as he assembled his museum, "leaf by leaf" (*I* 19). "Reading one leaf at a time"—this was Peale's hope—we might come to enjoy not only a life but, bit by bit, the whole world (*DI* 33).

Charles Willson Peale's autobiography merely "collects" where overarching syntheses and retrospective summaries might be expected; like his collection, it is not more than the sum of its parts. Peale's careful balancing of the particular against the general was based on the premise that a good museum collection should serve as the meeting point of two processes which for Peale were not yet opposed: the humanization of nature and the naturalization of humanity.

CHAPTER 3

Collecting Human Nature

P. T. BARNUM

Collecting as Storekeeping

Accounts of the history of the Philadelphia Museum or its immediate offshoots in Baltimore and New York frequently end on a somber note. Fighting against declining revenues and rising expenses, Peale's sons, so hopefully named Rembrandt, Rubens, and Titian, were finding it increasingly difficult to preserve their father's dream of a museum that would entertain as well as educate urban audiences. When Rubens Peale became the "conductor" of the Peale Museum in Baltimore, founded by his brother Rembrandt in 1814, he announced that the museum's galleries would be lighted every night, that additional seats would be installed in the lecture room, and that a band would play "three times a week." He paid $650 for the exclusive exhibition of John Warren's Egyptian mummy, which he displayed along with a tattooed head from New Zealand, allowed "learned dogs" to bark their way through silly tricks, and hired bands of Native Americans for performances of "traditional" dances and "scalping maneuvers."[1] When he finally launched his own museum on Broadway, auspiciously called the Parthenon, Rubens devoted his lecture room "more and more to purely theatrical attractions," preferably those that required audience participation, mesmerism among them.[2]

A good example of the new emphasis in curatorial interests is the Western Museum in Cincinnati, whose deterioration from a solid natural history collection (in which John James Audubon had been responsible for the mounting of birds and fish) to popular entertainment has been documented by Daniel Aaron.[3] In 1823, Joseph Dorfeuille, one of the collection's curators, was appointed as the new director. He soon began to offer a variety of entertainments, the most enduringly successful of which became widely known as "Dorfeuille's Hell": a wax-figure tableau featured Lucifer amidst his demons, surrounded by fountains of flames and columns of icicles. The wax figures were animated by hidden machinery, while Dorfeuille's employees shrieked, snorted, grunted, and groaned in the wings.[4]

In the Philadelphia Museum, Titian, the most accomplished naturalist among Peale's sons, still strove to maintain a more or less consistent level of quality. He saw his father's collection through two major moves, first to the Philadelphia Arcade in 1827 and then, in 1838, to an entirely new building at Ninth and George streets. Here an English traveler, otherwise full of disdain for the "worthless and trashy articles" accumulated in most museums in the United States, found much to praise about Peale's collection, calling it "the very finest, if not indeed the only really good Museum in the United States . . . quite equal to many of the best in Europe."[5] Such approval notwithstanding, the financial situation of this "really good Museum" was becoming more and more precarious. At a time when one quarter to one third of the federal government's budget went into the subsidy of the scientific exploration of the West, there was less money left than ever for the support of already existing collections.[6] Quantity often took the place of quality: Wilkes's "Great United States Exploring Expedition," one of whose members had been Titian Peale himself, returned with a staggering 160,000 specimens, many of which were subsequently lost or destroyed. The Philadelphia Museum as planned and implemented by Charles Willson Peale would have hardly been equipped to cope with the massive influx of information that came with these and similar expeditions. As Paul Farber notes with reference to Buffon, "in the middle of the eighteenth century an ambitious, disciplined, and well-placed individual . . . could seriously consider writing a complete natural history, and could indeed publish a thirty-six volume one covering a theory of the world, man, the quadrupeds, the birds, and the minerals. By the middle of the nineteenth century, even the contemplation of a serious project of Buffon's scope would have been hubris or folly."[7]

On 1 May 1843, the building that housed Peale's museum was sold at a sheriff's auction. Peale's grandson Edmund acquired what was left of the original collection and, two years later, moved the exhibits to the Old Masonic Hall, where he became increasingly desperate in his attempts to stay afloat amidst a competition for audiences that had become truly fierce. Peale's "pet" exhibit, the mammoth, was soon sold to the Hessisches Landesmuseum in Darmstadt.

It had been Peale's fate to be a kind of institutional nationalist before the political climate had been ripe for large national institutions capable of processing the new wealth of data that was becoming available. However, the ultimate blow to Peale's enterprise came not with the establishment in the 1840s of the Smithsonian Institution in Washington but with the arrival on the scene of the ultimate "showman"—Phineas

Taylor Barnum, a.k.a. "the Prince of Humbugs" or "the Show Prince" (titles he invented and happily used himself). Through a clever ruse, Barnum, in December 1841, became the owner of Scudder's American Museum in New York City. His new establishment became so successful that within a short period of time the walls of Rubens Peale's tawdry temple came tumbling down, too. Barnum merged Scudder's enterprise and the collections from Peale's Parthenon to form his own American Museum. In 1849, acting jointly with his friend and accomplice Moses Kimball, the proprietor of the Boston Museum, Barnum also acquired the remnants of the Philadelphia Museum once owned by Peale pere: among the items listed in the catalogue of the sheriff's auction were 1,624 birds, 214 fish, 11 cases of minerals, 194 mounted quadrupeds, and sundry Native American, Chinese, Northwest Coast, and South American curiosities.[8] Barnum sold the entire collection again just a few years later; it is anyone's guess how many of the specimens survived the fire that, in December 1851, destroyed the Swaim Building in Philadelphia, the last home of Peale's displays.

Phineas Taylor Barnum was a wily storekeeper from Bethel, Connecticut, whose spectacular success as a museum-keeper quickly added a new word to the English language. When Charles Sherwood Stratton, better known as Barnum's dwarf "General Tom Thumb,"[9] returned to his hometown of Bridgeport after his first tour through Europe with his new manager Barnum, one of Stratton's neighbors exclaimed: "We never thought Charlie much of a phenomenon when he lived among us . . . but now that he has become 'Barnumized,' he is a rare curiosity" (*L* 292).

Curiosities are *made*, this statement implies, not *born*. And as he was making and remaking his curiosities, Barnum made and remade himself. Like Peale, he attributed his success to perseverance or, as he preferred to call it, "Yankee-stick-to-it-iveness," which he said had always been "a noted feature in my character" (*ST* 2: 574).

Barnum's patient investment in his own admirable character traits paid off, literally as well as metaphorically. As early as 1850, a guidebook to his American Museum, sold at the door for twelve and a half cents, hailed Barnum as the "NAPOLEON of his profession" (*AMI* 1). Thirty years later, in 1882, the Smithsonian, in a letter signed by Spencer Fullerton Baird, requested the favor of being allowed to produce a mask from Barnum's face "to be placed in our series of representations of men who have distinguished themselves for what they have done as promoters of the natural sciences" (*ST* 2: 798).[10]

This chapter concerns itself mainly with Barnum's American Museum and thus with the first three decades of "Barnumization." Beginning in December 1841, Barnum during this period devoted most of his considerable energies, with an interruption of five years during which he unsuccessfully engaged in other business ventures,[11] to the expansion of his museum collections. Although Horace Greeley encouraged him to take such a disaster as a sign that he should quit and go fishing, Barnum remained unfazed even when, in July 1865, a fire devastated what diarist George Templeton Strong had called Barnum's "eyesore" on Broadway.[12] Only after a second catastrophic fire in 1868 did the accident-stricken Barnum finally shift his attention to projects that allowed him and his performers greater flexibility. He invented the three-ring traveling circus, "The Greatest Show on Earth," and developed such gigantic displays as the "Great Roman Hippodrome" (1874–1875) and the "Grand Congress of Nations" (1884). Although these enterprises grew directly out of Barnum's earlier activities, the ever-increasing scale on which they were conceived and carried out makes a comparison with previous efforts such as Peale's museum difficult, if not impossible. Drawing mainly on Barnum's autobiographies, his posterbills, and his museum catalogues, I shall try to define some of the crucial similarities and differences between Peale's museum-keeping and Barnum's reinvention of the museum as "show-shop."[13]

Like Peale, Barnum had received no training that would have qualified him for his museum work. But if Peale's conservatorial efforts profited from the skills he had acquired as a saddler's apprentice, many of Barnum's later enterprises and achievements also directly derived from experiences in his childhood, notably his interactions with people in country stores. As a twelve-year-old, Barnum began work in his father's country store. Subsequently he clerked in other stores in Grassy Plain, Connecticut, and Brooklyn, New York. In 1828, he returned to Bethel to open his own store, selling fruit, confectionery, oysters, and toys. "There is something to be learned," Barnum reminisced in his autobiography, "even in a country store" (L 39). This was the place where "exceptions to the general rule of honesty" occurred with "sufficient frequency" (L 39), where "nearly every thing was different from what it was represented," since each party, the customers as well as the storekeeper, "expected to be cheated, if it was possible" (L 99). In a store that depended on cash as well as credit or barter, the customer as well as the storekeeper knew that neither could trust the other's words: "Our eyes, and not our ears, had to be our masters" (L 99). Trading in a place where "worthless trash" (L 93) could be found next to treasures seemed eminently suitable for someone with Barnum's "speculative character" (L 107), but he eventually decided he wanted to engage in business "faster than ordinary mercantile transactions would admit" (L 140). In 1835, he acquired his first "curiosity," Joice Heth, a former slave who looked infinitely older than she was and went on show for Barnum with the preposterous claim that she had once nursed little George Washington. Thus began Barnum's career as a "showman."

The formative environment during young Barnum's first visit to New York was, significantly, a store. The many curious toys displayed on its shelves kept him spellbound. Again and again, the boy would return to the shop, looking around and perceiving "many things I had not noticed previously." At night, young Barnum, his imagination on fire, went to bed only to dream of "all my possessions" (*L* 24). Soon his limited funds were depleted. Nevertheless, this is exactly what Barnum wanted his museum, his "show-shop," to be like—a place where people could go again and again and yet discover something new on each visit. Stores, for Barnum, encapsulated the shifting relations between appearance and reality, surface and essence, deception and discovery. In a sense, Barnum's American Museum was an attempt to replicate for his visitors and his own benefit that childish sense of wonder that had seized him on his first visit to New York—and combine it with an adult's keen sense of profit lest such an experience again deprive him, as it did so shatteringly during that visit, of all his finances. Have your cake and eat it, too: Barnum started out with hardly a dollar in his pocket and died fabulously rich and famous, "the most celebrated American of the nineteenth century," as one critic has summarized the showman's fairy-tale career.[14]

The period in which Barnum rose to wealth and fame has been called, with some justification, the "Age of the Storekeeper," and there is perhaps no one who better embodies the rapid changes of the era than Barnum himself. In the country, the storekeeper had been the "main player in all aspects of community anywhere," a powerful middleman between the demands of his local community and the economic realities of the city where he went to buy his supplies from the wholesalers.[15] The railroad and increasing cash trade led to changes in the economic landscape, too; in an expanding market the smaller stores soon had to compete with the mass retailers in the cities. Manufacturers now began to package their goods as effectively as possible to lure their customers into buying *their* products and not those of their competitors. Under the influence of new advertising techniques, the visual chaos, the hodgepodge appearance, of the old store gave way to a new order determined by images, brand names, and the needs not just of individual townspeople but of *all* potential customers.

In 1841, the same year that Barnum haggled his way into ownership of Scudder's American Museum, a fancy goods store on Broadway had opened its doors to the public, the "raree show" of Tiffany, Young, and Ellis, whose fascination is memorably captured in a description Edgar Allan Poe sent to the *Columbia Spy*. These warehouses were, rhapsodized Poe, "beyond doubt the most richly filled of any in America; forming one immense *knicknackatory* of *virtu*." Walking along overflowing shelves stacked with imported goods, Poe noticed

a beautiful assortment of Swiss osier-work; chess-men—some sets costing five hundred dollars; paintings on rice-paper, in books and sheets, tiles for fencing ornamental grounds; fine old bronzes and curiosities from ancient temples; fillogram articles, in great variety; a vast display of bizarre fans; ranging, in price, from sixpence to seventy-five dollars; solid carved ebony and "landscape-marble" chairs, tables, sofas, & c.; apparatus for stamping initials on paper; Berlin iron and "*artistique*" candle-sticks, taper-stands, perfume-burners, *et cetera, et cetera*.[16]

Barnum's American Museum had different delights to offer. There were no fine bronzes, exotic fans, perfume burners, or fancy candlesticks. Still, as a minstrel song published in 1849 and probably performed much earlier, observed:

> Barnum's Museum can't be beat:
> De Fat Boys dar am quite a treat.
> Dar's a Big Snake too, wid a rousing stinger;
> Likewise, Pete Morris, de Comic Singer.[17]

Unlike Tiffany, Young, and Ellis's store, Barnum's own version of the "raree show" catered to the masses, allowing them, through the purchase of the countless catalogues, pamphlets, and fifteen-cent photographs, the symbolic (and inexpensive) appropriation of the wealth of products Barnum had accumulated.[18] Barnum's visitors could even take parts of the museum home with them: "Several Hundred Tame Black, and Grey Squirrels from Ohio, may be seen here," announced a posterbill for Barnum's museum in 1842. "They are beautiful pets, and will be sold VERY CHEAP, singly or otherwise." And those patrons of Barnum's museum who wanted to have their furry or feathery friends stuffed also knew where to turn for expert help: "An Experienced Taxydermist!" cried another posterbill, "is engaged at the Museum, and persons having Pet Birds or Quadrupeds they wish preserved, can have them mounted in the best style" (27 March 1843).

Cheerfully Barnum invited "Mechanics, Tradesmen, Manufacturers, Inventors, Artists" to exhibit their goods and "cards of Business" at a "perpetual fair" inside the American Museum, the "most efficient and cheap mode of ADVERTISING that can be adopted" (28 October 1843). If Barnum offered to help his visitors advertise their goods, he also, unabashedly, asked them to help him advertise *his*: "Mr. Barnum will be glad at all times to supply, gratuitously, pictures of curiosities to such visitors as will put them up in conspicuous places" (2 July 1860). In 1862, Barnum introduced his own "Improved Mercantile Advertising Drop Scene," which he, with his usual modesty, described as "the most brilliant plan for advertisers ever yet suggested." He promised to display the names of advertisers on the curtain in the museum's theater and lecture hall to attract the attention of those seated in the house. Their names would also appear on "all of his

Programmes and Bills, which are circulated in Steamboats, Railways and Cars . . . daily amounting to no less than 5,000 and making no less than 390,000 bills in 3 months" (7 July 1862).

However, Barnum's patrons were not only supposed to buy, they were also encouraged to sell and deliver: "1,000 Living Mice" were always wanted for the museum's anacondas (19/20 April 1860), and the owner of the museum, who liked to assure his visitors that he himself was "in daily attendance . . . and ready at all times to purchase curiosities" (3 December 1860), offered to pay twenty-five dollars per pound to whomever brought him a "live speckled brook-trout, in good condition" (9 May 1860). As impossible as it sounds (given the gigantic scale on which he soon operated), Barnum tried to maintain in his museum some of the colorful diversity of the old country store, which had been just as much a "community meeting place" as a commercial outlet. Barnum's American Museum came crammed full of items procured by a whole network of collecting agencies and soon functioned like a modern corporate enterprise. And yet, as will become evident in the course of this chapter, it still harked back to a time when "keeping store was a game of guessing just where something might be found."[19]

The story told in the following pages is a surprisingly complex one, given the reputation of Barnum's American Museum (which, in spite of recent rescue operations, is still treated with fastidious disgust by some historians of nineteenth-century culture). Barnum never raised his twenty-five-cent admission fee; only those eager to see special exhibits such as the "Beautiful Anatomical Venus" were asked to throw a little more into the coffer (31 August 1847). The condition was, of course, that patrons knew how to behave themselves. There was, threatened Barnum, "No Admittance for Improper Persons" (3/4 October 1845) or, more specifically, for "FEMALES OF KNOWN BAD CHARACTER." "Ladies and Families," Barnum stated, "will be perfectly safe, and no more exposed to evil companions than in their own Parlors" (25 June 1845). The American Museum was "the NICEST, CLEANEST, MOST ACCEPTABLE, and MOST ATTRACTIVE place of public amusement" in the world (9 April 1860), where "TEN TIMES MORE" attractions could be found than "at any other establishment" (27 May 1861). As indicated by the popular song quoted earlier, Barnum had succeeded in transcending the class barriers that some historians think had remained more or less in place even in Peale's museum.

But the American Museum *was* comparable to the Philadelphia Museum in that it also reflected a coherent program on the part of the collector himself. The story of Barnum's museum is a story of boundary crossings: boundaries between people small and large, rich and poor, black and white, able and supposedly "disabled"—and, perhaps most important, between those who really *are* people and those who are *not*, that is, between humans and animals. Barnum's strategy was so phenomenally successful because these transgressions remained transient experiences, safely temporary violations that would help to reinforce, if not create, the reassuring "normalcy" of the visitor's view of "himself."[20] The aim of Barnum's collections was to establish a view of human *collectivity,* but this togetherness was in fact even more *selective* than the one aimed at by Peale's museum, since it was based on the eventual expulsion of what Barnum's exhibition playfully purported to include. However, as we shall see, one of Barnum's transgressions was far from temporary: the flaunting of "common sense" against the natural scientist's allegedly superior insight and expertise. The paradigm of classical "natural history" could no longer contain what humans began to see once they had stopped to take a long look at themselves.

"The Long and the Short of It"

Throughout his adult life, Barnum liked to surround himself with images of childishness.[21] For all his professional cunning and business acumen, he retained a childlike frame of mind, which became most evident in the pleasure he took in things exceedingly small (such as his "dwarfs" Tom Thumb and Lavinia Warren) or toweringly large. Notable among Barnum's exhibits were the rhinoceros, one of the "most gigantic" animals found in "nature's whole collection" (*AMI* 18), and, Barnum's answer to Peale's mammoth, Jumbo the elephant, "the largest and noblest animal on earth," "THE ONLY MASTODON ON EARTH."[22]

His true love was reserved for persons meekly and mildly diminutive. In September 1855, Barnum hosted a "beautiful baby contest" at the American Museum, "the first exhibition of the kind ever held in New York." Premiums went to "the finest baby" in each of the designated age groups, and there were additional awards for twins, triplets, quadruplets, and the "fattest child" (21 March 1855). Over all this hovered the huge shadow of benevolent Barnum himself: "One hundred picked babies, under five years of age, all in one room," trumpeted a posterbill for 5, 6, and 8 June 1855, "is a sight never yet witnessed on this continent." A Barnum catalogue published later offers the following appropriate assessment of Barnum's "childlike" disposition: "Wordsworth's fancy 'the child is father to the man' was never more thoroughly illustrated than in the life of Barnum."[23] In 1843, poster-bills for Barnum's museum emphatically encouraged visitors to go and see for themselves what Barnum's "dwarf" Tom Thumb looked like, "this perfect Specimen of MAN," "this extraordinary and captivating piece of humanity," "the Smallest Dwarf ever heard of," "the most perfect Dwarf in the World," "The Smallest Person that ever walked alone," the "most SURPRISING and DELIGHTFUL CURIOSITY the world has ever produced" (26 June 1843; 28 October 1843). A picture included in one of the bills tantalized patrons with the sight awaiting them: a miniature Tom Thumb is seen surrounded by a group of normal-sized people. His excessive smallness, deliberately exaggerated in the picture, is enhanced by the presence of the children in the group of gawking visitors, all of whom tower above him. The display cases of the museum, just roughly sketched out in the background, indicate that in Barnum's museum "natural history" has become a more or less accidental backdrop to the real spectacle—that of humanity itself (ill. 14; 5 October 1843). Not coincidentally, Barnum invariably used the language of natural history to describe his human exhibits, calling them "specimens" or "curiosities" and identifying them as "belonging to the genus homo" (10 February 1862).

The 1850 guidebook to Barnum's American Museum gives us a fairly good impression of what the museum looked like after several years of "Bar-

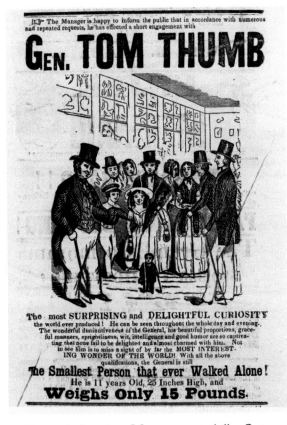

14. Barnum's American Museum, posterbill, 5 October 1843. The Harvard Theatre Collection, Houghton Library, Harvard University.

numization." It also reveals that one of Barnum's main obsessions as a collector was, not surprisingly, *size*. Enticed by the music of bands playing outside the museum, the visitors entered Barnum's establishment through a hall of wax figures, among which the guidebook singles out the life-sized replicas of the two giants Mr. Robert Hales and Miss Eliza Simpson, who were married on the "stage" of the museum in 1847. Ascending through a "handsome flight of stairs" to the museum's picture gallery on the second floor, visitors would walk past a daguerreotype of the museum owner himself and suddenly find themselves confronted with a mirror, a "huge looking-glass of magnifying power," a source "of considerable amusement." Several years before Lewis Carroll's Alice witnessed herself shutting up and then opening out "like a telescope,"[24] Barnum placed his museum visitor in front of a looking glass in which, "with the wish of ascertaining what it is," he "beholds himself all at once transformed into a giant.

His eyes and the other features of his face are enlarged ten-fold—even every pore in his skin is visible. However, he soon discovers that he is of the same size as usual, and enjoys a hearty laugh at his momentary marvel" (*AMI* 4).

Mirrors and other shimmering surfaces had been standard devices in museums of earlier periods, too: they suggested to visitors that seeing was a complicated activity that crucially involved and affected their own sense of self. For example, in Fernando Cospi's museum in Bologna, which straddled Renaissance curiosity and Baroque wonder, a mirror was placed directly overhead so that in its reflection the visitors could picture themselves as part of the collection, merged with the objects on display.[25] But the purpose of Barnum's reflecting device is more clearly defined: it is a feat of cunning engineering rather than an aesthetic subversion of the meaning of the museum as a whole. Encouraging people to take a look at themselves, Barnum would shock them with the brief reflection of the distorted face of a monster, only to reassure them, quickly, that everything was "as usual."

Inside the other exhibition rooms on the second floor, the play with shifting sizes continued. Barnum's guide reminds the visitor not only of Tom Thumb, whose tiny costume is displayed here, but also of "Titania, the Queen of Fairies" ("only twenty-four inches high") whom Barnum had also introduced to the public. The guide invites visitors to seat themselves in "the Giant's Chair," where they could compare their "own proportions with those of the 'great' personage for whom that chair was made" (*AMI* 8). And if confused by the measurements given for the birds and quadrupeds also on display in this room, visitors could go ahead and measure their own "inches"; one of the rooms was, according to the guidebook, "furnished with a machine expressly for that purpose" (*AMI* 11).

Interestingly, Barnum's obsession with size persisted even after his treasures had gone up in flames. A pocket-sized, in fact "little," guidebook to Barnum's Traveling Museum and Menagerie published in 1871 offered biographical sketches of Admiral Dot ("the smallest man in miniature that walks alive") as well as of Monsieur Goliath, "the great physical wonder of the nineteenth century." The portrait of M. Goliath included in the booklet shows him holding Admiral Dot in his hand—"an illustration of the comparative size of the *largest and smallest men on earth!*" The writer of the guidebook feels compelled to add that he himself is "five feet nine inches high, and yet, with his hat on, . . . can walk under Mons. Goliath's upraised arm without even touching it."[26]

Barnum's interest in people excessively small or hyperbolically tall, in "colossal giants and diminutive dwarfs" (6 June 1864), offers some striking parallels to Swift's *Gulliver's Travels* (1726), a book replete with references to seventeenth- and early eighteenth-century collections and popular fairs.[27] The living exhibits at the American Museum included an attraction billed as the "Lilliputian King" (whom Barnum liked to display seated on the enormous palm of the Nova Scotia giantess Anna Swan), as well as Miss Reed, "the Queen of Lilliput."[28] Like the diminutive Tom Thumb touring the cities of Europe with his manager Barnum, little Gulliver too was ferried around in Brobding-nag by his first master: "My Master finding how profitable I was like to be, resolved to carry me to the most considerable Cities of the Kingdom."[29] And as Barnum, more than a hundred years after the publication of Swift's novel, did with his giants, Gulliv-er's owner prepared posters describing Gulliver's "Person and Parts" so that he could better market him as a natural curiosity.

As has often been noted, the first two books of *Gulliver's Travels* turn a human be-ing into an often helpless animal that, depending on its size, either has the potential to destroy or runs the danger of being destroyed itself. When, in the second book, the macroscopic point of view becomes microscopic, images of animality appear even more frequently. The giants of Brobdingnag look at Lemuel as if he were "a strange Animal in the Field," even though they also notice with surprise that he walks "erect upon two Legs" and is "exactly shaped in every Part like a human Creature." If the six-inch Lilliputians seemed like insects to Gulliver, he now finds himself in the position of an insect. Smaller even than the king's dwarf, Gulliver is regularly attacked by bigger animals, and the king's naturalists wonder how such a small "carnivorous Animal" could possibly obtain its food. Gulliver's induction into the animal world is highlighted in a richly significant scene, in which he is adopted, fed masticated morsels of food, and patted appreciatively on the head by a monkey with a maternity complex who takes him for a "young one of his own Species." But the enforced change of perspective also makes Gulliver see humans in a different light. Swift's book is filled with terrifying images of distorted, distended parts of the human body: Gulliver watches with horror as a nurse gives her "monstrous" breast to the greedy mouth of his master's baby and is disgusted almost beyond words when one of the Brobdingnag queen's maids places him on her grossly inflated nipple. From Gulliver's microscopic perspective, the minute imperfections of people's skins assume the dimensions of craters and ravines. People, as Gulliver summarizes his predicament, appear beautiful to us only when "they are of our own Size."[30]

Barnum's pervasive, playful interest in the small and the tall deliberately shunned the biting satire of Swift's novel. Lemuel Gulliver's environment changes, and therefore, to his horror, so does he, although objectively he stays the same. Barnum's gentle giants and dapper "dwarfs," his Tom Thumbs and Anna Swans, Admiral Dots and Goliaths, never ask the visitor to readjust her entire frame of reference. Swift's Gulliver changes from a "Man-Mountain," with a brief interlude in normal life, into a mere "Manikin" and realizes that "nothing is great or little otherwise than by Comparison." And while

Gulliver understandably does not want to "look in a Glass . . . because the Comparison gave me so despicable a Conceit of my self," the mirror in Barnum's museum playfully suggests to visitors that all these distortions of scale will not lastingly affect their own sense of measurement. In Swift's novel, it is the human observer, though he remains the "normal" person he has always been, who becomes a *lusus naturae*.[31] Barnum, however, in the American Museum as well as in his traveling shows, worked hard to establish the statistical average—as represented by the ordinary visitor—as the norm. Significantly, it was at about this time, in the 1840s and 1850s, that the word "normal" and its cognates entered the English language.[32]

Barnum's ceaseless play with deviance and normalcy is evident also in the military uniforms in which he liked to dress up his giants and dwarfs. They were a reminder of the standard of ordinary "manliness," which in Barnum's "living curiosities" would appear either exaggerated or reduced. (It is revealing that Barnum's midget Stratton was a "general" whereas his giant Martin Bates, who had actually served in the Confederate Army, merely held the rank of "captain.")[33] And just as the king of Brobdingnag was adamant that Gulliver should find a mate "his own size," Barnum delightedly hosted the (well-publicized) wedding of Charles Sherwood Stratton ("Tom Thumb") and thirty-two-inch-high Lavinia Warren Bump in 1862. Ten years later, the giantess Anna Swan married Captain Bates. Though in both cases there is no evidence that Barnum had acted as matchmaker, the pleasure he took in such weddings was obvious.[34] If these occasions were the ultimate assurance that Barnum's "freaks" were somehow "normal," they insinuated that the excessively tall or small also "normally" tended to stay on *their* side of the divide.[35] As President Lincoln is supposed to have explained to his bewildered son when welcoming "General" Tom Thumb and his new wife to the White House in 1863: "'Dame Nature sometimes delights in doing funny things; you need not seek for any other reason, for here you have the *long* and the *short* of it' (pointing to himself and the General)."[36]

The Collector as Collectible

As hard as Barnum might have worked to make the eccentricities in his museum cause for a celebration of normalcy, it is clear that for *himself* he wished to reserve a status that was both ordinary and extraordinary, "both in and out of the game, and watching and wondering at it," to quote words used in a different context by Walt Whitman, who had interviewed Barnum on 21 May 1848 for his newspaper, the *Brooklyn Daily Eagle*.[37] Over the years, Barnum cheerfully established and deliberately exaggerated his "personal connection" (*ST* 2: 495) with his ever-increasing collection, just as Peale, whose portrait was included in the picture gallery of the American Museum, had emphasized *his* ownership of the Philadelphia Museum. But he was never just *one* of the many sights included in his collection: though this self-portrait does not actually exist, we can at least imagine Barnum pointing at himself rather than at his collection. And if he didn't do the pointing himself, others would. In his autobiography, once among the most widely read American books of the nineteenth century, Barnum shrewdly pretends to be amazed that he himself should have become "a curiosity": "If I showed myself about the Museum or wherever else I was known, I found eyes peering and fingers pointing at me, and could frequently overhear the remark, 'There's Barnum,' 'That's old Barnum,' etc." (*L* 292–293).

When Mark Twain inspected Barnum's museum in 1865, he thought it was little more than a glorified "peanut stand," with a few cases of dried frogs and some other novelties thrown in for good measure.[38] However, *pace* Twain and others, Barnum's collection, as he had conceived it, had many continuities with Peale's conception of the natural history museum. The introduction to the catalogue for Barnum & Van Amburgh's Museum and Menagerie, written after the main period of Barnum's museum work was already over, skillfully imitates some of Peale's very own rhetoric. There is the same appeal to the didactic value of natural history. No study, the catalogue states, "is more important to the growth of a rising generation, or to adult age." Natural history will teach museum visitors to recognize the hand of God in "brute" creation, since it instills in man's bosom a "knowledge of wisdom and goodness and omnipresence of a supreme and All-wise Creator." And we find similar discontent with the lack of national, federal support for the collector's selfless work, although the invocation of "individual capital" at the end of the following passage definitely introduces a new, ironical note into the discussion:

> That nation of Europe which has not its Natural Historical Institute is regarded as on that backward course that leads in the end to semi-barbaric ignorance and folly. In this country, in consequence of our peculiar form of government, based as it is on the assent of confederated will, nothing has so far been done, and probably nothing ever will be done, to promote the cause of natural history, through governmental influences and patronage. Thus situated, the whole of a work so glorious, as the one we contemplate and feebly discuss, has depended, and must ever depend, upon individual capital and enterprise.[39]

Barnum was not just a loud-mouthed, shallow, and self-promoting entrepreneur whose prime goal it was to put money in his purse while he was putting junk in his museum (*L*

400).[40] Although he himself often encouraged similar simplifications, we must remind ourselves that the American Museum also had, in the words of Joel Orosz, "considerable wheat along with the chaff."[41] Even George Templeton Strong allowed himself to be impressed by Barnum's aquaria."[42] After a fire had ravaged his $400,000 collection in 1865, Barnum seriously contemplated the creation of a *"National Free Museum"* that would contain "collections of natural history and . . . all specimens of *everything* presented by our govt. in any of its departments & everything presented by *anybody* in this or any other country." This institution, Barnum told Bayard Taylor, should be erected alongside the reconstructed American Museum with all the curiosities he had acquired himself, the "giants, dwarfs, fat women, bearded ladies, [and] baby show."[43]

However, as the spatial arrangement proposed here already indicates, it is hardly possible, as has been suggested, to look at Barnum's natural history collection *minus* "the freaks" so that we can better understand "the true value" of his efforts.[44] Against a new tendency to celebrate Barnum's "homely luster,"[45] it needs to be said that the showman's "true value" (a slippery category under the best of circumstances when applied to Barnum's work) is probably to be found somewhere in the middle, between the seemingly opposed identities of the cynical showman and the devoted museum curator. With Barnum, the collecting of specimens of human and animal nature had turned into happy, if often haphazard, hoarding.[46] "My organ of acquisitiveness must be large," Barnum observes in the first chapter of his authobiography, "or else my parents commenced its cultivation at an early period" (*L* 20).

For the tentacular Barnum, the chance "find," the good deal, the great purchase became more important than the hunt for and happy possession of the long-coveted item. Barnum appreciated the singularity of each of his purchases, but he did so chiefly in terms of their marketability. Like Peale, Barnum was a "monomaniac" when it came to his museum, but with him the activity of collecting whatever . . . money would buy or enterprise secure" (*ST* 1: 198) was curiously dissociated from the objects that he or his agents collected. In Peale's museum, accompanying texts had carefully identified the specimens on display, cocooning them with curatorial care; Barnum's use of such specifying verbal accoutrements was considerably more cavalier, and not only where the facts of natural history were concerned,"[47] In Barnum's 1850 guidebook, we find serious and solidly descriptive passages ("The Shell is turreted or fusiform; generally ribbed or striated transversely; the aperture oval, terminating anteriorly in an elongated canal"; *AMI* 22) side by side with carelessly impressionistic and almost nonsensical statements ("The black and white stripes of the Zebra, alternating with each other, are much admired. . . . The Leopard is also a fine animal"; *AMI* 7).

Museums are, as Robert Harbison has said, overindexed places, existing in "a perpetual day of inventory," a fact reflected in the "threateningly suitable form" that has arisen so that they can be documented, the catalogue.[48] Peale tried, as we have seen, to circumvent the inherent boredom of the list by narrativizing his catalogue entries as best he could. For Barnum, however, the dilemma posed by the inventory again became the occasion for a joke. Offering himself as security for one of his friends involved in a lawsuit, Barnum was asked by the suspicious lawyer of the plaintiff to be more specific about his property. "Do you desire a list of it?" Barnum inquires. "I do, sir, and I insist upon your giving it before you are accepted as further security," the lawyer responds. Casually, Barnum invites him to mark down the items on his list in the order in which he would call them off;

> "I will, sir," he answered, taking a sheet of paper and dipping his pen in the ink for that purpose. "One preserved elephant, $1000," said I. He looked a little surprised but marked it down. "One stuffed monkey skin, and two gander skins, good as new—$15 for the lot." "What does this mean? What are you doing, sir?" said he, starting to his feet in indignation. "I am giving you an inventory of my Museum. It contains only five hundred thousand different articles," I replied with due gravity. (*L* 357)

The frustrated plaintiff's lawyer finally decides to accept Barnum's affidavit "without going further into the 'bill of particulars'" (*L* 358).

The 1850 guidebook does not even attempt to list all the "particulars" amassed in Barnum's museum. Barnum's "Miscellaneous Specimens" numbered in the "tens of thousands," and each of them seemed "worthy of the detail our space will not permit us to give" (*AMI* 6). From the pages of his catalogue, Barnum emerges as a true Master of the Miscellaneous: "A Snuff Box, made out of the pulpit of John Knox, a Sword, used by a trooper of Oliver Cromwell at the Battle of Marston Moor, and a suit of Armor, of the reign of Henry the Eighth, make us think of three very different persons, and are illustrative of the universal character of Mr. Barnum's American Museum, which may justly be called an epitome of all time and men" (*AMI* 23–24). What the catalogue seems to be telling the visitor here, a bit tautologically, is that different things are different things. In Barnum's monstrous "cabinet of curiosities" extravagance appeared to be an end in itself, not an incentive for the sense of boundless wonder at the multiplicity of the world previous collectors had tried to instill in their visitors.

The route through the collections suggested by Barnum's guidebook is not characterized by a slowly emerging sense of a divinely ordered nature, but is instead punctuated by sudden, jarring transitions: "Having observed some varieties of the porcupine, we proceed in our survey. A human body . . ." (*AMI* 22). Within one or two paragraphs, the visitor breathlessly moves from "beautiful gems of shells" to a "very curious specimen, and one quite opposite in its nature," a tiger. Then she views

> in succession, a Chinese Umbrella Hat, a South American Saddle, a fine specimen of a Durham Cow, a Wolf, and several specimens of the Squirrel, and other animals, inclosed in the same case.
>
> Above the cases, the visitor may also see ranged around the room, the following specimens: the head of a Hippopotamus from the Nile, the head of a Fossil Elk, and a truly magnificent specimen of Quartz from Upper Missouri. Then follow the Lower Jaw of a Grampus Whale, an Indian Bowl, and a Chinese Music Pipe, each worthy of the attention it receives, and the enquiries it elicits. (*AMI* 23)

The principle of Barnum's museum is the profusion of sights, not the achievement of insight into a predetermined order of things. Even recurrent themes such as size or, as we shall see in the following section, "the connecting link between man and brute nature" point less to an order intrinsic to the collection than to a meaning superimposed on it by the avuncular collector, whose face and biography were everywhere present in the museum—on the cover of the catalogue the visitors purchased, in the pamphlets sold by his "living curiosities," on posters and daguerreotypes.

One of Barnum's most popular exhibits was the "Happy Family," a metaphorical slap in the face of the contextualizing curatorial philosophy that had informed Peale's "habitat dioramas." In Barnum's display, predator and prey were supposed to coexist peacefully in one cage, "a collection of Birds, Beasts, Reptiles, &c. of antagonistic natures, all living together in harmony" (6 October 1854). The reality seemed different to at least one caustic observer, Mark Twain again: "A poor, spiritless old bear—sixteen monkeys—half a dozen sorrowful raccoons—two mangy puppies—two unhappy rabbits—and two meek Tom cats, that have had half the hair snatched out of them by the monkeys, compose the Happy Family—and certainly it was the most subjugated-looking party I ever saw. The entire Happy Family is bossed and bullied by a monkey that any one of the victims could whip, only that they lack the courage to try it." Twain concluded: "The world is full of families as happy as that."[49]

The "order" experienced by the visitors to Barnum's museum was the order of the American Museum itself, a space shaped so that they could stroll around in it and, at their convenience, at least visually assemble their own, equally random collections:

> We will now, with the permission of the reader, accompany him through this truly splendid establishment. . . . Having paid the admission money, we pro-

15. "View of the Wax-work Room." From *Barnum's American Museum Illustrated* (1850). Harvard Theatre Collection.

16. "View of the Room on the Right of the Picture Gallery." From *Barnum's American Museum Illustrated*. Harvard Theatre Collection.

ceed, and presently find ourselves in a handsome new saloon on the ground floor, with a mellowed and subdued light, well adapted to the exhibition of wax figures . . . After contemplating them for a time, we turn to. . . . Before leaving this room, we pause some minutes. . . . We are then arrested on our onward way by the Carriage of General Tom Thumb. . . . Our next glance falls an. . . . We next see a portion of the bed on which Robert Burns was born. . . . The visitor can now, if he pleases, ascend another staircase. (*AMI* 2, 6, 8, 9, 24, 29)

In Barnum's expert hands, the museum of natural history had become a place where people would go not because they wanted to learn about nature but because this was where they could reassure themselves about themselves. This principle is particularly evident in the arrangement and sequence of the rooms as they were reflected in the 1850 guidebook. Visitors would begin their tour in a hall with wax figures (ill. 15), where they would look at various "illustrations of human life," displays extolling the "cultivation of domestic virtues" (*AMI* 4). From there they proceeded, past the famous mirror, into the portrait gallery and then into the exhibition halls. In one of the illustrations included in the catalogue, the animals in their display cases seem blurred, virtually indistinguishable, whereas the visitors' faces are sharply rendered (ill. 16). The museum itself now was little more than a background for the visitors moving around and enjoying themselves in its spacious halls.

Having toured the exhibits, Barnum's visitors would descend again to take their seats in the lecture hall—a place where humans could watch other humans (ill. 17). The theatrical metaphor incidental to Peale's self-representation in *The Artist in His Museum* had thus become central to Barnum's concept: on the museum's stage Barnum regularly presented plays, "sanitized" for the visitors' consumption.[50] Even architecturally, the theater had become the centerpiece of the museum.

Barnum's guidebook itself, instead of lecturing the visitors from the vantage point of superior knowledge, very shrewdly adopts the perspective of the uninformed, "normal" visitor. Barnum's deliberate eschewal of the "Voice of Institutional Authority"[51] is particularly evident in the frequent use of the passive verb form: "In this room we are also shown . . ." (*AMI* 5); "At the upper end of this room we are shown several specimens . . ." (*AMI* 12); "Our notice is attracted by a stick on which some Oysters are embedded" (*AMI* 22). The "peace" so frequently evoked in the guidebook is not the peace emanating from the patient perusal of the pages of the Book of Nature opened for the benefit of the curious visitor; it is the peace of propriety and good behavior. Barnum, who had the premises of the American Museum unobtrusively policed by plainclothes detectives, gently assured his visitors, especially those "prevented" from visiting theaters, on "account of the vulgarisms and immorality which are sometimes permitted therein" (*AMI* 2) that in his institution the "spirit . . . of order" reigned triumphant (*AMI* 28). Thousands of visitors were "daily passing in and out of the saloons," but, as Barnum was pleased to point out, there was "not the least confusion or noise" (*AMI* 28).

Peale's national aspirations survived, but in severely attenuated form. Barnum delightedly discovered that "America" too was a good sales pitch. On one Fourth of July, Barnum hoisted a flag outside his museum on Broadway because he felt confident that a conspicuous display of national flags would arrest the patriotic attention of "people passing the Museum with leisure and pocket-money" and bring "many of them within my walls" (*ST* 1: 216). The distance between Peale's nationalist zeal and Barnum's tireless touring of America *and* Europe can best be gauged by the peaceful juxtaposition of two items in Barnum's museum, documents signed by, respectively, George Washington and George III (for whom, we might recall, Peale never tipped his hat): "Notwithstanding the vast difference in the character and qualities of the two personages," states the guidebook, "the signatures are remarkably similar—both free and noble" (*AMI* 14). Barnum is, as the guidebook proudly points out, quintessentially American, not because of what he has (i.e., the intrinsic value of his collection) but because of what—or, rather, *who*—he is and continues to be.[52]

17. "View of the Stage of the Lecture Room." From *Barnum's American Museum Illustrated*. Harvard Theatre Collection.

Barnum continued to be who he was not just through his own expanding collection but also through the ceaseless revisions of his autobiography. Responding to the demands of his many fans and, as he also claimed, the bids made by "fifty seven different publishers,"[53] Barnum first bared his self to an eager public in December 1854. *The Life of P. T. Barnum* appeared a few months after Thoreau had published *his* autobiography, *Walden; or, Life in the Woods*, and a few months before Whitman announced to his readers, not on "leaves of grass" but on pages he had typeset himself, that he was "an American, one of the roughs, a kosmos."[54] The frankness with which Barnum talked about his own pranks, deceptions, and publicity stunts elicited a storm of unfavorable responses ("We have not read, for a long time, a more trashy or offensive book than this," exclaimed *Blackwood's Edinburgh Magazine*),[55] but the negative press helped to boost rather than depress sales.

A parody followed suit, in the form of a little book facetiously declaring itself to be *The Autobiography of Petite Bunkum, the Showman . . . Written by Himself.* The spoof successfully imitated and exaggerated the hands-on, bare-knuckled bravado of its model: "I became the proprietor of an extensive Museum in New York—how, or by what means, is none of the reader's business."[56] However, since the anonymous author of the playful parody was at such pains to assert, in a postscript, that he actually "liked" the "great Yankee showman," "one of the greatest men of the age," it has been suggested that the great self-advertiser himself perhaps had more than just a hand in the composition of his parody.[57]

Whether or not he helped compose *Petite Bunkum*, Barnum went on to mimic himself in the many editions of his *Life* he would put out over the following years. "Advertising," he said, "is to the genuine article what manure is to the land,—it largely increases the product."[58] His shameless self-promotion and self-inflation in his protean autobiography is still a cause for condescension among historians of the genre, who have largely ignored it.[59] But Barnum's *Life* was an ambitious, serious project, though it remained, like Whitman's *Leaves of Grass,* eternally and intentionally unfinished. It was a "collector's autobiography," like Peale's, if more geared to an audience's expectations of good fun. If Peale's episodical autobiography had been held together by the picaresque presence of "poor Peale," surviving life's multiple adversities, Barnum's chapters were a string of jokes, sometimes only tenuously linked by the mere chronology of the narrative. Consider the chapter titles: chapter 2 is entitled "Clerk in a Store—Anecdotes," and chapter 4 is called, even more vaguely, "Anecdotes with an Episode." Chapter 5 announces itself simply as a "Batch of Incidents," and the following chapter modifies this blunt heading only slightly: "Incidents and Schemes."

In the first six of his fourteen chapters, Barnum never tires of varying one standard situation: the meeting of "wags," or "joke-loving" people, in the store. "One afternoon," begins a fairly typical section of chapter 6, "the usual number of customers being gathered together in my little store, one of our joke-loving neighbors asked . . ." (*L* 120). Barnum's perseverance, which he also identifies as "the propensity of keeping out of harm's way" (*L* 12), consists in the autobiographical subject's ability to joke better, more efficiently, with more lasting consequences than the people around him, who often (which increases the challenge as well as the glory) happen to be competent jokers themselves. It was in the "associations" of his youth, declares Barnum, that his "natural bias was developed and strengthened" (*L* 105).

The later chapters of Barnum's *Life* strive for greater narrative continuity, but the *episode* remains the underlying structural unit. Chapter 8, for example, begins with Barnum's account of his association with Aaron Turner, proprietor of a traveling circus and a good "*practical joker*" himself. It continues with a detailed account of a joke played on Barnum by Turner, moves on to a joke played by both of them on a stingy landlord in Virginia, and then describes a trick Barnum pulled on Turner. In the next episode Barnum deceives his audience by impersonating a black singer after his own African-American performer, James Sandford, has walked out on him. Barnum does this so well that he is nearly killed by an enraged visitor before he identifies himself as white. Sudden interruptions or abrupt transitions are always possible, and Barnum rarely bothers to justify them adequately. "This is not exactly the place to introduce a newspaper," admits Barnum in chapter 12, after having just summarized his attempt to stage a buffalo hunt at Hoboken(!). But then he goes on to talk about the newspaper anyway: "The incidental mention of Mr. West suggests 'The Sunday Atlas,' which was always a favorite of mine" (*L* 356). This is indeed how Barnum's *Life* presents itself—as an *incidental narrative of incidents.*

The obvious advantage of such a loose structure is its open-endedness. If Whitman, in the opening lines of "Song of Myself," declared that he hoped to "cease not till death" and then went on to revise his forever growing *Leaves of Grass* until he died,[60] Barnum kept a similar promise, continually rewriting and expanding his autobiography (while physically he also, as photographs show, expanded). Indefatigably he tacked on yet further and further chapters to his book, now called *Struggles and Triumphs*. Barnum's written *Life*, comments Neil Harris, had a career almost "as complicated" as the life he lived.[61]

Because Barnum's autobiography was so relentlessly cast as a work-in-progress whose temporary manifestations in print just failed to tell it all, every one of its versions became obsolete the minute it appeared. Whatever copy of Barnum's autobiography readers had bought, the book would always taunt them to wait for the next and updated version. No wonder that Barnum's audience felt, as George Haines put it in 1874, that the author of this autobiography might very well live to attend his own funeral.[62] In a prefatory note attached to the 1875 edition of *Struggles and Triumphs*, Barnum took delight in the confusion caused by his autobiographical practice, according to which writing about his life had become well-nigh simultaneous to living it: "During the six years which have elapsed since the first publication of this volume numerous striking incidents in my life have occurred, which have induced the addition of several chapters, as well as the revision of certain portions of the text which even the lapse of so few years seemed to render necessary." If every visitor to Peale's museum received a copy of the printed "Guide," Barnum was adamant that his admirers, too, should not go without the book in which the collector (and not the collection) was on relentless display: "I wish to have five million or more of the inhabitants of the United States read that book for themselves" (27 October 1855).

Whitman hoped, in a poem that ended *Leaves of Grass* beginning with the book's third edition (1860), that whoever touched his work was touching not a book but a man,[63] and the readers of Barnum's immensely successful autobiography obviously were meant to derive a similar feeling from the showman's ever-changing memoirs. A popular anecdote related by Barnum's publicity agent George Haines (and since Barnum was his own best advertiser, this one might have originated with him, too) also plays on such a notion. While Barnum's traveling show was exhibited in Cleveland, a country butcher from Cuyahoga County, clutching several newly purchased, shiny butcher's knives, walked up to Barnum and in no uncertain terms demanded his "life." Or so it seemed:

"Do you intend to kill me, sir?" asked Barnum, in a half tremulous tone.
    "Oh! no, Mr. Barnum," said he, "I mean your great 900 page book, I have hearn so much tell about, called *The Life of P. T. Barnum, Written By Himself,*

which you advertise to sell for $1.50, and give in a free ticket to the show, be-
sides," said he.

By this time a large crowd had gathered around, who heard the above conversation,
and understanding fully the situation, and seeing how completely the "Old Showman
himself" had been done for by the country butcher, joined in the roar of laughter.[64]

Peale's autobiography flaunted detachment; Barnum's account of his "charmed" life reveled in its own forever expanding egotism: "All autobiographies are necessarily ego-
tistical. If my pages are plentifully sprinkled with 'I's,' . . . I can only say, that the 'I's'
are essential to the story I have told. It has been my purpose to narrate, not the life of
another, but that career in which I was the principal actor."[65] However, those readers
expecting revelatory insights into the life led offstage by America's greatest showman
were inevitably disappointed by Barnum's book. He barely mentions his wife, Charity,
and his children. On the last pages of *The Life of P. T. Barnum,* Barnum hurriedly ticks
off the birth dates of his daughters and busily announces that to him "home" and "family" are "the highest and most expressive symbols of the kingdom of heaven" (*L* 404). In
a work remarkably devoid of peaceful domestic scenes and private musings, crammed
full instead with anecdotes, jokes, and descriptions of the author's peregrinations, this
conventional conclusion, coming as it does in a chapter pacifically entitled "My Family,"
is a surprise. If family members appear at all in the autobiography, they are frequently
part of a joke. At the time that Barnum was managing the singer Jenny Lind Barnum's
daughter Caroline, while singing in a church, is mistaken by some gullible churchgoers
for the "Swedish nightingale," simply because she has been seen walking in the streets
in her father's company. A true Barnum, Caroline "did not undeceive" the churchgoers,
and "many persons that afternoon boasted, in good faith, that they had listened to the
extraordinary singing of the great Swedish songstress" (*L* 321).

All in all, the autobiographical self in Barnum's text appears strangely devoid of
personal significance. In Barnum's reading of his own life, the self becomes a cipher in
an interminable charade of people playing jokes on each other, alternately deceiving
others and, if they are not careful, being deceived themselves. If Barnum's private self
does not figure prominently in the autobiography, this of course should not be taken to
mean that Barnum had no identity other than his public one. Granted, "failure to find
something does not necessarily prove its non-existence,"[66] but it is certainly legitimate
to look at the kind of identity Barnum carved out in the many editions of his autobiography and to notice that some aspects are emphasized while others are consciously
downplayed. Barnum's autobiography, like his American Museum, was a planned exhibition, not a "truthful" confession.[67]

Barnum's "Connecting Links"

Just as he happily perverted the taxonomic sense of the term "family" by lumping to-gether many different "antagonistic" species in one cage, Barnum would also merge classes, orders, and genera to create composite specimens—what he called "connect-ing links." As Benjamin Franklin already knew, such "links" were eminently suited for jokes. Those spoofs were easy to pull off for as long as the "New World" still served as a convenient repository for hitherto unheard-of natural "curiosities." Corresponding with a French naturalist, Franklin bragged he had discovered a bird that was a cross be-tween the animal and vegetable kingdoms. This imaginary "bird" carried two tubercles at the joints of its wings, which after its death, so Franklin claimed, would turn into "the sprouts of two vegetable stalks which grow out of the juice from its cadaver and which subsequently attach themselves to the earth in order to live in the manner of plants and trees."[68]

Recent apologists have warned us, as Barnum sometimes did himself, not to pay too much attention to such bizarre creations of his as the "Fejee Mermaid" or the "Woolly Horse."[69] The standard argument is that these were advertising stunts meant to induce visitors to go and see the many other "bona fide" objects on display in other departments of the American Museum.[70] However, without denying that thousands of genuine speci-mens in Barnum's collection were of real value even for serious-minded naturalists such as Henry David Thoreau, there cannot be any doubt that a large portion of Barnum's fame rested on the relentless propagation of fakes. Most of these fakes, though, had a common theme and a message that, on closer inspection, was far from funny.

The human body had always figured prominently in Barnum's collection. Visitors be-held, for example, in "close vicinage" with a stuffed seal, the severed dried hand and arm of the pirate appropriately called "Tom Trouble"; a blackened mummy from Thebes; a hu-man body found in Glasgow, Kentucky, "in a salt-petrous cave, nine feet under ground"; a body, "in a state of extraordinary preservation," recovered from "a copperas cave on Cum-berland River, Warren County, Tennessee" (*AMI* 22). The real crowd-pleasers, however, were Barnum's living exhibits, humans in various states and stages of oddness, dispro-portion, disfigurement, or deprivation, often hidden behind or between the showcases, where they would surprise the visitors and sell them pamphlets with their life stories or photographs. At one time or another, Barnum's collections sported people of uncom-mon height and size (his dwarfs and giants, "Fat Boys," "Mammoth Girls," and "Living Skeletons"), "Animal People" ("Bearded Ladies," "Jo-Jo the Dog-Faced Boy," and "Lionel the Lion-Faced Man"), "Wild Men" (such as the "What-Is-It" and the retarded Ohio siblings whom Barnum passed off as the "Wild Australian Children"), "Half People," and, last but not least, genetic "oddities" such as the "Albino Lady," the "Celebrated African United Twins," and the famous "Siamese" twins Eng and Chang Bunker, whose wax figures Barnum owned and who began to appear personally in the Museum in the 1860s. These "living curiosities" were all, in varying degrees, a reflection of Barnum's preference for "anomalies" that threatened neat categories. Some of them, more precisely, suggested a temporary suspension of the imaginary line dividing humans from animals.[71]

At about the same time that Barnum hosted performances by "NEGRO-EXTRAVA-GANZISTS" such as the "Ethiopian Minstrels" or the "Sable Brothers" ("who personate the Southern plantation Negroes"), he was also exhibiting his enormously popular orangutans, all of whom he declared to be "connecting links" between humankind and the animal kingdom. The metaphor of the "connecting link," a hypothetical type combining two life forms, harks back to the earlier concept of the Great Chain of Being, which generations of natural philosophers in the wake of Aristotle had imagined as rising continuously from the minerals through the plants and animals up to man and, finally, God Himself. As Stephen Jay Gould has shown, the Chain of Being posed a substantial empirical problem by drawing attention to the "large and apparent gaps between major units . . . minerals and plants, plants and animals, monkeys and humans."[72] It was these gaps that Barnum deliberately addressed in his American Museum. Arthur O. Lovejoy, the author of the seminal book *The Great Chain of Being*, claimed that Aristotle, had he been alive in the 1840s, would "have made haste to visit Barnum's Museum."[73] Whether or not we agree with Lovejoy's exuberant assessment, one thing is certain: the "Great American Showman," as numerous of his posterbills prove, consistently exploited the notion of the "connecting link" for his own purposes.

In a posterbill for 6 January 1845, the orangutan "Mad'lle Fanny," named after the Austrian dancer Fanny Elssler, is praised as "the largest animal of the kind ever brought to his Country" (ill. 18). Mlle. Fanny was, so the poster said,

18. Barnum's American Museum, posterbill, 6 January 1845. Harvard Theatre Collection.

universally allowed to be the nearest
approach to humanity of

ANY ANIMAL

ever yet discovered. She is indeed the

**Connecting link between Man and
Brute!!!!**

possessing as many of the characteristics of the one as the other. She will
be seen at all hours of the day and
evening.

Fanny was portrayed sitting on a chair, wearing a prim
dress with a ruff collar, fastidiously eating soup out of a
little bowl precariously balanced in her left hand. In October 1845, Barnum featured a male orangutan, "the most
perfect specimen of this strange animal ever seen in this
country." In a reinforcement of traditional gender stereo-
types, this "wild man of the woods" looked just a little
"wilder" than Fanny (ill. 19). His apparel was restricted
to a long sheet casually draped over his thin legs. But
his body, with the exception of the overlong arms, was
that of a child, while his face wore the benign expression
of a well-groomed, balding adult. Twenty-five thousand
people had already gone to inspect this ape,

every one of whom expressed the most
perfect astonishment
*to find an animal deprived of speech so
nearly resembling the*

HUMAN RACE

And so perfect a personation of the
Wild Man of the Woods!!

He is the greatest of Living Wonders
ever seen in this or any other country—
and no man, woman or child should
fail seeing him.

19. Barnum's American Museum,
posterbill, 3 and 4 October 1845.
Harvard Theatre Collection.

A year later, when, amidst a great deal of publicity, Barnum and Tom Thumb returned from their tour of Europe, they were accompanied by "Mlle. Jane," an orangutan instantly hailed, by the *New York Daily Tribune,* as a "curiosity worthy of all lovers of natural history."[74] And in October 1847, Barnum's Museum advertised yet another anthropoid victim, recently purchased, in terms that had now become familiar to the public—namely as the

> finest specimen of this SEMI-HUMAN RACE of animals ever seen; her intelligence and sagacity approaching very nearly the cultivated intellect of the human family, while her other characteristics blended therewith, constitute the STRANGEST ANOMALY, and the most interesting feature ever witnessed in any age or country. She is as FULL OF THE PLAY AND MISCHIEF, ss [*sic*] the rudest country school boy. She may be seen at all hours of the day and evening, in conjunction with all the other wonders of the museum.

In the following years, the apes came and went, a bizarre parade of humanlike hybrids intended to illustrate "the nearest approach of irrational creation to HUMANITY AND REASON" (5 July 1852). Charles Darwin couldn't have agreed more. Watching orangutans at play in the London Zoo, he exclaimed: "Let man visit Ourang-outang in domestication . . . & then let him dare to boast of his proud preeminence." He suspected that "we may be all netted together."[75]

Barnum's advertisements suggest such amazing closeness even by means of their very layout. In a kind of pictorial joke, an announcement published on 31 August 1847 stuns the hurried visitor who would at first catch just the two terms printed in bold type: "orang-utans" and "organs of speech." The fact that the noun "organ" is an anagram of "orang" adds to the effectiveness of the message, which would have taunted potential visitors with the prospect of talking apes to be encountered in the museum halls (ill. 20). As we saw in the previous chapter, it was exactly the conjunction of these two terms—"orang" and "organs of speech"—that had been a source of considerable worry for eighteenth-century naturalists.

And those worries were not over yet. For Darwin, reflections on man's place in nature had been intimately bound up with the question of race, as the notes taken on his voyage with the HMS *Beagle* (1831–1836) prove. Seized as he was by the urgent wish to "congratulate myself that I was born an Englishman," Darwin's notion of different races as representing different stages of "improvement" did not affect his general conviction that man was, at

20. Barnum's American Museum, posterbill, 31 August 1847 (detail). Harvard Theatre Collection.

least in principle, not exempt from the laws of the animal world. Within this general framework, even an Englishman was nothing else but a highly domesticated and supremely adaptable animal, one that has used his "greater power of improvement" to the maximum: "The varieties of man seem to act on each other the same way as different species of animals—the stronger always extirpating the weaker."[76]

Faced with their own racial dilemmas, white American naturalists had been less generously disposed. Barnum's orangutans, tirelessly advertised as possessing "an equal share" of human and animal features, conjured up the specter of hybridity that had already made Jefferson shudder when, in *Notes on the State of Virginia* (1785), he luridly evoked the "preference of the Oranootan for the black women over those of his own species."[77] Since according to him black men in turn preferred white women, Jefferson's ascending scale of skin colors—culminating in the pleasant hues, the mixture of white and red, on the Caucasian woman's face—was not endangered, though it was clear that the "negro's" desire for the white woman somehow had to be contained. In 1839, the Philadelphia physician Samuel G. Morton published his *Crania Americana*, a skewed tabulation of the cranial capacities of Native Americans, blacks, and Caucasians, arguing emphatically (and on a woefully inaccurate basis of evidence) that there *were* several species of mankind, created at various stages in the earth's history, which resulted in a clearly discernible hierarchy of races. The Charleston naturalist and minister John Bachman feared that Morton's model did away with a good Christian's belief in the single act of divine creation that had made the entire world what it was today. He pointed out that, while cases of hybridity between distinct species of animals existed, they had never led to the creation of new "races." Unions between the "different varieties" of the human species, however, obviously did produce offspring. If all of these were hybrids, why, then, "the whole world must by this time be made up of hybrids."[78] Morton retaliated with a detailed documentation of fertile animal hybrids, arguing that hybridity itself had nothing to do with the definition of species. The "mere fact," wrote Morton, that, like animals, "the several races of mankind produce with each other, a more or less fertile progeny, constitutes, in itself, no proof of the unity of human species."[79] However, a sort of "natural repugnance" prevented the random intermixture of species and kept nature from lapsing into complete chaos.

What seemed to be at stake in this discussion was a proper definition of "humanity." Bachman, though he apparently shared Morton's twisted interest in the collection of human skulls, nevertheless campaigned for a complete separation of mankind from the animal world in order to be able to defend his paternalistic, "family-oriented" vision of southern slavery. For Bachman, "the African" was "an inferior variety of our species," incapable of running his own life: "Our child that we lead by the hand, and who looks to us for protection and support is still of our own blood notwithstanding his weakness and ignorance."[80] Morton, allowing for the division of humankind into inferior and superior species, was more radically suggesting that there might in fact be connecting links between the animal kingdom and man, "the most domestic animal." Whatever the theories were that American naturalists offered in the 1840s, whether or not they identified themselves as advocates of polygenism, most of them finally found a way to underwrite the inferiority of blacks: if genetic distinctness did not help to move some humans closer to animals than others, "natural repugnance" had to confirm that some humans were more human than others. Josiah Nott and George R. Gliddon in *Types of Mankind* (1854) included drastic figures illustrating the "palpable analogies" between the skulls of blacks and those of a "superior type of monkey."[81]

While Barnum was showing off his "semi-human" orangutans at the museum, his contemporary Edgar Allan Poe let real or apparent apes drift through the pages of some of his most terrifying works.[82] The last of Poe's stories to feature apes, which is incidentally also the last story published in Poe's lifetime, seems strangely relevant for our purpose here. "Hop-Frog; or, The Eight Chained Ourang-Outangs," was written in February 1849.[83] The story is unique in terms of both the intensity of the conflict that is portrayed and the narrative resolution that is suggested. In "Hop-Frog," the tyrannical ruler of an unnamed kingdom and his seven equally ruthless ministers seem—as some people thought Barnum did—"to live only for joking": "To tell a good story of the joke kind, and to tell it well, was the surest road" to the king's favor. As it happens, the king is especially fond of jokes at the expense of Hop-Frog, his fool, who, conveniently enough, is also a "dwarf and a cripple"—in other words, a "triplicate treasure" (900).[84] Hop-Frog cannot walk "as other men do": he propels himself forward "by a sort of interjectional gait—something between a leap and a wriggle." For this deficiency he tries to compensate by the use of his well-developed muscular arms, which enable him "to perform many feats of wonderful dexterity, where trees or ropes were in question, or anything else to climb" (900). Small wonder, then, that in such moments, as Poe's narrator adds, "he much more resembled a squirrel, or a small monkey, than a frog" and small wonder, too, that the king finds his fool so funny—not because of the jokes he makes but because of the way he looks.

Poe's story takes place on the day chosen for one of the king's masquerades. With the festivities imminent, the fat king and his equally obese ministers still haven't found suitable costumes for themselves. They summon their fool and demand his help. More precisely, they ask him to write a story for them in which they can play a part: "Characters, my fine fellow; we stand in need of characters—all of us—ha! ha! ha!" (902). The king forces Hop-Frog to drink wine, although he knows that this will have a terrible effect on the dwarf's "excitable brain." As tears are streaming down Hop-Frog's face, the king maliciously compliments him on his "shining eyes" and suggests that he perhaps wants another drink. Hop-Frog's only ally, a "dwarfish" but "well-proportioned" young girl called Trippetta,[85] intervenes on his behalf, and the enraged king throws a goblet of wine into her face. Inwardly seething, the king's fool now comes up with a suggestion for the king and his ministers. Dressed up and chained together as orangutans who have escaped from their keepers, they are to surprise and frighten the other masqueraders. "The resemblance shall be so striking, that the company of masqueraders will take you for real beasts" (904). With their delighted consent, Hop-Frog dresses the king and his ministers in tight-fitting shirts and then coats them with tar and flax. When the disguises are completed, Hop-Frog chains them all intricately together "after the fashion adopted, at the present day, by those who capture Chimpanzees, or other large apes, in Borneo" (905). At midnight, eight giggling orangutans, their chains jangling, traipse into the ballroom and, sure enough, frighten the guests.

Unbeknownst to them, the joke-loving king and his companions are now themselves the targets of a terrible joke, Hop-Frog's "*last jest.*" Ostensibly worried about the safety of the masqueraders, Hop-Frog has asked for the chandelier to be removed from the ballroom. Amidst the confusion occasioned by the spectacular entry of his chained "apes," Hop-Frog quickly lowers the empty chandelier chain. When his orangutans have arrived in the exact center of the circular ballroom, he attaches the hook of the chandelier's chain to the chain that is fastened around the bodies of his apes and draws them together "*in close connection,*" While everybody in the room, including the apes, is still "convulsed with laughter, " Hop-Frog climbs up the chandelier chain "with the agility of a monkey" (907) and then lets the chain fly up again, "dragging with it the dismayed and struggling ourang-outangs, and leaving them suspended in mid-air between the sky-light and the floor" (907), Baring his sharp teeth, Hop-Frog now puts a torch to the flax-and-tar coats of the helpless king and his ministers and watches his grotesquely dangling, human/animal candles burn down. His "joke" played out, the king's jester makes his final exit through the ballroom's skylight, where Trippetta is already waiting for him: "Neither was seen again" (908).

The gratuitous violence of Poe's story has surprised many readers.[86] There is nothing in his source, a chapter from the fourteenth-century *Chronicles* of Jean Froissart, that would have anticipated the extraordinary conclusion of Poe's text. Froissart reports a disastrous incident that took place in fourteenth-century France at a wedding at the court of Charles VI. However, the masqueraders who burned to death then were victims of an unfortunate accident, whereas in Poe's story the king and his ministers die as the result of a carefully wrought plan. Besides, Froissart's courtiers were dressed up as "savages," not as apes, which emphasizes the originality of Hop-Frog's idea. Poe's narrator points out that orangutans, at the time in which the story takes place, had "very rarely been seen in any part of the civilized world." Rather, we might add, they had not been seen or even heard of *at all*. In the United States, the first orangutan made its appearance in the halls of Charles Willson Peale's Philadelphia Museum in 1799: "How like an old Negro?" Peale asked his visitors (*W* 7). Fifty years later, when Poe wrote "Hop-Frog," apes were a more frequent, if not exactly common, spectacle in America, not the least thanks to Barnum's museum, which became the place where, as a poem included in Barnum's autobiography suggested, "the orang-outang / Or ape salutes thee with his strange grimace" (*L* 311).[87]

But Poe's apes are a far cry from Barnum's friendly orangutans, and Poe himself thought that the subject of his text was "a terrible one."[88] Consider, for instance, the many conspicuously precise descriptions of, and references to, chains in his story, especially the symbolically resonant chain that ties the king and his ministers together. The traditional idea of the Chain of Being, of a hierarchical order of beings on a scale rising from the lowest to the highest forms of life, had not been rendered fully obsolete by the new narratives of progress and development in nature that were increasingly cherished by nineteenth-century scientists. Tennyson's *In Memoriam* (1850) still recommended that humans get rid of the animal in themselves and "arise and fly," "working out the beast, / And let the ape and tiger die."[89] In Poe's story, thanks to the unsightly *lusus naturae* Hop-Frog, all such Models through which humans had tried to render themselves part of, and yet apart from, the rest of nature are exposed in their full, deadly artificiality, and humans arise and fly only to descend into hell.

With lethal irony, Poe's narrator remarks that Hop-Frog's artful "chaining arrangement" was intended to "make all things appear natural." But in Poe's story, chains have become fetters: "A long chain was now procured. First, it was passed about the waist of the king, *and tied*; then about another of the party, and also tied; then about all successively, in the same manner. When this chaining arrangement was complete, and the party stood as far apart from each other as possible, they formed a circle; and to make all things appear natural, Hop-Frog passed the residue of the chain, in two diameters, at right angles, across the circle" (905). The second chain in the story, the chandelier's chain, lifts the king and his ministers up into the air and, instead of elevating them to godlike status, causes their downfall. The chandelier's chain has turned into a hangman's noose.

Hop-Frog's repeated promise ("*I* can soon tell who they are!"; "*I* shall soon find out who they are!"; "*I* begin to see who these people *are*, now!") sounds indeed like an echo of the question Barnum had been exploring since 1845: "What-Is-It?" But if Barnum's play with the difference between humans and animals first jeopardized and then ultimately maintained natural distinctions, Poe's story erases them completely: "The eight corpses swung in their chains, a fetid, blackened, hideous, and *indistinguishable* mass" (908; my emphasis). At the end of "Hop-Frog," the human/animal divide has disappeared to the point of indistinctness, or, rather, *extinction*. The full impact of the story's conclusion is brought home in Hop-Frog's last speech: "I now see *distinctly* . . . what manner of people these maskers are" (908). The distinct insight the story "Hop-Frog" provides is that ultimately there are no distinctions. Poe's "connecting links" display their interconnectedness in the most literal sense possible. They are connec*ted* links: chained together, a bundle of animal costumes and burning human flesh, struggling in a grotesque net formed by their own ambiguous bodies. And Poe's fool Hop-Frog becomes a connecting link, too, in that he literally *connects, enchains, welds together* his victims. As Poe's "Hop-Frog" indicates, connecting links do not connect, they collapse. At the end of Poe's story, Hop-Frog, the agile man-monkey, climbs up the "scale of nature" from which his clumsy man-apes dangle and vanishes through the symbolic "skylight" to his home, "some barbarous region."[90] Poe's "Hop-Frog" drastically evokes a completely "blackened" world, a world in which gradations, be they natural, racial, or political, have ceased to matter. To translate Poe's allegory into terms that better capture the anxieties of readers in antebellum America: in a spectacle created and carefully arranged by a cunning, ugly slave, the masters themselves become slaves and expire. In Poe's text, man is no longer, as the eighteenth-century poet Edward Young thought, the "distinguished link in being's endless chain."

In late 1838, Poe assisted Thomas Wyatt in preparing an adaptation of Céran Lemmonnier's *Synopsis of Natural History* (which he also reviewed in July 1839 in *Burlington's Magazine*). "Of all animals," wrote Wyatt, "the ourang is considered as approaching most nearly to man in the form of his head, height of forehead and volume of brain."[91] If orangs had already come uncomfortably close, one of Barnum's exhibits, the "What-Is-It," suggested that there might be something even closer. "Is it an animal? Is it human?" asked the advertisement in the *London Times* on 29 August 1846 that accompanied the first exhibition of such a mysterious "connecting link" in London. "Or is it the long sought for link between man and the Ourang-Outang, which naturalists have for years decided does exist, but which has hitherto been undiscovered?" The article went on to describe not just a composite body but also a form of composite behavior: the features, hands, and the upper portion of the "What-Is-it" were "to all appearances human," while the lower part of its body, "the hind legs" and the "haunches" of the mysterious being, which was covered with long hair and "larger than an ordinary sized man, but not quite so tall," were "decidedly animal." Barnum reminded his associate Moses Kimball that "the thing" was not "to be called *anything* by the exhibitor. We know not & therefore do not assert whether it is human or animal. We leave that all to the sagacious public to decide."[92]

The "sagacious public" in Barnum's America realized, of course, that questions about the line dividing humans from animals intersected with questions about race.[93] While for Linnaeus the Chain of Being had been a simple connective, other naturalists were quick to use it in a more hierarchical way—as a vertical scale—to rationalize the worrying association of humans with beasts and bestiality. Here, African-Americans proved handy indeed; as Winthrop Jordan has pointed out, "the association of the Negro with the ape ordered men's deep, unconscious drives into a tightly controlled hierarchy."[94] Barnum stunned his visitors into sudden recognition of their closeness to the animal kingdom and then allowed them to detach themselves from the disturbing implications of his "connecting links," and he did this most effectively through the enduringly famous embodiment of the "What-Is-It" by an African-American, William Henry Johnson. Johnson, who remained on the "freak circuit" for more than sixty years and was known to contemporary audiences as "Zip,"[95] began to appear at the American Museum even before Barnum had officially become the museum's proprietor again. Advertisements published in the *New York Times* in February and March 1860 visually played with the idea of connection and separation that had found its perfect expression in the "What-Is-It":[96]

THE WHAT IS IT?

THE WHAT IS IT?

THE WHAT IS IT?

CONNECTING LINK BETWEEN MAN AND MONKEY

CONNECTING LINK BETWEEN MAN AND MONKEY

CONNECTING LINK BETWEEN MAN AND MONKEY

LOOKS LIKE A MAN!   ACTS LIKE A MONKEY!

LOOKS LIKE A MAN!   ACTS LIKE A MONKEY!

LOOKS LIKE A MAN!   ACTS LIKE A MONKEY!

BUT CAN'T TALK!

BUT CAN'T TALK!

BUT CAN'T TALK!

Zip appeared at the museum dressed up in a furry smock. His head was shaved, apart from a small topknot, to exaggerate what visitors would have regarded as a distortion of a normally shaped human skull, and this is how he still appears in a publicity shot entitled "The Original What-Is-It" that was produced around 1903 (ill. 21). A posterbill of Barnum's Museum for 17 March 1860, which advertised "Zip" as the "MOST MARVELLOUS CREATURE LIVING," asked: "Is it a lower order of **man**? Or is it a higher development of the Monkey ?" This "very unique specimen of BRUTE-HUMANITY—this most wonderful of NATURE'S FREAKS—this most insolvable of PHISIOLOGICAL RIDDLES" had been discovered near the source of the Gambia River in Africa by a party of explorers in search of the "Gorillia."[97] At the time of its capture, the "What-Is-It" had not been alone, proof that he was not just an individual aberration but the representative of an entire species:

There were five of these singular creatures in company; but, despite the efforts of their pursuers, two of them escaped. The remaining three were captured and shipped to this country. Unfortunately, however, two of them (one of each sex) died on the passage out, and the one now presented to the public, (a male,) is the sole survivor. He possesses the skull, limbs, and GENERAL ANATOMY OF THE ORANG-OUTANG, with the actual COUNTENANCE OF A HUMAN BEING! He is, probably, about 10 years old; is four feet high, weighs 50 pounds; and is intelligent, docile, active and PLAYFUL AS A KITTEN.

21. "Zip, Original What Is It." Photographer unknown. Philadelphia, 1903. Harvard Theatre Collection.

The accompanying illustration shows the "What-Is-It" (ill. 22) wearing a hairshirt and holding a stick in a kind of diorama-like setting, complete with exotic ferns, rocks, and palm trees in the background. Barnum's comical being featured "the skull" and "general anatomy" of an orangutan and the "countenance" of a man—just the countenance, mind you, not the brains: the profile of Zip's head emphasizes the medical condition from which Johnson allegedly suffered, microcephalism. Too human to be an ape, Zip appeared too simian to be fully human.

The description offered in the posterbill for 31 March 1861, which also celebrated Barnum's return as the proprietor of the American Museum, threw in some more details as to the natural history of this mysterious "nondescript." But this was natural history of a peculiar sort, describing less what was there than what was not. Engaging his curious visitor's imaginative participation, Barnum tries to influence her through the repetition of the phrase "instead of" and by using such adverbs as "too" or "now" and the auxiliary "should," all of which insinuate that the "What-Is-It" is some kind of a provisional creature:

22. Barnum's American Museum, posterbill, 17 March 1860. Harvard Theatre Collection.

You will observe that it is something very peculiar indeed. The formation of the head and face combines both that of the native African and the Orang Outang. The upper part of the head, and the forehead in particular, *instead of* being four or five inches broad, as it *should* be, to resemble a human being, is LESS THAN TWO INCHES! The HEAD OF THE WHAT IS IT? is very small. The ears are set back about an inch *too far* for humanity, and about three-fourths of an inch *too high* up. They *should* form a line with the ridge of the nose to be like that of a human being. As they are *now* placed, they constitute the perfect head and skull of the Orang Outang, while the lower part of the face is that of a Native African. In the next place the teeth, *instead of* standing erect, occupy a slanting position, like those of the horse or sheep, slanting to a great distance under the tongue, and into the roof of the mouth, (my emphasis)

Barnum's poster also suggested that, although the "distortions" of the body of the "What-Is-It" were to be considered natural or "animal-like," their effect, to a human viewer's mind, was disabling. Specifically, the "What-Is-It" *cannot* close his *mouth, cannot* unbend his legs, *cannot* straighten his arms:[98]

The teeth are double nearly all around, and the creature *is not able* to close its mouth entirely, owing to the formation of the jaws, which are crooked instead of straight, thus leaving the front of the mouth open about half an inch. THE ARMS OF THE WHAT IS IT are much too long in proportion to its height at least some three inches. They are also crooked like those of the Orang Outang, and it is *not able* to straighten them. He has great strength in his hands and arms. Anything he can get hold of he will cling to for quite a length of time. There is apparently more strength in his hands and arms than in all the rest of his body combined. In the next place, his legs are crooked, like those of the Orang Outang. He *cannot* make them straighter than you see them now. He has no calf to his leg, but exhibits a gradual taper from the knee to the ankle joint, THE WHAT IS IT'S FOOT is narrow, slim, and flat, and has a long heel like that of the Native African. The large toe is more like a man's thumb. The others are bent under, and the distortion seems to be natural.

The feet of the "What-Is-It" are formed like hands—a reminder that in nineteenth-century taxonomy the term "quadrumana" often replaced "primates" to help exclude humans.[99] The long stick in Zip's hand further convinced Barnum's visitors how precariously this "man-monkey" maintained an erect posture. They "climb trees with the greatest facility," wrote Thomas Wyatt in his *Synopsis* about the "quadrumana," "while it is only with pain and difficulty they can stand or walk upright; their foot then resting on its outward edge only."[100] In spite of such tough impediments, further emphasized by the crutch-like staff, Zip's huge smile indicated that he was basically a happy being.

Zip's crutch also appears in one of the lithographs in Currier & Ives's *Barnum's Gallery of Wonders* (ill. 23), which repeats the allegation that this intermediate specimen should be considered the true "*connecting link* between THE WILD NATIVE AFRICAN, AND THE ORANG-OUTAN."[101] (Notice that wherever Darwin, who harbored his own distinct racial prejudices, still would have said "man," these texts pointedly limit the extent of the "connection" to the "Native African.") Invidiously, the lithograph's caption suggests that at the time it was caught, the "What-Is-It" ran on all fours; only "with difficulty" had it been prevailed on "to stand as nearly erect as here represented." The similarity between this print and the posterbill illustration featuring Tom Thumb amidst a circle of bemused visitors is striking (ill. 14). The faintly delineated background in the Currier & Ives print takes Zip out of his habitat of ferns and palm trees and places him firmly into Barnum's museum, more specifically in front of the famous aquaria. This time he is wearing not his fur but a red frock, short brown pants, and green slippers. Two couples are shown to be spellbound by the unusual spectacle, and through the inclusion of the child, which like Zip is turned toward the viewer, the "What-Is-It" seems even smaller, more fragile, more domesticated than in other contemporary illustrations—a childlike being, as pleasant, if not exactly as attractive, as little Tom Thumb.

Although "Zip" Johnson's "simian features" impressed the fastidious diarist George Templeton Strong, who also joked that the "What-Is-It" was "a great fact for Darwin," he did not for a moment believe that the "What-Is-It" was anything other than a "nigger boy."[102] But it did not require the sophistication of a Templeton Strong to see through Barnum's rather ham-fisted hoax. Yes, the "What-Is-It" seemed to tell every visitor, endowed as it was with (partial) "HUMANITY AND REASON" and (partial) bes-

tiality, we are all "netted together." But, no not to worry,
Zip was obviously black. As
a nice, pleasant, playful, black
"connecting link," he easily
qualified for Bachman's definition of the African slave
as a "child that we lead by
the hand." Barnum cast his
net and yet kept the meshes
wide enough to allow his
visitors to escape. By carefully designing the context
in which his "connecting
link" appeared and by offering what was in fact a great
deal of description to accompany a specimen supposed to
be a "nondescript," Barnum
succeeded in simultaneously
proposing and ridiculing the
threat of "connectedness."[103]
An important aid in this
strategy was Zip's seeming
disability: it helped Barnum
to steer a midway course between the display of naturally

23. "What Is It?—Or 'Man Monkey.'" From *Barnum's
Gallery of Wonders*, no. 26. Lithograph by Currier & Ives.
Ca. 1865.

preordained differences and the suggestion of accidental but harmless aberrancy. The
"What-Is-It" thus inhabited a strange twilight zone between assumed typological generality and "pathological" peculiarity.[104]

If Zip still proved too "bewildering," the frustrated white visitor could always, as the
New-York Daily Tribune suggested, walk on to the animal exhibits, where he would find
"repose and certainty in a Giraffe, a Whale, or a Rhinoceros."[105] The vertigo that "nor-
mals" are said to experience when they see a "freak" (which Leslie Fiedler thinks is "like
that experienced by Narcissus when he beheld his image in the reflecting waters"[106]) did
not affect Barnum's visitors. Encouraged by the general context of the Barnum museum,
they would gravitate between the surprised recognition of similarities and the relieved
acknowledgment of differences. As embodied in Barnum's "connecting links," Darwin-

ism (or the inter-connectedness of living beings that it stood for) became harmless fun. Barnum's handbills deliberately stressed how like a "kitten," like a pet, the "What-Is-It" really was. Indeed, there was "NOTHING REPULSIVE IN HIS APPEARANCE."

As if he wanted to top his own performance in the construction and exhibition of the "What-Is-It," Barnum for a brief time confronted his visitors with a display titled "What Can They Be?" In September 1869, Barnum facetiously encouraged "naturalists and other persons" to identify two "wonderful," "indescribable" animals, found "in a singularly perfect state of preservation" in "a CAVE in the hitherto unexplored Wilds of Africa." He offered a thousand dollars to whomever could "correctly classify the

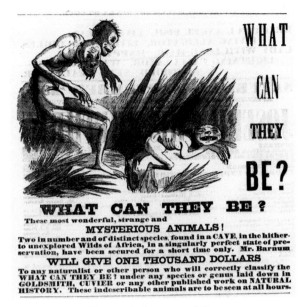

WHAT CAN THEY BE?

WHAT CAN THEY BE ?
These most wonderful, strange and
MYSTERIOUS ANIMALS!
Two in number and of distinct species, found in a CAVE, in the hitherto unexplored Wilds of Africa, in a singularly perfect state of preservation, have been secured for a short time only. Mr. Barnum
WILL GIVE ONE THOUSAND DOLLARS
To any naturalist or other person who will correctly classify the WHAT CAN THEY BE? under any species or genus laid down in GOLDSMITH, CUVIER or any other published work on NATURAL HISTORY. These indescribable animals are to be seen at all hours.

24. Barnum's American Museum, posterbill, 29 September 1860. Harvard Theatre Collection.

WHAT CAN THEY BE?" according to the established rules of natural history (29 September 1860). The poster illustration shows two monkey-like animals, one of which sports two heads (ill. 24). This time, Barnum might have strained the credulity of his visitors too much: the "What Can They Be?" appeared for the last time on the bill for 12 November 1860 and thereafter was consigned to oblivion.[107]

Apart from Zip, Barnum's most enduringly famous "connecting link" (and actually the first he invented) was the "Fejee Mermaid," a creature that briefly appears in Joyce's *Finnegans Wake*—as a "Feejeean grafted ape on merfish, surrounded by obscurity."[108] Barnum had borrowed the sickly-looking specimen from his accomplice Moses Kimball, the owner of the Boston Museum, and was pleased to discover that, within a few weeks, it had helped him triple his monthly profits. Even his own tolerant museum naturalist Guillaudeu had warned him that mermaids didn't exist, which to Barnum characteristically was "no reason at all" (*L* 231). As the first step in his elaborate scheme "to modify general incredulity in the existence of mermaids" (*L* 231–232), Barnum invited Levi Lyman, who had already successfully assisted him earlier in the exhibition of Joice Heth, to pose as the aptly named "Dr. Griffin," a representative of the Lyceum of Natural History in London. Soon letters recommending Dr. Griffin's "mermaid" began pouring in from different parts of the United States to the offices of newspaper editors in New York. All written by Barnum himself, they excitedly confirmed that Dr. Griffin was transporting a "genuine" mermaid to the London Lyceum. But Barnum also spread the word that Dr. Griffin, forever in the service of hard science and not disposed to play up to the expectations of the public, was reluctant to display his treasure in public (*L* 237). While the "mermaid fever," as Barnum termed it, "was . . . getting pretty well up," Barnum sold ten thousand "mermaid pamphlets" at a penny each in hotels and stores. Then, his name not yet officially implicated in the stunt, he rented the Concert Hall on Broadway and announced the opening of his long-awaited exhibition of "the mermaid and other wonderful specimens of the animal creation."

Alongside the mermaid, Barnum displayed other "connecting links in the great chain of Animated Nature"—a platypus, "being the connecting link between the Seal and the Duck," a flying fish (the connection of "the Bird with the Fish"), and a siren, "an intermediate animal between the Reptile and the Fish" (*L* 238). Barnum was aware of the discrepancy between his prurient advertising of a bare-breasted mermaid on large transparencies outside the exhibition hall and the rather disappointing appearance of his "black-looking specimen of dried monkey and fish that a boy a few years old could easily run away with under his arm" (*L* 239; see ill. 25), but he used it to his advantage. This, to be sure, was not the "sea-maid" Shakespeare's Oberon heard "Uttering such dulcet and harmonious breath, / That the rude sea grew civil at her song" (*A Midsummer Night's Dream* II, i). If the "reality" of this mermaid's body did not match the popular image, what better proof could there be that, after all, the mermaid *was* real?[109]

Discussing his mermaid in his autobiography, Barnum gave his readers an impressive sample of his abilities as a natural historian, pretending that he had in fact taken great care with his analysis of the weird specimen:

> The monkey and fish were so nicely conjoined that no human eye could detect the point where the junction was formed. The spine of the fish proceeded in a straight and apparently unbroken line to the base of the skull—the hair of the animal was found growing several inches down on the shoulders of the fish, and the application of a microscope absolutely revealed what seemed to be minute fish scales lying in myriads amidst the hair. The teeth and formation of the fingers and hands differed materially from those of any monkey or orang-outang ever discovered, while the location of the fins was different from those of any species of the fish tribe known to naturalists. (*L* 234)

Barnum shrewdly used the verbal trappings of the language of natural history (the references to the "formation" of the mermaid's teeth and fingers; the assertion of her difference from "any other species . . . known," which normally precedes the establishment of a new type specimen) in order to mock his own doubtful credentials as a naturalist as well as his readers' gullibility. He would have been the last admit that he had heard the mermaids singing.

What seems even more important within the economy of Barnum's mermaid narrative is that the shameless exhibitors of the hoax at least on one occasion ended up being deceived themselves, Barnum treats his readers to a funny description of frock-coated Levi Lyman's carefully prepared appearances in the Broadway Concert Hall, where he would stand in front of the "hideous-looking mermaid" in her glass vase and, "surrounded by numerous connecting links in nature," learnedly hold forth to his audiences about "the works of nature in general, and mermaids in particular" (*L* 238). At one point, Lyman leaves the room for a few minutes, and some visitors, seizing this opportunity for a "joke," reach into the mermaid's glass case, stick a half-smoked cigar into her mouth, and disappear quickly. No wonder that those in the next group, seeing "a little, black, dried mermaid with a cigar in her mouth," immediately feel that "the whole thing was an imposition" (*L* 240). Lyman, unaware of the dramatic change in the prize exhibit, approaches the stage "with a dignity which

A CORRECT LIKENESS OF THE FEJEE MERMAID.
Reduced in size from Sunday Herald.

25. "A Correct Likeness of the Fejee Mermaid." Woodcut from *The Life of P. T. Barnum* (1855).

no man could assume better than himself" and resumes his harangue about the miraculous mermaid, which included the information that, having been caught in a fisherman's net in the "Fejee Islands," it lived "upwards of three hours after its capture." Feigning genuine interest, one of the visitors asks Lyman: "Was her ladyship smoking the same cigar when she was captured that she is enjoying at present?" (*L* 240), Turning around and discovering the unauthorized alteration in his "critter," Lyman, completely nonplussed, perspires as if he had been "mowing a heavy crop of grass" (*L* 240).[110]

This passage shows Barnum in full possession of his considerable talents as a writer, which should secure him a place, even if only a minor one, in the history of nineteenth-century American literature. More to the point, it also reveals the crucial premise of Barnum's autobiographical narrative as well as, more generally, the basis of his "showmanship." Remember the country store: "Each party expected to be cheated, if it was possible" (*L* 99). Barnum's jokes, his tricks and hoaxes, often depended on the *conscious* cooperation of his audiences. Even when they were taken in, they accepted the deception as part of a system that would allow them, given the opportunity, to retaliate.

When Barnum's "Fejee Mermaid"—called the "Fudge Mermaid" in "Petite Bunkum's" parody—toured the South in the winter of 1843, it met with a rather unfavorable reception in South Carolina. Here the aforementioned Reverend John Bachman, D.D., defender of human uniqueness and Caucasian superiority, took Barnum's hoax seriously enough to launch, on 20 January 1843 and under the pseudonym "No Humbug," a vitriolic attack in the willing pages of the *Charleston Mercury*. Bachman, Audubon's correspondent and collaborator on *The Viviparous Quadrupeds of North America* (1846–1854), was, of course, more than just a "local" naturalist,[111] and he knew his science better than to doubt the veracity of the stuffed platypus that patiently accompanied the mermaid on her trip south.[112] It was the other "connecting link" he was concerned about. He proposed to take it out of its glass case so that it could be shown to be the "lamentable hoax" it was. Bachman cringed to see how the public would pay to inspect a fish-tail "attached to the head and shoulders of a Baboon." Above all, he saw a larger political significance behind the scam: "Our Yankee neighbors usually show more ingenuity, and they ought to have recollected that although we poor simpletons are a long way off from the Banks of New Foundland, we are not to be imposed on by a tail of a Codfish." Barnum's accomplice, his uncle Alanson Taylor, defended himself and the mermaid rather feebly the next day, criticizing his assailant for seeking the shelter of anonymity. Strangely, this did not prevent him from signing his own piece, rather inconsistently, as "the Man Who Exhibits the Mermaid."

The true reason for Bachman's agitation became evident in a longer piece printed in the *Mercury* on 29 March 1843, long after Alanson and the mermaid had fled the scene of their imminent humiliation. Bachman was worried about the status and public relevance of natural history itself. He and his fellow naturalists in Charleston—among them John Edwards Holbrook, about to release the second and improved edition of his *North American Herpetology*—were "somewhat anxious for the sake of our reputations to afford the public an opportunity of deciding on the accuracy of our judgments as Naturalists." Bachman reminded his readers that "opinions in matters of science are not hastily formed" and ended his diatribe on a somewhat whining note, addressing no longer Barnum and his helpmates but the public at large:

> If now an indulgent public consider our studies and our labors of no value to them; if they enjoy the joke, whilst men who are ashamed of their names hurl their missiles at them from all quarters; if they chuckle and say, "I see what another lashing these naturals have received"—let it be so—we will bear it as well as we

are able; but be not surprised if under all these provocations we evidence a little of the spirit of human nature; and that when you again call upon us to explain to you the mysteries of nature, or the humbugs of art, our reply should be: "thank you gentlemen, you have long since taken these matters into your own hands; why call upon Hercules when your own shoulders are already at the wheel."

Bachman had accurately perceived the subversive principle behind Barnum's hoaxes and dreaded the consequences: Barnum would turn each of his visitors into a potential "expert," making his jokes credible enough so that people would pay to see the result and sufficiently incredible so that they would still feel in a good position to question what they had inspected. What Bachman had not grasped was that Barnum was in fact making a point similar to his own. The mermaid was the bizarre living proof against the existence of "connecting links," the ultimate assertion, in the guise of a joke, that there really was no serious connection between humankind and nonhuman nature. This conclusion was even easier to establish since the mermaid's upper body was so noticeably ape-like, composed as it was of orangutan's skull and the jaw and teeth of a baboon. Unrepentingly, Barnum displayed his "connecting link" again in 1855, celebrating that

the renowned and inexplicable

FEJEE MERMAID

about which there has been so much, and such an animated controversy, has been added to the Exhibition. (3 April 1855)

Barnum's Candor

As his biographers have emphasized, Barnum liked to call his own bluff and would, where appropriate, speak candidly to his visitors. This was also the primary aim of one of his books, ambitiously titled *The Humbugs of the World: An Account of Humbugs, Delusions, Impositions, Quackeries, Deceits, and Deceivers Generally, In All Ages* (1865). But there are significant problems with Barnum's candor, as one episode in *Humbugs of the World* illustrates especially well.

Having purchased a live beluga whale in 1861, Barnum wanted to create a suitable environment for his "sea-monster." He constructed a special glass tank, a "MINIATURE OCEAN" (2 December 1861), in the basement of the American Museum. But in his new gas-lit environment the poor whale became so frightened that he mostly kept at the bottom of the tank "except when he was compelled to stick his nose above the surface to breathe." In what seems like an (unintended) parody of Melville's invocation of the great Leviathan, Barnum describes how his disappointed visitors would stand for half an hour waiting to "sight" the whale—here (s)he didn't blow![113]

Distressed by the whale's stubbornness "in not calmly floating on the surface,"[114] Barnum himself appeared beside the tank, assuring his visitors that the whale was indeed in residence, to be seen "at considerably less trouble than it would be to go to Labrador expressly for that purpose." Unimpressed, one of Barnum's visitors, a "sharp Yankee lady," concluded that Barnum was again "humbugging" his visitors with a fake whale made of "india-rubber," operated by "steam and machinery, by means of which he was made to rise to the surface at short intervals, and puff with the regularity of a pair of bellows." Barnum, seizing the opportunity, "very candidly" acknowledged to the lady that she was "quite too sharp" for him: "I must plead guilty to the imposition; but I begged her not to expose me, for I assured her that she was the only person who had discovered the trick. It was worth more than a dollar to see with what a smile of satisfaction she received the assurance that nobody else was as shrewd as herself." A gratified customer, the Yankee lady had "received double her money's worth in the happy reflection that she could not be humbugged, and that I was terribly humiliated in being detected through her marvelous powers of discrimination!" Whenever afterwards he met the good lady, Barnum would try hard "to look a little sheepish."[115]

Obviously, even Barnum's "candor" came with a limited warranty. The pinnacle of Barnum's humbug was to let people believe that they had successfully uncovered a humbug. Barnum reserved the right to make the discovery of his joke an integral part of yet another joke—here perpetuated by the "sheepish" attitude he adopted toward his skeptical visitor whenever he ran into her. Catering to the fancies of his patrons, he knew how to keep his counsel, how to make sure that the smooth surface of his changing deceptions remained intact. In Maureen Howard's recent novel *Natural History,* which describes the interlocking fates of a family and the city of Bridgeport, where Barnum resided, the narrator warns the reader: "Mind your attitude towards the great showman. . . . Best not to belittle our hero as the quintessential American trickster for he'll disarm you with full disclosure."[116] As Howard suggests, with Barnum the "candid" look behind the scenes could be yet another stage device, a new strategy for deception, and could lead to the "discovery" of another, better joke.

Whether living or dead, whether animal or human, almost all of Barnum's exhibits—when taken on their own terms and without the surrounding Barnumesque ballyhoo—were not, it seems, particularly original. Elizabethan "wonder cabinets" contained mermaids, mummies, and embalmed bodies of human babies, as well as the bones and skulls of apes.[117] In his visits with the "CROWNED HEADS of Europe" (22 December 1852), Tom Thumb echoed the dwarf John Wormberg, thirty-one inches tall, who paraded before James II and the nobility in Whitehall.[118] Even Barnum's gentle giants had their predecessors. For example, when the writer of Barnum's catalogue of "Living Curiosities" proudly claims that he could easily stand upright under Monsieur Goliath's outstretched arm, we remember that Samuel Pepys once made his own measurement of a giant's height at a fair by standing under him "with my hat on."[119]

However, in spite of all it contained that seemed secondhand, as a concept Barnum's collection was unique. In Barnum's American Museum, form had finally superseded content. Barnum's museum is best understood as an ongoing conversation between the collector and his *visitors*. If one of the main themes of Barnum's exhibition was size, his collection, in all its different guises, was a vast experiment in size. Barnum's autobiography expanded; his museum, fires and all, expanded; his circus became "the Greatest Show on Earth."[120] A color poster created for Barnum in 1881 by the Strobridge Lithograph Company of Cincinnati, Ohio, summarized Barnum's career: "Born in the town of Bethel, Conn., July 5th, 1810, started as a Showman in 1835, has conceived & exhibited more gigantic amusement enterprises than any other showman that ever lived."[121]

Metonymically (through his various gigantic enterprises) as well as metaphorically (in his constant self-advertising), Barnum became a curiosity himself, a kind of gentle giant, larger than life, more impressive than anything he ever displayed. The cover of the 1850 catalogue shows Barnum's huge face hovering godlike over the cute little figure of one of his most profitable exhibits, General Tom Thumb, a kind of alter ego as well as a substitute son. Tom Thumb stands in the center of his adoptive parent's collection, just as Peale's Noah had rested amidst his animals (ill. 26). In February 1862, Barnum displayed a likeness of himself next to Commodore Nutt, his most recent "wonderful prodigy" (a great tautology!), described as a "mere pigmy of the genus homo" (10 February 1862; ill. 27). Barnum had reputedly paid thirty thousand dollars for the exclusive right to exhibit Nutt over a period of three years—more than any of Barnum's competitors had been able to offer. Barnum's pride in his authorship of an "entertainment unprecedented in the annals of amusement" is evident. Interestingly, while there are several familiar illustrations showing Barnum with a "dwarf," there is not a single well-known image of Barnum standing next to one of his giants.

In 1888, sculptor Thomas Ball had a photograph taken of himself cowering on the pedestal of his completed monumental stature of a seated Barnum—a dwarf crouching on the steps of a king's throne. But Barnum himself ordered that the statue should remain hidden in a warehouse until after his death (when it was erected in Seaside Park, Bridgeport). P. T. Barnum was a cuddly, not petrified, colossus, a gentle giant the people

26. *Barnum's American Museum Illustrated*, cover (1850). Harvard Theatre Collection.

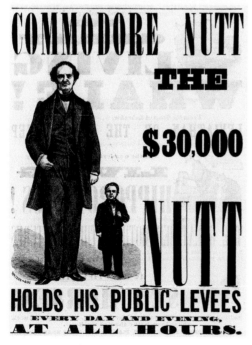

27. Barnum's American Museum, posterbill, 10 February 1862. Harvard Theatre Collection.

could love and occasionally play with, a giant who would, on selected occasions, allow his audiences to reduce him back to size again: "Mr. Barnum, you look much like other common folks," he appreciatively quotes a surprised museum visitor in his autobiography (*ST* 1: 231).

Barnum tantalized his audience, "the common folks," by suggesting how the lines between humans and animals, deviance and normalcy, smallness and largeness, ugliness and beauty, could become blurred. Then he would relieve them again from their state of uncertainty by showing how they themselves could become active participants in the process and thus make it seem harmless, Barnum's visitors could stick a cigar in a mermaid's mouth, or they could argue that the whale in the aquarium was really a dummy. Fun was an integral part of the show; the best example was Barnum's shabby mermaid, a display that was effective enough to dupe some visitors and crude enough to make others laugh. Barnum's crafty reaffirmation of white superiority—independent of class lines—was all the more effective since it did not rely on patronizing lecturing but required the museum-goers' creative, if skeptical, participation. Since we all live by our imaginations, most people, when given a choice, "had rather be deceived than not," sighed Ralph Waldo Emerson, summarizing what he regarded as the cause and the curse of Barnum's museum.[122]

However, in spite of his folksy manner, the "giant" Phineas Barnum remained, as the story of the Yankee lady and the whale shows, ever in control, making sure that his visitors would always stand under *his* outstretched arm and not vice versa. Barnum turned himself into a product and insisted that he retain the copyright. Even before John Wanamaker established the first "department store" in New York in 1876, Barnum had successfully engineered the transition from the museum-as-country-store into the museum-as-department-store. In the 1860s, the largest building in downtown New York City was Alexander Turney Stewart's "Marble Palace," a five-story cast-iron retail dry-goods store, illuminated by gaslight. This, rather than the Smithsonian, was the real competition for Barnum's "Marble Building" (27 March 1843). But although there can be no doubt that Barnum's American Museum took part in the creation of "a new powerful universe of consumer enticements,"[123] its owner still held fast to the traditional identity of the country storekeeper who retains absolute control over his inventory. Barnum offered his own products, not somebody else's merchandise, and his avuncular "blatant face" loomed large on all the museum's advertisements.

Decades later, Henry James first recalled Barnum's face when describing childhood visits to the American Museum on his cherished "non-dental Sundays" (those Sundays not reserved for visits to the dentist). Outside, lurid images of "spurious relics and catchpenny monsters in effigy" would tempt passersby with the promise of the "more . . . abnormal" living originals inside the museum. James memorably describes his sense of wonder and anticipation while waiting "in the dusty halls of humbug, amid bottled mermaids, 'bearded ladies' and chill dioramas" for the doors of the lecture room—Barnum's theater hall, the "true seat of joy"—to open.[124] James recognized that the theater hall was the imaginative center of Barnum's museum enterprise: the American Museum depended on the clever and careful modeling of audience expectations, a purpose to which the device of "natural history" had finally become incidental. Barnum called himself a student not of natural history but of "human nature."[125]

For those distanced from the smoothly running mainstream of Victorian society, Barnum's concept did not always exert its normalizing influence. In December 1889, twenty years after the last of Barnum's "show-shops" had burned to the ground, Henry James's invalid sister Alice was looking for words to describe how her body was becoming alien to herself as well as to others ("There are some half a dozen people who have come to see me once and who have never come again"). She confided to her diary that she had become like "a Barnum Monstrosity which had missed fire."[126]

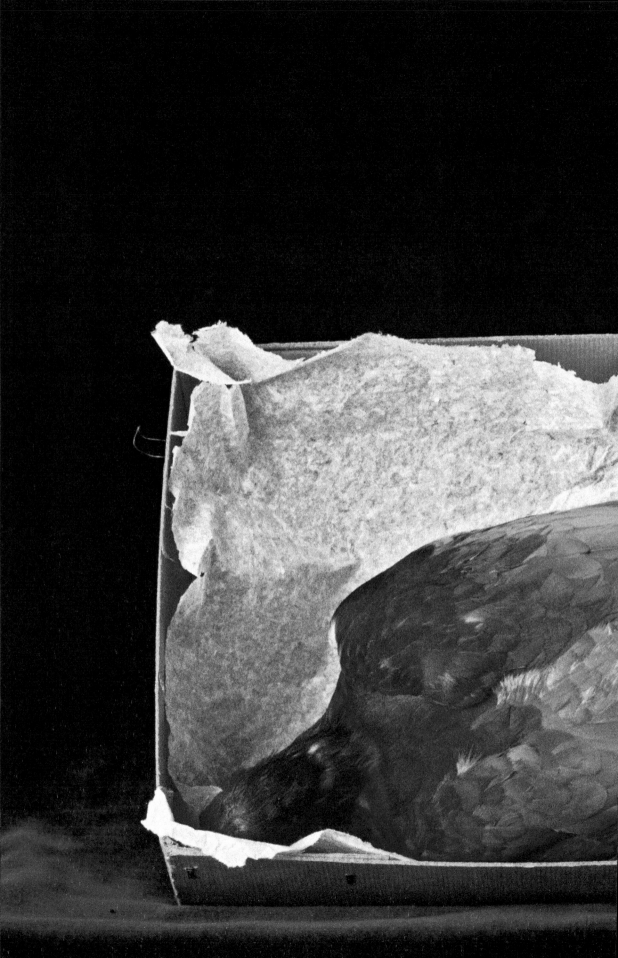

PART TWO

REPRESENTING

CHAPTER 4

The Power of Fascination

*When charmed by the beauty of that viper, did it never occur to you to change
personalities with him? to feel what it was to be a snake? to glide unsuspected in grass? to
sting, to kill at a touch; your whole beautiful body one iridescent scabbard of death?*
Herman Melville, *The Confidence-Man*

Serpents in the Garden

In the comfort of his London home, Peter Collinson would tremble when he imagined the dangers that John Bartram and his son William encountered on their arduous collecting trips. His inner eye beheld large poisonous snakes wrapped around the limbs of his innocent American correspondent, as if he were a kind of New World Laocoön. In the crowded halls of Charles Willson Peale's Philadelphia Museum, the writhing snakes confined in fragile wire cages sent shivers down the spines of impressionable visitors. And P. T. Barnum's lurid advertisements for the American Museum baited passersby with the prospect of "monster snakes" within. "Twelve feet long and Ten Inches in Circumference," these creatures spent their days in constant anticipation of the live mice that were fed to them while horrified visitors were allowed to look on (17 September 1849; 3 December 1860). Somehow, the idea of the New World, in the susceptible imaginations of generations of observers, had come to be intimately connected with the specter of ravenous, rattling reptiles spewing poison as their luminescent bodies disappeared into the dark.

It is easy to see why the members of God's people, dispatched on an errand that was to be its own reward, should have been so obsessed with the idea of deadly snakes hiding in the leafy undergrowth of the precarious paradise they had entered unbidden. What is remarkable, though, is just how long such fears persisted. At a meeting of the Royal Society in London on 17 February 1736, the assembled gentlemen, after listening wearily to yet another letter on the mysterious powers of snakes sent to them by a zealous American correspondent, agreed that "the people of that Country" were "strangely possessed" with "that dreadfull Reptile, inasmuch they entertain faith in the most wonderfull things concerning it."[1]

Such a strange obsession became visible in *Images or Shadows of Divine Things*, a collection of loose notes and typological reflections the Puritan theologian Jonathan Edwards had begun to compile in 1727. Edwards, one of the most brilliant minds in colonial America, had demonstrated a remarkable gift for entomology in his early essay on "flying spiders." Now he apparently had developed an interest in herpetology, too. The "serpent's charming of birds and other animals into their mouths," he wrote in his notebook, was like "the spider's taking and sucking the blood of the fly in his snare"; both, to Edwards, represented the workings of the devil attempting to catch "our souls by his temptations."[2]

A later passage in Edwards's text elaborates the allegory of the "charming" snake and details its applicability to the life of the human sinner, promising redemption through Christ, who, reassuringly, will arrive to "bruise the serpent's head":

> The animal that is charmed by the serpent seems to be in great exercise and fear, screams and makes ado, but yet don't flee away. It comes nearer to the serpent, and then seems to have its distress increased, and goes a little back again, but then comes still nearer than ever, and then appears as if greatly affrighted, and runs or flies back again a little way, but yet don't flee quite away, and soon comes a little nearer and a little nearer with seeming fear and distress . . . until at length they come so [near] that the serpent can lay hold of them and so they become their prey. Just thus often times sinners under the Gospel are bewitched by their lusts. . . . Whatever warnings they have and whatever checks of conscience that may exercise them and make them go back a little and stand off for a while, yet they will keep their beloved sin in sight . . . but will return to it again and again, and go a little further until Satan remedilessly makes a prey of them. But if any one comes and kills the serpent, the animal immediately escapes.[3]

The narrative embedded in Edwards's tension-packed exercise in typology (notice how the unfortunate animal comes nearer, moves back again, comes "nearer than ever," goes back, then comes "a little nearer and a little nearer") is probably less readily recognizable today than the more conventionalized reading of the "serpent" as one of the forms in which Satan makes his appearance.[4] What is interesting is not just the equation of the snake's unwitting prey with easily tempted, wavering, fallible humanity, standing back "a while" yet gradually if reluctantly giving in to the centripetal force of sin.

Perhaps even more striking is Edwards's description of the execrable reptile's preferred mode of capture, through "charming" or, as naturalists would soon call it, "enchantment" or "fascination."

Oliver Wendell Holmes would later sardonically remark about Edwards's prose, perhaps not unfairly, that, thickened as it was with scriptural references, it was more heavy than holy.[5] In this excerpt, however, Edwards presents his case quite skillfully. His story is predicated on the opposition of *good* (here represented by the checks and warnings of conscience) and *evil*, but it derives its special effect from the association of these two theological terms with another set of opposites, *attraction* and *repulsion*. In the "Queries" appended to his *Opticks* (1704), Isaac Newton had suggested that many phenomena at the atomic level could only be explained by assuming the work of "attractive and repelling powers which intercede the particles."[6] In a minutely ordered universe ruled by a supreme God, physical laws were those of the spirit, too, and it is possible to argue that Edwards's story of the "animal" and the snake was so disturbing because the relationship between good and evil, attraction and repulsion, through the intervention of the force of "fascination," appears dangerously skewed. Equilibrium, so important in Newton's and Edwards's universe, was threatened: what should repel the sinner because it is evil seems attractive to him, charming, even enchanting, a fact captured in Edwards's description of the sinner keeping his "beloved sin" not just in mind but "in sight." And immediately a third set of terms, not mentioned by Edwards but implicit in his story, enters into the debate, complicating it even further: we begin to suspect that the dichotomy of good and evil, as endangered as it seems here, cannot be equated with the opposition of "beauty" and "ugliness"—that, horrible to think, something repulsive, because it is so attractive, might also seem "beautiful" (in a sense quite different from the symmetrical "sweet mutual consent" Edwards attributes elsewhere in *Divine Things* to the experience of beauty).

This suspicion was made explicit by Edwards's British contemporary William Hogarth. In his *Analysis of Beauty* (1753), Hogarth cheerfully ignores the religious dimensions of the topic, evincing sympathy for Eve and taking unashamed delight in serpents and serpentine lines, which to him represent the hallmark of good art—"variety." The title page of Hogarth's book includes a succinct epigraph from Milton's *Paradise Lost*, in which the serpent, chiefly because of its pleasing form, "fascinates" humans, "curl[ing] many a wanton wreath, in sight of Eve, to lure her eye." "Divest," Hogarth ecstatically invites his readers, "one of the best antique statues of its serpentine winding parts, and it becomes from an exquisite piece of art, a figure of such ordinary lines and unvaried contents, that a common stonemason . . . might carve out an exact imitation of it."[7]

Edwards's text belongs to a tradition of narratives about the dangerous powers of snakes that, as early as 1794, induced Benjamin Smith Barton to refer condescendingly to "volumes of tales."[8] Snakes are not, to modify Emily Dickinson's phrase about the rat, God's concisest tenants. Legless and limbless, their scaly skins glittering in manifold hues, they can be strangely beautiful. More frequently, however, they evoke terror and disgust in the human observer, and this not just because one of them was said to have seduced our first mother. Emerson watched with distaste several snakes moving up and down a hollow, with no apparent purpose whatsoever, "gliding . . . not to eat, not for love, but only gliding."[9] Their slightly scandalous mode of locomotion as well as the fact that, like other reptiles, they have "neither hair, feathers nor mammae" (as the naturalist John Edwards Holbrook put it) and that they shed their skins—shift, as it were, their shapes—has helped to relegate snakes to a more than questionable position among the vertebrates. In the orderly train of animals obediently entering Noah's barnlike ark, as represented in Edward Hicks's *Noah's Ark* (1846; Philadelphia Museum of Art), snakes are conspicuously absent. And the tiny but obviously determined spotted snake in Erastus Salisbury Field's idealized representation of the Garden of Eden (1865) appears a disturbing, solitary outsider in an otherwise quiet, dreamlike exotic paradise, the only being on the move while most other animals remain transfixed in orderly pairs, arrested in their symmetrical positions, in silent contemplation of themselves and their beautiful environment (ill. 28). Behind the snake, we see Eve reaching for the apple.

Apart from its exegetical relevance, Edwards's charming snake had a more local habitation and a name. Edwards was firmly at home in the landscape of New England, "a child of the wilderness as well as of Puritanism," according to Perry Miller.[10] He was an acute observer of his natural environment and thus of a world that would have included reptiles. (*Images or Shadows of Divine Things* also contains, for example, a description of how snakes swallow their prey—"they draw them down little by little.")[11] And he would have been one of the first to know that, among venomous snakes, the American rattlesnake was said to have the ability of "fascinating," "charming," or "enthralling" its prey.

The rattlesnake was the serpent in the American Garden, defining its beauty as well as defying those who would dare walk around in it too carelessly. Captain John Smith listed it, in his *Advertisements for the Unexperienced Planters of New England* (1630), among the reasons why some of the new settlers would have eventually decided to return to England: some, he says, complain that "the Country is all Woods," some talk "of the danger of the rattell Snake."[12] When, nearly a hundred years later, William Byrd II set out to investigate the "dividing line" between Virginia and Carolina, he did not forget to pack a "Strong Antidote against the bite of it," the Rattlesnake Root, which

28. *The Garden of Eden*, by Erastus Salisbury Field. Ca. 1865. Oil on canvas, 63.5 X 105.41 cm. Collection of the Shelburne Museum, Shelburne, Vermont. Gift of Mr. and Mrs. Dunbar W. Bostwick. 1959–266. Photography by Bruce Schwarz.

he had "made into Doses in case of Need." Amusingly, the rattlesnakes his band runs into, or rather steps onto, prove to be just as polite as his men appear to be dumb: "Will Pool trod upon one of them without receiving any hurt, & 2 of the Chain Carriers had march't over the other, but he was so civil as to bite neither of them."[13] And Alexander Hamilton, traveling from Annapolis to New York, Albany, and Boston in 1744, even regretted, somewhat facetiously, that a snake he more or less accidentally killed in the Hudson River Valley was harmless: "Going on board again at 4 o'clock I killed a snake, which I had almost trod upon as I clambered down the steep. Had it been a rattlesnake I should have been entitled to a colonel's commission, for it is a common saying here that a man has no title to that dignity until he has killed a rattlesnake."[14]

Most of the early colonists' efforts understandably were focused on possible antidotes against the rattler's poison, and in this they often seemed helpless to the point of being comical. In 1720, Captain Hall, an amateur scientist in Carolina, selected a number of local pets for instant extermination by a tied-up rattlesnake, with haphazard results.[15] Hall's unfortunate victims, mostly "Curr-dogs," were held right up to the jaws of the securely fastened "fine healthful" snake, were inevitably bitten, and then, right on schedule, died their miserable and not-so-experimental deaths while the assembled gentleman naturalists measured the duration of the wretched animals' agonies with their pocket watches. When "we could get no more Dogs," a cat was substituted; after the fatal bite, it accidentally escaped and was found only days later—so grotesquely disfigured that none of the gentlemen felt capable of examining its body. To the experimenters' amazement, a hen survived and had to be killed, so that a hasty inquest could be conducted, while a bullfrog passed away within a mere two minutes. Increasing complaints in the neighborhood forced the gentlemen to suspend their investigations: "Cats and dogs were not to be had; for the good Women, whose Dogs had been killed, exclaimed so much, that I durst not meddle with one afterwards." When Hall's hapless team later tested a remedy provided by a doctor from Charleston, the only dog to survive was the one to which the antidote had not been administered.

A considerably more panic-stricken account reached the Royal Society from Philadelphia, where Joseph Breintnall had been bitten by a rattlesnake. The man whom Benjamin Franklin movingly described in his *Autobiography* as a "good-natur'd" and friendly lover of poetry now all of a sudden experienced, as he told Peter Collinson, the most extraordinarily tormenting nightmares: "In all Sicknesses before, if I could but sleep and dream, I was happy so long; being ever in some pleasing Scenes of Heaven, Earth, or Air. . . . Now if I slept, so sure I dreamed of horrid Places, on Earth only; and very often rolling among old Logs. Sometimes I was a white Oak cut in Pieces; and frequently my Feet would be growing into two Hickeries. This cast a sort of Damp upon my waking Thoughts." Worse still, infused with the poison of the monster, Breintnall became a kind of walking reptile himself. All summer long, he stumbled around with his arm and hand "spotted like a Snake."[16]

This astonishing narrative, sent to London on 10 February 1746, becomes even more poignant when we realize that the man who had been "happy so long" and had then experienced nameless terror died merely a month after he had written his letter. On 18 March 1746, Joseph Breintnall's body washed up on the Jersey shore—an apparent suicide. "When once Religion is banished the mind, I know not what Can Relieve a Man in deep distress, but death, & the mean, Low hope of Annihilation therein," the pious John Smith of Philadelphia observed when he heard of Breintnall's death.[17]

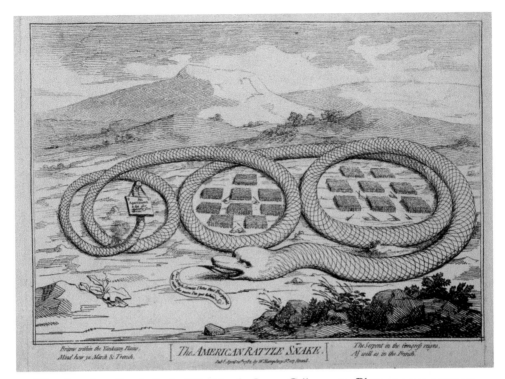

29. "The American Rattle Snake." Etching by James Gillray. 1782. Plate 25.4 x 33.2 cm; on sheet 26.3 x 36.1 cm. Published by W. Humphrey, London. The Library of Congress.

Freethinkers—Breintnall had become one—were even less equipped than their pious contemporaries to deal with the "dreadfull Reptile." Or so it seemed. Drawing on these and other reports about the effects of the rattlesnake's poison, Breintnall's friend Benjamin Franklin made his own "modest proposal" in the *Pennsylvania Gazette* of 9 May 1751. Facetiously he suggested that, for each ex-convict deported from England to America, the fledgling colonies send a rattlesnake back to their mother country: "I would propose to have them carefully distributed in *St. James's Park,* in the *Spring-Gardens* and other Places of Pleasure about *London.* . . . *Rattle-Snakes* seem the most *suitable Returns* for the *Hitman Serpents* sent us by our *Mother* Country."[18]

The threatened American colonies, in the War of Independence, could not have found a better emblem than this dangerous animal to represent their cause: "Don't tread on me."[19] A British cartoon from 1782 depicts a rattlesnake of truly hyperbolic proportions enveloping in its gigantic folds the armies of Burgoyne at Saratoga and Cornwallis at Yorktown, generously offering its third, still empty loop as an "apartment for rent" by yet another unlucky army: "The British Armies I have thus Burgoyn'd," sneers the animal, "And room for more I've got behind" (ill. 29). As late as 1836, a

30. "The Spirit of the Times." Lithograph, artist unknown. Published by P. Desobry, New York, February 1836. Photograph by Gerda Hellmer.

rattlesnake appears, alongside an American eagle with Andrew Jackson's bony face, in a political cartoon titled "The Spirit of the Times" (ill. 30).

Ostensibly dealing with outstanding claims against France for the alleged illegal seizure of American ships and the bumbling British attempt to establish a compromise, the cartoon pits British clumsiness (with John Bull holding a rifle) and French pride (represented by puffed-up birds and a cock with a grotesquely inflated chest) against the passion of the Americans and their very real potential to harm: "Let them come a'shore; I'll rattle them."

For better or worse, the rattlesnake, metonymically associated with the American landscape, had also become a kind of metaphor for the extraordinary challenge, the danger as well as the promise, of the New World itself, On the eve of the French revolution, M. le comte de Lacépède, the Keeper of the King's Cabinet, stood wonderingly before a case of neatly arranged snake specimens in his collections. As a chapter in his *Histoire des quadrupèdes ovipares et des serpents* (1788–1789) reveals, monsieur le comte shuddered to think that behind the beauties of America there should be hidden the sleek, undulating bodies and the sharp, rapid fangs of such hideous reptiles, waiting to prey on the unsuspecting traveler. Not realizing that the streets of Paris would soon look worse than anything he could have imagined about the wild woods of the New World, Lacépède expressed his relief that Europeans did not have to live in the forests of America, where the snake, "ce funeste Reptile," exerted the same reign of terror as the tiger (!) did in the deserts of Africa: "We do not miss their natural beauties or their climate so much warmer than ours, their more luxuriant trees, with their wonderful foliage, their sweeter and more beautiful flowers: these flowers, these leaves, these trees, alas, hide the abode of the dreadful rattlesnake."[20] Lacépède's rattlesnakes lie still, isolated from their natural habitat in the dangerous forests of the New World: their life seems to reside in the shadows tracing the contours of their immobile bodies on the whiteness of the page, as flat as the board in the Cabinet du Roi where they, as if they were archeological artifacts, have been rigged up for durable display (ill. 31).

The rich, open, indeterminate presence of America becomes tangible not only in the poems that would come to be written about it. The following narratives dealing with the rattlesnake's alleged "power of fascination" are offered

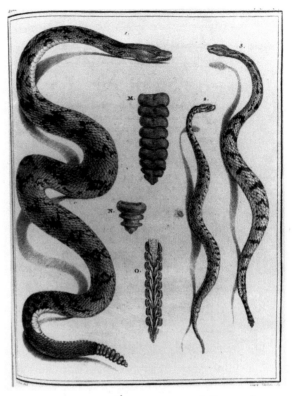

31. Bernard Germain Étienne, comte de Lacépède, *Histoire naturelle des quadrupèdes ovipares et des serpens* (1789), vol. 2, pl. 18. Museum of Comparative Zoology, Harvard University.

to the reader not as cautionary tales about what happens when a naturalist's own good sense falters and his power of judgment collapses. Rather, I present them here as products (some of them splendid) of the storytelling human imagination working its way through the wild, bewildering world of American nature.

Creatures from Hell

In 1683, Edward Tyson, member of the Royal College of Physicians, sliced a rattlesnake from Virginia into pieces, excitedly announcing to the readers of the London Royal Society's *Philosophical Transactions* that he lacked "Words to express what the *Pencil* could not imitate, much less can be represented in a *Print*."[21] Thus, in 1714, when Cotton Mather sent the first installment of his "Curiosa Americana" to the distinguished

Royal Society in London (hoping that some day in the not too distant future they would make him a fellow, too), he already knew that rattlesnake tales were eminently marketable.[22] In dead earnest, Mather presents as authentic several stories that, although a few realistic details have obviously been added for greater credibility, nevertheless sound as if they had been lifted straight from Pliny's *Historia naturalis.* Among them is a "Story . . . constantly affirmed by the *Indians,*" and it is a familiar one, too: "These Snakes frequently lie coiled at the Bottom of a great Tree, with their Eyes fixed on some Squirril above in the Tree; which tho' seeming by his cries and leaping about, to be in a Fright, yet at last runs down the Tree, and into the Jaws of the Devourer." In his *Natural History,* Pliny had warned his readers of a Medusa-like animal, the "basilisk," hardly larger than a foot in length and adorned with a "white spot like a star on its head" that looks like a diadem—a creature that terrifies humans as well as "other serpents," one whose mere breath kills trees and shrubs and breaks stones asunder. Like the Medusa, the basilisk also destroys through its gaze: "If he do but set his eye on a man, it is enough to take away his life."[23]

Mather's untidy mélange of supposedly new facts and ancient superstitions won him the coveted membership in the Royal Society (as he did not go to London to sign their constitution, he never achieved full status).[24] Another writer on rattlesnakes, who simply identified himself as "Captain Walduck," fared less well with the gentlemen in London: his notes ("most of this is Matter of Fact," he asserted rather charmingly in his conclusion) were apparently not deemed worthy of publication.[25] For Walduck, an officer stationed in the West Indies who had traveled widely in the British colonies, the rattlesnake was the "most pernicious Creature in ye English Empire upon ye Main of America . . . as well for their Number as Effect," and this even though there was "aboundance of Strange Reptiles in ye Woods in Virginia & Carolina . . . long more deadly than ye R.S." For Walduck, the rattlesnake's bite is "as mortall as Fate," "as deadly to themselves as to those they bite," a standard assertion in snake accounts of the time, to which Walduck adds the observation that the snakes themselves have to feed regularly on roots that serve as an antidote to their own deadly poison. Reveling in images of ugliness, the good captain tremblingly reports that in New England and Pennsylvania thousands of rattlesnakes would annually convene for the purposes of procreation in a "heap . . . some flying about others hissing & makeing a horrible noise, in a strange permiscuous manner," while "their Spirits" are "high & volatile & full of venom."

Walduck's almost hysterical images of wanton profligacy and sexual promiscuity reflect his general conviction that, by their very nature, rattlesnakes are *un*natural. The offspring of their parents' indiscriminate mass copulation, young rattlesnakes skip the phase of innocent childhood and slow growth; after merely a year, they are already as "bigg as a Mans thumb," ready to assume their pernicious pursuits. Since rattlesnakes are, of course, viviparous, the following story is definitely based upon hearsay. Walduck reports that when he once came across what he believed was a nearly hatched "rattle-snake egg" and hit it with a stick, "out ran a Snake all perfect, att least 7. inches long, & as big as a Goose-Quill." For Walduck, these snakes are "Slothfull" and "heavy" crea-tures; they live on "Squirrells & all sorts of Birds that they can catch, & by a wonderfull Facination they shall charm them . . . into their mouths, from yᵉ Top of a Tree 50 foot high." (In William Byrd's funnier version of a similar incident, the rattlesnake "ogles" its "poor little" prey, "till by force of the Charm he falls down Stupify'd and Senseless.")²⁶

Walduck claims that woodsmen recognize the sound of a rattler's tail "20 yards of"; however, even before humans hear, see, or smell a rattlesnake, they are, as Walduck notes with a touch of poetry, "seized with sorrow." No wonder, then, that Indians paint snakes on their bodies to defend themselves, by such "Diabolical Magick," against a creature whose poison is so potent that, when it chews on them, green switches turn "to a blackness all yᵉ way up to yᵉ Top." Walduck then enriches his fantasies of phallic threat and oral engulfment with a story that soon became the stock-in-trade of colonial and postcolonial rattlesnake lore, the story of the fatal boot. A farmer is bitten by a snake and dies; after his death, his boots are handed on to two successive husbands of his wife, who in turn pass away unexpectedly; finally, a surgeon ascertains the source of the inherited evil, a rattlesnake's deadly fang buried in the sole of one of the farmer's boots. The doctor then demonstrates the power of the ominous tooth on an unfortunate dog. (In one of its many later guises, the story reappears in letter 10 of J. Hector St. John de Crévecoeur's *Letters from an American Farmer* [1782].)²⁷

Having raised the specter of the lethal reptile, Walduck is confident that, as evil as such snakes might appear, they will prove insufficiently equipped to survive the onslaught of European settlement: "In some Ages a RSnake will be as great a Rarity in N. England as a Wolf is now in Old England, tho' they were once very populous in each Country."²⁸ In the meantime, however, the rattlesnake continued to hold captive the imaginations of American naturalists. For instance, in his contribution to colonial rattlesnake lore, Paul Dudley, the attorney general of the Province of Massachusetts Bay, excitedly confirmed that the eye of the rattlesnake "had something so singular and terrible, that there is no looking stedfastly on him; one is apt, almost, to think they are possest by some Demon."²⁹ Like Mather, Dudley had not actually *seen* any instances of charming, and this would hold true for most other "observers" of American reptile life at the time—with the notable exception perhaps of Robert Beverley, who, in the second edition of his *History and Present State of Virginia* (1722), coolly assured his readers that he himself had once observed a rattlesnake, its "Colours . . . more glorious and shining" than ever, fascinate and then devour a poor helpless hare in the seclusion of a cherry orchard. "I live in a Country where such things are said frequently to happen." Beverley swore that he was telling the truth: "This I have related of my own View, I aver (for the Satisfaction of the learned) to be punctually true, without inlarging or wavering in any respect, upon the Faith of a Christian."[30] But no matter how dangerous these rattlers were, as Beverely had previously informed his readers, they made "dainty food," too.[31]

Perhaps it was because they found themselves represented in such unlikely contexts that rattlesnakes would, as the Swedish traveler Peter Kalm (who also believed in "fascination," although he had never seen it) wrote in his diary, sometimes "look sharply" at human beings and, "as it were, marvel" at them.[32] The only victims ever "fascinated" by the rattlesnake, sneered the eminent Philadelphia naturalist Benjamin Smith Barton in 1794, were the credulous naturalists themselves.

So far, the fascinated victims in the more standard reports had been birds, squirrels, at best a lonely hare, but it seems only logical that humans should not have remained exempt. In November 1765, the *Gentleman's Magazine* in London printed a letter about "remarkable and authentic" cases of the involuntary paralysis of humans by rattlesnakes which Peter Collinson, F.R.S., had received from a correspondent in Philadelphia, identified only by the initials "J.B."[33] In view of Collinson's friendship with John Bartram, the subject of the first chapter of this book, it has sometimes been assumed that the author of this letter must have been his plant-collecting pen pal from Kingsessing.[34] We know that, prodded by Collinson, John Bartram had indeed developed an interest in fascination, Though generally skeptical ("What I have heard and remarked of the Rattle-Snake's Power of Charming is surprising"), he also acknowledged in one of his letters to Collinson that he had heard "from several persons of undoubted Credit," who had told him that "there is a surprizing Fascination in the looks of this Snake" and had affirmed that "there proceeds such subtle Emanations from the Eyes of this Creature beyond what we can comprehend" (*CJB* 39–40). Traveling along the Susquehanna in 1738, John Bartram came across a large rattlesnake himself and decided to try out its mysterious power here and now: *Hic Roma, hic salta.* Staring at the snake, he hoped the snake would stare back and, perchance, fascinate him. Bartram was disappointed, as he told Collinson: "I being just behind went round ye snake he drew up in a quoil & rattled then I stood staring at him to observe whether he would have any way afected mee by his looks but found no alteration" (*CJB* 91).

On the other hand, Franklin's snake-bitten friend Joseph Breintnall has also been proposed as a likely candidate for "J.B." From one of Collinson's letters to Bartram we know that in 1736 he had discussed the "long Received opinion of Charming" with Breintnall, too.[35] However, this would mean that he must have kept Breintnall's letter on file for decades before deciding to publish it, twenty years after Breintnall himself had died. A brief look at the Journal Book of the Royal Society clarifies the mystery and offers the "conclusive evidence" that David Scofield Wilson has demanded. Not only do the records of the meeting held on 17 February 1736 establish Breintnall beyond a doubt as the author of the ominous letter, they also explain why Collinson would have waited so long before making it public. "A Letter from Mr. Jos. Breintnall dated at Philadelphia Nov.ʳ 9 1735 to M.ʳ Collinson; on the subject of the fascinating Powers of Rattle Snakes was read," the Society's secretary entered into the journal Book in his neat handwriting. "The Author . . . seems to have taken no small pains in collecting together all the wonderfull Stories and accounts he could anyway gather relating to Rattle Snakes and black Snakes." Breintnall's stories apparently all centered around the same topic, namely the question of how a rattlesnake can "by its Eyes enchant a Man to that degree, that he shall not be able to remove from his place so long as they mutually beheld one another."

The distinguished members of the Society didn't believe what they heard. Why Breintnall had compiled these curious accounts wasn't clear to them, they snorted, "except it were merely for the Amusement of his Correspondent" and "to shew his own Talent and Address in giving the relation." His evidence was flimsy and relied on the reports of witnesses who were either not credible or "may drink a little too much." Almost all of his stories came "from second third & fourth hands," and the only event that Breintnall had witnessed himself, the mysterious disappearance of a blacksnake under his very eyes, was based on insinuation rather than factual demonstration. If one thing was evident from this letter, it was that not the rattlesnake but Americans themselves were "strangely possessed." This, indeed, was the letter quoted at the beginning of this chapter, the message that had made the savants of the Society shake their heads in disbelief at a country whose people were willing to entertain "faith in the most wonderfull things," In short, the meeting of the Royal Society on 17 February 1736 was a disaster for Joseph Breintnall and an embarrassment for Peter Collinson: "M.r Collinson however had thanks for communicating the Contents of this curious Letter," the secretary tersely ended his skeptical report.[36]

The gentlemen of the Royal Society, their harsh attacks on Breintnall's reliability aside, were correct about his talents as a writer. The text Collinson sent to the *Gentleman's Magazine* is a gem. The loose epistolary form allowed Breintnall to present his bizarre evidence as a series of mini-narratives, which, incidentally, are all set in the centers of the recent protest against the British Stamp Act mentioned in the same issue of the magazine—Maryland, New Jersey, and Pennsylvania. As usual, the witnesses are characterized as "persons of good credit," to be believed even when it turns out that they themselves have only heard the story and not seen the actual event:

Doctor *Chew* tells me, a man in *Maryland* was found fault with by his companion, that he did not come along; the companion stepping towards him, observed that his eyes were fixed upon a rattlesnake, which was gliding slowly towards him, with his head raised as if he was reaching up at him; the man was leaning towards the snake, and saying to himself, *he will bite me! he will bite me!*—Upon which his companion caught him by the shoulder and pulled him about, and cryed out, *What the devil ails you? He will bite you sure enough!* This man found himself very sick after his inchantment.

In a century that, in the words of the philosopher John Locke, wanted to define the "bounds between the enlightened and the dark parts of things" and at the same time showed an ever-increasing preference for marionettes, machine-men, and, finally, mesmerism, Breintnall's interest in the temporary loss of will-power does not seem at all outlandish. If the devil has been expelled from most areas of daily life, he might still be looking at us through the "large eyes of a glistening snake." Breintnall's stories present identifiable people, real human beings with a name, a profession, and in some cases even a direct relationship to the writer himself: "Mr Nicholas Scull a surveyor"; "*Joshua Humphreys* in *West Jersey*"; "*William Atkinson*, an honest man in *Bucks* County"; "*Thomas Hatton*, a merchant in this town." And they always describe how these people, faced with the imminent threat of fascination, finally regained, in the nick of time, their— indeed sorely needed—presence of mind: "then he had the resolution to push himself from the fence"; "He presently came to himself"; "He gave the snake a lash with his whip, and this taking off the snake's eye from his prey, the charm was broken." Even the skeptical members of the Royal Society had noticed how "particular and circumstantial" Breintnall was "in the names and characters of person & things."

The fear of snakebites, concedes Breintnall, is so great that some people "have suffered by them in imagination," and he adds an anecdote in which rational judgment and practical sense replace unwarranted fear and imagined sickness. A farmer believed that his body was swelling up as a consequence of such a snakebite, whereas in reality he had just mistakenly donned his son's somewhat tighter jacket: "He had presently more doctors than were good. . . . Yet he grew worse and worse, and had like to have died. At length came the son home, with a jacket too big for him. . . . This proved the best remedy of all." If this story inserts a note of skepticism into Breintnall's account, it also effectively heightens the authenticity of the other strange incidents reported, as it proves that this author, for one, is fully capable of distinguishing between real and imagined dangers, fact and fiction. Thus, at the end of Breintnall's account, the rattlesnake's alleged power of fascination remains as mysterious as before. In this context, the detailed description (singled out in the Royal Society's report as "the only Story" based on the author's "own knowledge") of how Breintnall himself once unsuccessfully pursued a blacksnake, a harmless reptile said to share the power of enchantment with its venomous brother, seems almost symbolic. A neighbor "noted for telling strange things" directed him to the spot where he had last seen the snake disappear and, sure enough,

> there was the snake within three steps of me, with his head raised above half a yard from the ground, and his neck curved like a goose's. The sun glittered on his head and breast, which offended my eyes, and made me the more resolute to kill what I naturally hate to see. I got over, and chose a stick among some bushes that were grubbed up hard by, one about five foot long and very fit, as I thought, to have cut a snake in two, or smote him to a considerable distance. He kept his posture, and I went near him, observing the grass to be short, and the place clear and plain enough; I viewed the length of my stick, and carefully set my left foot forward, to be within reach, and had a fair view of him until the moment I drew my arm from its extent, which I did suddenly, but I neither struck him, nor saw him again, tho' I searched diligently for about half an hour: Whether he darted away, or withdrew by a hole downright, I could not find out.

The snake, transported into the realm of fantasy and myth through the sun that glitters on its breast and blinds the human observer, retains its power to look without being looked at and mysteriously disappears from view, leaving the man to clutch his useless stick. We now understand why when, years later, Breintnall was bitten by a snake, he should have been so susceptible to nightmares that he would dream of "horrid Places, on Earth only."

Attractive and Enlightened Rattlesnakes

While Mather, Dudley, Kalm, and Breintnall were busily evoking the full horror of the American serpent, other writers, such as the British naturalist Mark Catesby, apparently felt more attracted than repelled by it. Catesby, the self-taught English naturalist so enamored with "the most wonderful Productions of *America*," was certain, as he said in the preface to his *Natural History of Carolina, Florida, and the Bahama Islands*, that he had described most, if not all, of the "Serpents" in Her Majesty's Dominion.[37] Like others, Catesby believed in the "attractive Power" of Rattlesnakes, although he, too, admits he never "saw the Action." He does not hide his horror when mentioning how he spotted one of these monsters "gliding into the House of Colonel *Blake*, of *Carolina*," where it aroused the domestic animals and instantly led to a strange alliance of "Hogs, Dogs and Poultry," all united in their "Hatred to him." The unimpressed snake, "regardless of their Threats, glided slowly along." The "most deadly venomous Serpent of any in these Parts of *America*" would, as Catesby discovered, not only enter the yards of people's houses and frighten their chickens but also creep into their beds, "a very extraordinary Instance of which" happened to himself in Colonel Blake's house in 1723. A servant was making the bed in the guest room on the ground floor "but a few Minutes" after Catesby had left it and on turning over the sheets, still warm from the unsuspecting naturalist's body, discovered an unlikely bedfellow—a "Rattle-Snake, lying coiled between the Sheets, in the middle of the Bed" (*NHC* 2: 41).

But Catesby s plates (nineteen of the hundred plates in the second volume of his *Natural History* are devoted to snakes) do not terrify. It is impossible to convey to a reader who has not seen one of the original editions of Catesby's masterpiece the full splendor of his hand-colored, meticulously detailed engravings, all of which he had cut himself. To Catesby's contemporaries, his snakes must have seemed like fairy-tale creatures out of a world quite different from their own. Their elegant, sinuous bodies range in color from mud-brown to metallic blue, a rich alphabet of interesting poses. The one potentially scary situation Catesby depicts (in plate 45, a "Brown Viper" is shown preying on a lizard during the time of a great flood, when snakes, ready to kill, were crouching on floating vegetable refuse, seizing in their jaws whatever happened to come their way) is considerably lessened in its impact on the viewer by the fact that Catesby was obviously more interested in the snake's elegantly curved body than in the violence of the scene.

Unlike earlier naturalists, Catesby not only distinguished between species but also carefully described their colors, allowing the aesthetic sense of his readers full play without sacrificing the precision of his descriptions. About the "American Rattlesnake" ("Vipera caudisona Americana"), the canebrake rattler, Catesby writes: "The Colour of the Head of this Rattlesnake is brown, the Eye red, the Upper Part of the Body a brownish yellow, transversely marked with irregular broad black Lists." The rattle is "of a brown Colour, composed of several horney, membranous Cells," which form, as Catesby beautifully puts it, an "undulated pyramidal Figure." His explanation of the construction and effect of the rattle combines accuracy with admiration for the subtlety of such an intricate mechanism: the horny rings of the rattle "are articulated one within the other, so that the Point of the first Cell reaches as far as the Basis or protuberant Ring of the third and so on, which Articulation being very loose, gives Liberty to the Parts of the Cells that are inclosed within the outward Rings, to strike against the Sides of them, and so to cause the rattling Noise, which is heard when the Snake shakes its Tail." Catesby's plate of a pygmy rattler ("Vipera caudisona minor," ill. 32) is especially captivating. It shows the animal not in its natural habitat but drifting weightlessly against a decorative background consisting of shrubs and tree leaves from the Bahamas that seems almost as disembodied as the reptile itself. What the drawing rather nicely underscores is the peculiar—indeed, serpentine—form of the snake's body, which is so unlike anything in else in nature.

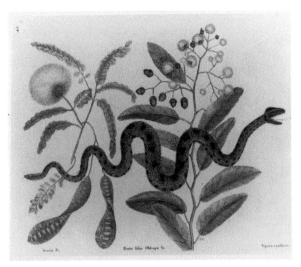

32. *"Vipera caudisona minor."* Hand-colored engraving by Mark Catesby. From Catesby's *Natural History of Carolina . . .*, vol. 2, 2nd ed. (1754), pl. 42. Courtesy, The Lilly Library, Indiana University, Bloomington.

Among the naturalists attracted by the rattlesnake was John Bartram's son William. In his *Travels through North & South Carolina, Georgia, East & West Florida*, Bartram wearily rejected "the power of fascination" as a likely factor in the rattlesnake's hunt for prey. "It is generally believed that they charm birds, rabbits, squirrels and other animals," he writes, his choice of words betraying studied detachment. He goes on to evoke a quasi-surreal, deliberately unbelievable scene, in which, under the stern, "stedfast" eyes of large snakes, hordes of "inthralled," helplessly "infatuated" little animals meekly march, single file, toward their inevitable doom, "creeping" obediently right into "the yawning jaws of their devourers" (*T* 224).

Bartram's extraordinary book shows its author to be a firm believer in "the benevolent and peaceable disposition of animal creation in general" (*T* 224). We understand his dismay when we read how he was once compelled to kill a rattlesnake that had been feasting on leftovers in a camp of Seminoles, leisurely "licking their platters" (*T* 220). Because of their "extraordinary veneration or dread of the rattle snake"—an attitude that is ironized almost as soon as it is asserted by the reference to the snake's rather mundane activity—the Seminoles had turned to a stranger to help them; "none of them had freedom or courage to expel" the intruder from their midst (*T* 219–220). When, after the deed is done, some Seminole men attempt to "scratch" Bartram as a punishment for having so recklessly exterminated the godlike animal, he carelessly identifies their unpredictable behavior as a "ludicrous farce" intended to "appease the manes [spirit] of the dead rattle snake" (*T* 222).

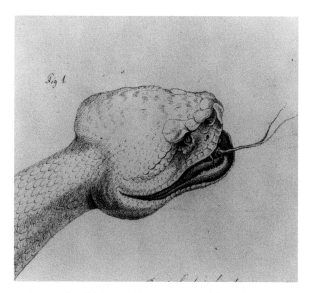

33. *Rattle Snake's Head.* Drawing by William Bartram. Ca. 1774. Natural History Museum, London.

Not coincidentally, "Indians" are a shadowy, vague presence in quite a few of these early accounts, where they appear either as anonymous witnesses vouching for the accuracy of stories Europeans would otherwise find hard to believe or, as is the case here, as fools with superstitious notions to be refuted and discarded by modern "scientific" insight. In 1793, for example, Benjamin Smith Barton pointed out that, because of the nature of their lives, "the rude savages of our continent" were generally "more liable to be injured by the bites of serpents than in the more polished stages of [man's] improvement."[38] In the present passage from *Travels*, Bertram's strategy is obvious: seen against the dense foil of supposedly native superstition and childlike fear, the white man's confidence in the essentially good disposition of even a venomous snake seems doubly enlightened and reasonable. Bartram informs us that he kept the snake's bleeding head for his collection of natural "curiosities," and this is probably what his drawing of a diamondback rattlesnake's head depicts or, as naturalists at the rime would have said, "figures" (ill. 33). At least at first sight, the album of drawings Bartram sent to Dr. Fothergill in London shows him to be remarkably less immune to stories about rattlesnakes and their ability to "fascinate" than he himself would have perhaps liked to think. In Bartram's rendition, the head of the snake, with its precisely drawn scales,[39] looks like an obscenely clenched fist or a grotesquely distorted skull, its one eye glaring, the thin forked tongue protruding menacingly from the partly open, weirdly grinning mouth. Confronted with this drawing, we might find it easier to understand why, quite like the rabbits or squirrels in the stories he had heard and disbelieved, even the enlightened Bartram in his chance encounters with rattlesnakes in the South (where they are "the largest, most numerous and supposed to be the most venomous and vindictive") would have sometimes felt, in his own words, "so shocked with surprise and horror as to be in a manner rivetted to the spot, for a short time not having strength to go away" (*T* 223).

Although Bartram continually reminds his readers that a rattlesnake never strikes "unless sure of his mark," his description of such an animal defending itself against an attacker at first reinforces existing fears rather than instilling confidence in the "magnanimity" of all of God's creatures. But what we finally hear in Bartram's lines, if we listen closely enough, is not so much the uncanny sound of the rattlesnake's bony rattle as the resonance of the writer's well-chosen words. According to Bartram, the rattlesnake, when surprised,

> instantly throws himself into the spiral coil; his tail by the rapidity of its motion appears like a vapour, making a quick tremulous sound; his whole body swells through rage, continually rising and falling as a bellows; his beautiful partico-loured skin becomes speckled and rough by dilatation; his head and neck are flattened, his cheeks swollen and his lips constricted, discovering his mortal fangs; his eyes red as burning coals, and his brandishing forked tongue of the colour of the hottest flame, continually menaces death and destruction. (*T* 223)

Bartram's powerful synesthetic images—right down to the heaving bellows of the snake's body, fanning, as it were, the fire of its flaming eyes—reek of the sulphurous pits of hell. His description of the rattlesnake enacts the drama of fascination, but the author is the stage manager, too, in full control of his own text. We realize that the horror in this passage—as was the case in his drawing—is in fact well orchestrated.

In the same chapter, Bartram almost shamefacedly admits his complicity in the death of yet another rattlesnake in East Florida, "about six feet in length, and as thick as an ordinary man's leg" (*T* 226). Rambling through swamps in the search of unknown flowers, his father, John, had suddenly noticed the venomous animal just before his unsuspecting son's feet. William, in the "fright and perturbation of his spirits" and annoyed by the unwelcome interruption of his important botanical research, arms himself with a stick and kills the animal: "I cut off a long tough withe or vine, which fastening around the neck of the slain serpent, I dragged him after me, his scaly body sounding over the ground, and entering the camp with him in triumph, was soon surrounded by the amazed multitude, both Indians and my countrymen" (*T* 226–227). The unfortunate snake is served up for dinner the same night, "in several dishes," at Governor James Grant's table, where Bartram himself, however, cannot swallow it, hoping sincerely he would never again have to be the accessory to the death of such a beautiful creature.

This passage is characteristic of Bartram's amazing book, in which long passages of seemingly dry description and extensive lists of species are continually intermixed with moments of dazzling narrative power and high dramatic intensity. It is significant that what could have turned into an emblematic narrative of man's triumphant victory over the serpent in Bartram's skillful hands now acquires a different meaning. The sound of the dead snake's scaly body trailing on the ground behind its human slayer provides a hollow comment on Bartram's supposed achievement and prepares us for his subsequent embarrassed admission that, if aimed precisely enough, a "stick no thicker than a man's thumb" will suffice to kill a creature that can only be called "wonderful . . . when we consider his form, nature and disposition" (*T* 227, 223).

A few years after Bartram's *Travels,* Charles Willson Peale published his reflections "On Collecting Reptiles and Insects" (1795), declaring that to "overcome our groundless fears of certain reptiles and insects" would mean "a real increase to our hapiness" and that "to cure ourselves of illfounded antipathies, is truly philosophic" (*PP* 2: 124). During collecting trips to New Jersey, which he neatly recorded in his diaries, Peale, to the amazement of his companions, would handle rattlesnakes as carelessly as if they were earthworms. Peale's spectacular collection contained several live rattlesnakes who would "caution" visitors not to approach any further (*PP* 2: 69), and it was here that physicians should come, said Peale, who wanted to find "a cure for the bite of this viper" (*PP* 2: 125).

And so they did. French-born naturalist and Haitian emigré A.M.F.J. Palisot de Beauvois, for example, found in Peale's museum the evidence that led him to conclude that "it appears . . . repugnant to reason" to attribute the behavior of a rattlesnake's prey "to enchantment, giving to that expression the full latitude which it presents to the imagination. We are no longer in that barbarous age in which men gave credit to enchantments, witchcraft, and miracles." Since rattlesnakes obviously inspire "unfavourable and repulsive dispositions," human observers, in order to render them "still more hideous," had been only too quick to give an "absurd interpretation" to simple or accidental facts. Beauvois resolved that "we have almost every thing yet to learn relating to these extraordinary reptiles."[40] Benjamin Smith Barton likewise prepared his "Memoir Concerning the Fascinating Faculty Which Has Been Ascribed to the Rattle-Snake, and Other American Serpents" from observations he made in Peale's Museum. In his essay, Barton refuted, step by step, the arguments of all those who still believed in the animal's mysterious powers. With some exaggeration he described the question of how naturalists would handle the topic of "fascination" as crucial to the future of the discipline and saw himself as a lone prophet in a wilderness of thickening "error and credulity": "Their dominion is extensive. The belief in the fascinating faculty of serpents has spread through almost all the civilized parts of North-America. Nor is it confined to America. It has made its way into Europe." Even children at school were now taught to believe in a prejudice "so deeply and so powerfully rooted" as to mock "the light and sureness of facts, and all the strength of reasoning."

Specifically addressing Lacépède, Barton pointed out that American forests were inhabited not only by dangerous rattlesnakes but also by plenty of courageous birds, whose "charmed" behavior could easily be explained as a natural, instinctive attempt "to protect their nest or young," resulting from the close attachment of the mother bird to her offspring. Such a feeling, declared Barton, was not after all "peculiar to the human kind" but should be regarded as an "instinct which pervades the universe of animals"— a sure indication of the benevolence of the workings of a "first great cause, who regards with partial and parental, if not with equal, eyes the falling of a sparrow and the falling of an empire."[41] Apparently, though, the eyes of the first great cause rest less benevolently on the rattlesnake; even though this animal has now been stripped of its evil powers, it is referred to by Barton, in the last paragraph of his essay, as a "hideous reptile." Interestingly, Charles Willson Peale, who had received little credit from Barton for his own experiments and observations, maintained in a private letter he sent in 1797 to his illustrious French colleague Geoffroy Saint-Hilaire that his naturalist colleague had been lying through his teeth when he claimed to have had "intimate" contact with rattlesnakes. Peale had seen him at work in his museum, and Barton's antipathy to such reptiles, said Peale, was "so great that he is afraid to touch or come near to the most inofensive of them" (*PP* 2: 199–200).

Decades after the experiments of Peale and his colleagues, *Niles' Weekly Register* carried a story that might serve as an interesting and amusing postscript to these attempts to "enlighten" the rattlesnake. In 1822, two tame rattlesnakes were exhibited by their owner, Mr. Neal, at different locations in Virginia. Mr. Neal would stroke the backs of his strange pets "as if they were so many strings" and treat them as if they were his lovers, encouraging them to "crawl up his breast and face, caress and kiss him" (a spectacle the writer of the article earnestly considered a "valuable addition to natural history"). Mr. Neal assured his audiences, "especially the ladies," that "there is not the least danger in witnessing" his rattlers in action. In spite of Mr. Peale's and Dr. Barton's efforts, Neal apparently still believed in the snake's power of fascination, even though the event itself now took place in his backyard: "As soon as the eyes meet, he says the process of charming commences." The snake's victim will, Neal testified, "hop from bough to bough, and rock to rock, overcome with apprehension, until approaching each other, the snake seizes him." However, Mr. Neal obviously did not believe in the deleterious nature of the rattlesnake's breath, "for he has often kissed them, and in blowing their breath upon him, he has found it uncommonly sweet."[42]

Science from the Swamps

Nothing could differ more from such lively, if perhaps not always entirely truthful, accounts of interactions with various snakes than the disinterested prose of John Edwards Holbrook's *North American Herpetology; or, A Description of the Reptiles Inhabiting the United States,* which first came out in three volumes from 1836 to 1840. Dissatisfied with the quality of the illustrations and worried about the ceaseless influx of new specimens, Holbrook burned all the sets of the first edition that he could recover from the public. His *Herpetology* was republished in a lavish second edition in 1842–1843.[43] Holbrook's magnificent work, later praised by the Harvard naturalist Agassiz as "far above any previous work on the same subject,"[44] was also an effective rejoinder to contemporary northern assumptions that the South, in terms of cultural refinement and scientific sophistication, was still in the swamps and that the only way to pass the time there, as a character in John William De Forest's novel *Kate Beaumont* (1872) quipped, was by having drinks.[45]

Born in 1794 in Beaufort, South Carolina, Holbrook obtained a medical degree from the University of Pennsylvania in 1818 and went on to travel through Europe, where he was deeply impressed with the Jardin des Plantes in Paris. Upon his return to the South, he was instrumental in the founding of the Medical College of South Carolina in Charleston, which conveniently provided him with a professorship of anatomy, a position he would hold for thirty years. Most of his time, however, was devoted not to medicine but to the collection and careful description of every living reptile inhabiting the United States, "from the Atlantic Ocean to the Rocky Mountains, and from Canada to the confines of Texas"—an "immense mass of materials," as Holbrook writes, which he tackled "without libraries to refer to, and only defective museums for comparison" (*NAH* 1: xiv; [ix]). Holbrook, a member of the Royal Medical Society of Edinburgh and a corresponding member of the American Philosophical Society, the Academy of Natural Sciences of Philadelphia, the New York Lyceum of Natural History, and the Boston Natural History Society, completed his work before the acquisition of Texas and California by the United States, which more than doubled the number of American reptiles. Even so, of the twenty-nine species he designated, "a very large proportion" has survived with his specific names.[46]

Accompanied by dazzling illustrations, the chapters in Holbrook's *Herpetology* feature a tightly ritualized succession of careful descriptions of the skins, the colors, and the sizes of frogs, lizards, tortoises, and snakes. These are followed by brief comments on their habits and habitats and by often critical reflections on the more common names of the animals and the circumstances of their first discovery by European or, preferably, American naturalists. In no department of American zoology, the professor exclaims, "is there so much confusion as in Herpetology," a sorry state of affairs due partly to the ignorance of earlier naturalists,[47] partly to the established practice—not adopted in his work, of course—of "describing from specimens preserved in alcohol, or from prepared skins" (*NAH* 1: ix). This southern gentleman enters a strange world here,[48] and in order to leave it in its full strangeness, he rarely makes an appearance in it himself. There are exceptions, though: at one point, we are permitted to witness Holbrook's excitement while watching a little newt in winter, "under ice of an inch thickness, swimming with great vivacity" (*NAH* 5: 78). Sometimes, if the species he is dealing with seems tractable enough, we see him experimenting with the members of his reptilian universe, as, for example, when he wants to find out if a little, hitherto undescribed frog is capable of swimming or not: "An individual thrown into water floated, struggling with its limbs extended, as though altogether unacquainted with the art of swimming. I have never heard it produce any sound" (*NAH* 4: 105).

In his introduction to the first volume, Holbrook dramatically announces that the study of reptiles offers "difficulties more . . . insurmountable than those presented by any other class of vertebrated animals," since reptiles inhabit, "for the most part, deep and extensive swamps, infected with malaria, and abounding with diseases during the summer months," which is exactly when, alas, these animals are "most numerous" (*NAH* 1: 18). Holbrook's neatly printed pages are populated with such swamp creatures, animals that are, as the author frequently notes, "very voracious." If they do not live in "muddy waters" or "three feet or more deep in mud," they pass their days "in concealment, near old fences, or under the bark of fallen and decaying trees, emerging only towards evening and after heavy rains," or they "seek out holes in the earth" where they spend their winters (*NAH* 5: 91, 24; 2: 59). Others hide their sinuous bodies "on the low branches of trees that overhang the water" in order to seize their prey—often other, smaller reptiles (*NAH* 4: 31). And while the plantation owner Holbrook dutifully notes the various encroachments of reptiles upon the world of humans (in his opinion it is, significantly, mostly in the "cabins of the negroes" that we find snakes; *NAH* 3: 90), he leaves no doubt that reptiles usually live in their own secluded areas.

Only rarely does Holbrook point out how much trouble he has taken to establish accurately the identity of a species he believes other, earlier naturalists have misjudged. No sooner has he done so than he needs to invoke the authority of one of his countless naturalist friends who happens to agree with him: "After twelve years' search, both in Carolina and Virginia . . . I have never seen any animal bearing the least resemblance to Catesby's figure . . . and my friend Professor Geddings . . . is of the same opinion" (*NAH* 4: 28). On occasion, where matters of principle are concerned, Holbrook can be stern, and a note of suppressed nationalism becomes audible: "I am not disposed," he declares at one point, as if somebody had tried to force him in this direction, "to make an exception in favour of this animal to the general rule, that there is not a reptile in Europe identical with any one of the United States" (*NAH* 4: 101).

It is when he attempts to catch the reptiles' beauty on the page that Dr. Holbrook's language finally begins to soar. When the chameleon is basking in the sun, its coloring has, he tells us, the "liquid brilliancy of the emerald" (*NAH* 2: 71). The blacksnake's upper-body surface is, he exclaims, "of beautiful bluish-black; the abdomen and tail are bluish-slate, while the chin and throat are pure silver-white" (*NAH* 3: 57). Holbrook truly admires the "alternate folds" of his snakes' "long and slender bodies" (*NAH* 3: 3), an enthusiasm that shines through even in his long, strenuously scientific, painstakingly meticulous descriptions of the structure of their skins, especially of their colors and characteristic patternings. His discussion of the family "Crotaloidea," "the most remarkable of the Serpent tribe," opens with observations on the head of the *Crotalus durissus* (the timber rattlesnake),[49] assessing in strenuously Latinate diction first its general shape, then the composition of the skin, the geometrical form of the frontal plates, the nasal plates, the plates around the eye, the labial plates, and the characteristic pit, before again returning to a more general impression of the snake's head. One of the recurrent motifs in this passage rife with minute detail, which can be cited here only in abbreviated form, is, interestingly, largeness:

> The head is enormously large, triangular, but broad and truncate anteriorly, covered with plates only in front, and with minute scales on the vertex and occiput; the rostral plate is large and triangular, with its basis downwards and its apex upwards and truncate; the frontal plates are also triangular, with their bases directed backwards. There are two nasal plates; the anterior is quadrilateral and excavated behind; the posterior is lunated in front to *complete the nostril.* The superior orbital plates are regularly oval. . . . Their outer margin forms a strongly marked projection over the eye. There are thirteen labial plates to the upper jaw. . . . Above the labial range is a row of small scales, that form the lower walls of a deep pit, *completed above by a large lunated plate.* . . . The nostrils are large, and very near the snout, but open laterally. The eyes are large, and extremely brilliant when the animal is enraged; the pupil is dark, oval and vertical; the iris flame colour. The mouth is large, the jaws strong. (*NAH* 3: 10; my emphasis)

What seems striking to the general reader is not Holbrook's technical language (no longer rare in texts of this sort); it is the extent to which he looks at the animal's head as if it were a work of art, its individual parts smoothly interacting to create and, indeed, *complete* a harmonious whole. This is particularly obvious in the detailed description of the snake's body, which is of a "pale ash colour," marked "with a triple series of dark irregular blotches and bars along the back." In front, "the blotches of the vertebral series are oblong," but, adds Holbrook, they vary in shape "near the middle of the body; they resemble chevrons, with an acute angle towards the head; beneath the terminations of these spots on the flanks is a row of sub-quadrate dark spots; near the tail the vertebral and lateral series unite to form a band, and between these there is another row of . . . grey spots" (*NAH* 3: 11).

After these descriptive details, the narrative section of the chapter seems lackluster, and it is not at all unexpected that Holbrook, who has read his Pliny, only has a wan smile left for those stories about the rattlesnake's alleged power of fascination: "I have every reason to believe it a fable; a modification of that of the basilisk of the ancients" (*NAH* 3: 12). One paragraph suffices to relegate such myths, along with the tree-climbing rattlesnake pictured in Audubon's *Birds of America* ("For this his high organization . . . seems ill adapted"), to the dustbin of scientific error. "If the Rattle-

snake has other 'charming powers,' they lay in the horror of its appearance, or in the instinctive sense of danger that seizes a feeble animal fallen suddenly into the presence of an enemy of such a threatening aspect—rather than to any mysterious influence not possessed by all venomous or ferocious animals upon their weak, timid, and defenceless prey" (*NAH* 3: 13).

Holbrook's work is particularly impressive because of the drawings prepared by two of his principal artists, J. Sera (about whom we know very little, except that he was an Italian painter who settled in Charleston) and John H. Richard, who later worked for the Smithsonian Institution and contributed illustrations to the official reports of western surveying expeditions.[50] The plates the Philadelphia lithographer Peter S. Duval and his associates produced from Sera's and Richard's drawings show the snakes in splendid isolation and in their full aloofness, as creatures

34. "Der Schauer Klapperer. Der schleuderschwaenzige Klapperer." From J. M. Bechstein, *Herrn de la Cepede's Naturgeschichte der Amphibien oder der eyerlegenden Thiere und der Schlangen* (1801–1802), vol. 5, pl. 9. Universitäts und Landesbibliothek Bonn.

inhabiting their own wonderful worlds, which consist of nothing but themselves and their own inimitably as well as illimitably beautiful bodies. This is even more evident when we compare the illustrations in Holbrook's *Herpetology* with those in a more eclectic work, such as Johann Matthäus Bechstein's German version of Lacépède's *Histoire naturelle* (1801–1802), which collects illustrations from earlier herpetological works and arranges the various species, together with anatomical details, on the page in bewildering profusion, their overlapping bodies bent so that they all fit (ill. 34).[51] By contrast, Holbrook's artists use the snakes' sinuous forms to full advantage, representing them in the most intricately coiled positions. Thus, they create a decorative, abstract visual alphabet, the components of which are all complex variants of one and the same winding S-shape.

In a painting by Sera, the *Crotalus durissus* (ill. 35) at first sight appears rolled into a compact mass, but closer inspection reveals the carefully contrived impression of weightlessness created through the deliberate balancing-out of two competing centers. Resting on the thick basis of the solid first coil (the middle part of the snake's body), the snake's expansive second coil is, in the right part of the composition, counterbalanced by its apparently weightless, slightly tilted angular head, with its single, precisely contoured eye. The snake's coiled body, as the viewer's eye travels from the head downward, loops out and, in the upper left part of the composition, seems to encircle blank space, while it is in fact stabilized by the axis of the erect, tapering tail, which protrudes in the upper left corner, constituting the diagonal head's compositional counterpart. The snake's body, an image of fierce energy, looks as if suspended in midair, an impression also evoked in Sera's representation of the eastern diamondback rattlesnake (*Crotalus adamanteus*), where head and tail appear as ornamental appendices to the reversed S-shape of the basic pattern (ill. 36).

A more experimental design is Sera's depiction of a pygmy rattler (today regarded as belonging to a different genus of rattlesnakes), in which the artist's use of blank space as part of his composition is particularly evident: the rattlesnake's circling shape delineates the white silhouette of another phantom snake, thus making the page part of the expressive design (ill. 37). Even more effective is the representation, by John H. Richard, of a species called Kirtland's Rattlesnake (the eastern massasauga; ill. 38). The animal's body resembles a horizontally arranged figure eight, vertically offset by the beacon-like tail, which seems to anchor the snake in weightless space. Holbrook's reptiles do not creep; they seem to drift. There is no attempt at all to present the snakes as part of a living natural world, interacting with other members of their species, confronting other animals or, for that matter, human beings. These bodies, examples of concentrated, elegantly distributed energy, have their purpose in themselves: it is not important that

35. *"Crotalus durismus."* Lithograph by Lehman and Duval from a watercolor by J. Sera, for J. E. Holbrook's *North American Herpetology*, 2nd ed. (1842–1843), vol. 3, pl. 1. Courtesy, The Lilly Library, Indiana University, Bloomington.

the snakes see, but that *we* see them. The purveyors of fascination have turned into endlessly fascinating objects to be known as fully as possible. In Holbrook's work, the old interest in the "fascinating" faculty of the rattlesnake finally gives way to the fascinated look of scientific spectatorship itself.

More than ten years later, in 1859, Charles Darwin published his thoughts on *The Origin of Species*. Feeling that it was time to rid nature of the foolish human analogies naturalists liked to impose on it, he mocked the belief held by some that the rattlesnake had been equipped not only with poisonous fangs to destroy its victims but also with a rattle "for its own injury," namely so that it could warn its prey (and food) to escape.[52] "I would almost as soon believe that the cat curls the end of its tail when preparing to spring, in order to warn the doomed mouse."[53] It is not recorded how Professor Holbrook, friend of Darwin's archenemy, Louis Agassiz, responded to *The Origin of Species,* but he certainly would have shared Darwin's scorn for the interpreting and interfering presence of the naturalist. In my collection of loose leaves from the "volumes of tales" about American rattlesnakes, we still find the observer's presence writ large. Even at their most bizarre, these tales of fascination articulate the anxieties Americans had about themselves and their own precarious existence in a land that seemed to be theirs before they were the land's. It was only partly a joke when *Niles' Weekly Register,* as late as 1824, reported the killing on Bullard's Plains of a large rattlesnake, purportedly older than the oldest inhabitants of that section of Louisiana. The snake, concluded the writer of the article, "was perhaps entitled to a pre-emption right from the land commissioners at St. Helena court house, as the first settler of Bullard's plains."[54]

36. "*Crotalus adamanteus.*" Lithograph by
Lehman and Duval from a watercolor by
J. Sera, for J. E. Holbrook's *North American
Herpetology*, vol. 3, pl. 2. Courtesy, The Lilly
Library, Indiana University, Bloomington.

37. "*Crotalophorus miliarius.*" Lithograph by
Lehman and Duval from a watercolor by
J. Sera, for J. E. Holbrook's *North American
Herpetology*, vol. 3, pl. 4. Courtesy, The Lilly
Library, Indiana University, Bloomington.

38. "*Crotalophorus Kirtlandi.*" Lithograph
by Lehman and Duval from a watercolor
by J. H. Richard, for J. E. Holbrook's *North
American Herpetology*, vol. 3, pl. 6. Courtesy, The Lilly Library, Indiana University,
Bloomington.

The philosopher Hans Blumenberg has brilliantly described natural history of the
Baconian tradition as a massive attempt to regain a lost paradise. It became the dream
of naturalists that, by reading the Book of Nature attentively and by (re)calling the
forgotten names of things contained therein, "man" would be able to recover the power
over nature that his earliest ancestor allegedly possessed.[55] Claiming dominion over the

"New World," some of the early colonists felt that they had—if only tentatively—regained "Paradise." Is it a coincidence that the first generations of American observers of nature from Jonathan Edwards and Cotton Mather to Benjamin Smith Barton would be so obsessed and *fascinated* by the possibility of losing it yet again? And that this possibility would naturally be enacted in countless confrontations with the *fascinating* figure of the "American serpent"?

As we have seen, making sense of rattlesnakes came to be of pivotal importance in early American natural history. With Holbrook's book, this fascination ceases to be the *subject* of scientific discourse and instead becomes one of its *qualities,* at least in the best examples of such prose. It is now almost a commonplace that science is not only a way of knowing but also a way of writing.[56] Science is, as Oliver Sacks puts it, a "human enterprise through and through," characterized by "sudden spurts" and "strange deviations."[57] Some of the strangest deviations from the straight and narrow path toward "the truth" have been the subject of my considerations here. But as John Edwards Holbrook's sober and beautiful book about American reptiles shows, even at its narrowest this path has its little twists and turns—or, to honor the real subjects of this chapter, its coils.

Fascinating Elsie

The only illustration of the actual process of fascination I have been able to find appeared long after serious naturalists had ceased to believe in it. Captioned "Rattlesnake Charming a Rabbit" (ill. 39), it appeared in 1855 in *Harper's New Monthly Magazine,* one of several lurid engravings accompanying an article by Thomas Bang Thorpe, pretentiously titled "The Rattlesnake and Its Congeners."[58] The scene takes place in a clearing, against a background of lush, vaguely southern vegetation, complete with ferns and palm trees. The snake's fangs, forked tongue, and upright rattle are clearly visible, but what is most striking is the human-like, half erect position in which the unfortunate rabbit is represented, its front legs extended pleadingly.

39. "Rattlesnake Charming a Rabbit." *Harper's New Monthly Magazine* 10 (1854–1855).

In the context of such illustrations, the promise of the article's author to dispense with the "fables of times past" in the light of the new and "interesting facts brought forth by the naturalist" appears mealy-mouthed. Not surprisingly, the body of the article restates, in only superficially updated and supposedly more "scientific" form, the belief in the rattlesnake's secret powers: "The rattlesnake . . . as certainly has an eye of command as had Napoleon; and the power of the reptile's gaze is not only acknowledged by the humbler class of animals, but man, with all his superior powers, has felt the thrill of helplessness pass through his soul, as he beheld that mysterious eye glaring full upon him" (478–479). Thorpe repeats all the stock-in-trade anecdotes about the dangerousness of rattlesnakes (among them Walduck's fatal-boot story, which he regards as "authentic") but adds one that gives an interesting twist to the familiar combination of beauty and danger associated with the rattlesnake. A painter with a soft spot for natural history has for a long time wanted to portray the "gorgeous beauties" of the rattlesnake. Just as he is painting a lovely girl (in fact, he is just adding "the rich carnations to the lips, the azure to her eyes, and the sunny auburn to a profusion of glowing locks"), he learns that a freshly killed rattlesnake has become available. He instantly switches sitters:

> In place of all that was beautiful—in place of intelligence—in place of woman in her loveliest estate, there was reared the form, according to our instinctive ideas, of the most repulsive of created things. . . . Soon by the aid of chalk the spiral and expressive form was produced upon the canvas; next was seized the pallet, the colors all beaming in flesh tints, the pencils still glowing with the imitating hues of Hebe; but all in their purity found a place upon the serpent's form. The delicate pink, the softened carmine, the Tyrian blue, the deep auburn, were not only necessary, but just as they appeared in their clearness and purity, so did they most approach the original, and aid in starting forth a picture of horrid fascination.

Within just an hour the "enthusiastic painter" has finished his snake portrait, the positively "magical" effect of which now teaches viewers the "vivid, glowing" lesson that "the things we love and worship in nature, or shrink from with terror, are but combinations of the same charming effects of form and color."

Forget the pseudo-philosophical explanation offered at the end of Thorpe's story. The association of ophidian terror and feminine allure stands, and the sparingly applied veneer of scientific seriousness makes the anecdotes offered in the article even more "plausible." A few years later, Oliver Wendell Holmes, the first Parkman Professor of Anatomy and Physiology at Harvard Medical School, published *his* contribution to the natural history of fascination—influenced, it seems, in more than one way by some of the ideas previously offered, in cruder form, in *Harper's New Monthly Magazine*. Under the title "The Professor's Story," Holmes's novel was serialized in the *Atlantic Monthly* before it came out in book form in 1861, now called *Elsie Venner: A Romance of Destiny*.

Holmes's reputation as a writer has changed greatly since the nineteenth century: once regarded as one of America's foremost poets and lionized by Boston society as the most brilliant conversationalist in "the Hub of the Universe," he has since been relegated to the margins of literary history. Holmes, wrote Daniel Aaron in 1973, "seldom rose high enough to see beyond the Boston horizon or ventured very far beyond the precincts of his class."[59] These limitations are painfully evident also in *Elsie Venner*, the first of Holmes's three forgotten novels. However, in the context of this chapter Holmes's novel is of considerable interest, perhaps not so much for what it says but for what it fails to say.

Elsie Venner features as its central female character an eighteen-year-old girl with an "odd," snake-like "sort of fascination about her" (*EV* 1: 162). Holmes's heroine repels and yet attracts her teacher, the anemic male protagonist of the novel, Bernard Langdon, the descendant of an eminent Boston Brahmin family that has fallen on hard days. "Master Langdon" is an aspiring young scientist forced to earn his bread at a school for girls in Rockland, the "Apollinean Female Institute" (a barely disguised jibe at Mount Holyoke Female Seminary), headed by the despicable Mr. Silas Peckham. We are guided through the novel by the avuncular voice of "the Professor," Bernard Langdon's former teacher, who appears unconcerned about his responsibility to spin a good yarn and occasionally needs to remind himself that "this is a narrative, and not a disquisition" (*EV* 2: 53). After the manner of the extraordinarily popular "Autocrat of the Breakfast Table" invented by Holmes, this narrator freely indulges in epigram, comment, repetition, and excessive interpretation and allegory, often to the detriment of probability and narrative immediacy.[60] In many ways, *Elsie Venner* seems to have been conceived not as a good read but as an experiment-become-narrative.[61]

For all his narrative antics, Holmes's professorial narrator makes no secret of the fact that he personally prefers the "trim gardens" of Boston to the "nervous" wild woods of western Massachusetts: "Nature, when left to her own freaks in the forest, is grotesque and fanciful to the verge of license" (*EV* 1: 233; 2: 45). It is with considerable trepidation and a sense of dark foreboding that the Professor says he sent out his promising but penniless student Bernard to teach at a country school: "Any one who looked at this young man could not fail to see that he was capable of *fascinating* and being *fascinated*" (*EV* 1: 32; my emphasis). As it turns out, "Rockland," Massachusetts, is a place for frustration as well as fascination, distinguished mainly by its population of rattlesnakes, which have replaced the previous inhabitants, the Indians. While it had been "easy enough" to drive away the latter, the former proved to have more staying power, "nobler" creatures that they were—infinitely preferable to "mean ophidians . . . poor crawling creatures, whom Nature would not trust with a poison-bag" (*EV* 1: 76).

As the Professor suggests at the beginning of the novel, Bernard Langdon is "marked out" for future greatness (*EV* 1: 34). However, once Bernard is out of the classroom, the first serious object of study he comes up against is very nearly his last one. Elsie, "an apparition of wild beauty" (*EV* 1: 129), is literally a snake become a woman. She has sharp white teeth; around her neck, she always wears a gold chain, which she dreamily coils and uncoils with her slender fingers; on her wrist glistens a bracelet, which looks "as if it might have been Cleopatra's asp, with its body turned to gold and its eyes to emeralds" (*EV* 1:192, 103, 129).

Elsie's cold, glittering "diamond eyes" (*EV* 2: 204) are reptilian eyes, as Langdon confirms when he inspects the caged snakes that he has obtained for curiosity's sake: "Their eyes did not flash, but shone with a cold still light. They were of a pale-golden or straw color, horrible to look into, with their stony calmness, their pitiless indifference, hardly enlivened by the almost vertical slit of the pupil, through which Death seemed to be looking out like the archer behind the long narrow loop-hole in a blank turret-wall" (*EV* 1: 260). Like the eyes of the rattlesnake, Elsie's eyes, as repulsive as they are, draw others to her and make them faint (*EV* 2: 234).[62]

Not only her eyes betray Elsie's "utter isolation" (*EV* 1: 230). She hisses, too, although this is, like everything else about her, only evident to particularly attentive observers: "Not a lisp, certainly, but the least possible imperfection in articulating some of the lingual sounds,—just enough to be noticed at first, and quite forgotten after being a few times heard" (*EV* 1: 231). Her moods quickly change from "paroxysms" of passion to "dull languor" (*EV* 2: 191), and it is in those moments of despondency that she retires to "the Mountain" behind her father's stately mansion, seeking refuge on the notorious Rattlesnake Ledge, "shunned by all, unless it were now and then . . . a wandering naturalist who ventured to its edge in the hope of securing some infantile *Crotalus durissus,* who had not yet cut his poison-teeth" (*EV* 1: 179). Small wonder that Elsie's only significant friendship in the novel is with the animal-like Sophie,[63] an aged former slave, whose grandfather, as the narrator rarely neglects to remind us, was a "cannibal chief" and whose grandmother's teeth were, like Elsie's, "sharp as a shark's" (*EV* 2: 145). But even for old Sophie rattlesnakes are "Ugly Things," and the terrible knowledge that Elsie, the beautiful, strange girl under her care, is a human rattlesnake nearly breaks her heart ("It a'n't her fault. It a'n't her fault"; *EV* 2: 15).

Plainly, Elsie suffers from "ophiditis,"[64] and the rather outrageous explanation Holmes, a practicing physician, gives his readers is that Elsie's mother, during her pregnancy, was bitten by a rattlesnake, surviving her daughter's birth only by a few hours but passing on the poison to her. Elsie was born with the mark of a rattlesnake's teeth on her throat, now covered by a golden coil she always wears around her neck. The problem at the center of the novel's strained plot arises from the fact that, in spite of her inherited reptilian nature, Elsie aspires to an emotion normally experienced only by humans, the feeling of love, and it is Bernard she has selected as the object of her affection. Bernard's interest in Elsie, however, is characterized by scientific detachment, not by romantic infatuation. Ironically, Elsie saves his life when, throwing caution to the wind, Bernard Langdon hikes up the Mountain and finds himself confronted by a rattlesnake. It is here that suddenly Holmes's often pedestrian narrative begins to sparkle, indeed "glitter":

His look was met by the glitter of two diamond eyes, small, sharp, cold, shining out of the darkness, but gliding with a smooth, steady motion towards the light, and himself. He stood fixed, struck dumb, staring back into them with dilating pupils and sudden numbness of fear that cannot move, as in the terror of dreams. The two sparks of light came forward until they grew to circles of flame, and all at once lifted themselves up as if in angry surprise. Then for the first time thrilled in Mr. Bernard's ears the dreadful sound that nothing which breathes, be it man or brute, can hear unmoved,—the long, loud, stinging whirr, as the huge, thick-bodied reptile shook his many-jointed rattle and adjusted his loops for the fatal stroke. His eyes were drawn as with magnets toward the circles of flame. . . . Nature was before man with her anæsthetics: the cat's first shake stupefies the mouse; the lion's first shake deadens the man's fear and feeling; and the *crotalus* paralyzes before he strikes. (*EV* I: 238–239)

The scene takes a surprising turn when suddenly the snake's eyes lose their light and terror. The aggressive rattler changes from perpetrator to victim as Bernard beholds "the face of Elsie Venner, looking motionless into the reptile's eyes, which had shrunk and faded under the stronger enchantment of her own" (*EV* I: 239).

In an attempt to account for his "fascination," Bernard turns to the Professor as well as to the work of the Scottish doctor James Braid (1795?–1860), whose theory of "hypnotism" had dealt, as Jonathan Miller notes, "a death blow to the pseudo-scientific theory of animal magnetism."[65] Braid stripped the process, which had been "discovered" and then widely advertised by Franz Anton Mesmer, of its occult implications, vigorously denying that some kind of secret, magnetic power passed from the operator into the patient.[66] The hypnotic state required neither the assumption of an operative substance—such as Mesmer's mysterious "magnetic fluid"—nor the presence of an operator; it could also be self-induced.[67] Bernard opens his notebook and discovers a relevant passage from Braid's work: "By fixing the eyes on *a bright object* so placed as *to produce a strain* upon the eyes and eyelids, and to maintain *a steady fixed stare,* there comes on in a few seconds a very singular condition, characterized by *muscular rigidity* and *inability to move,* with a strange *exaltation of most of the senses,* and *generally* a closure of the eyelids,—this condition being followed by *torpor*" (*EV* I: 256).

This passage, overlooked by all critics of Holmes's novel, is derived, almost verbatim, from the beginning of the second part of Braid's most important book, *Neurypnology; or, The Rationale of Nervous Sleep* (1843). Braid's interest in the hypnotic trance, explained by him as "merely a simple, speedy, and certain mode of throwing the nervous system into a new condition, which may be rendered eminently available in the cure of certain disorders,"[68] was based on a redefinition of the old concept of fascination so central to the doctrine of mesmerism. In effect, Braid argued that even the attraction of the bird to the snake could easily be explained as a process in which astonishment and the apprehension of danger send an impulse through the nerves and muscles, causing a corresponding movement without any control of the will.[69]

Braid's criticism seems all the more pertinent when we look at publications such as the New York physician John Newman's popular *Fascination; or, The Philosophy of Charming* (1847), a series of ten conversations between a "Doctor" and a "Lady." In the first of these strange colloquies, ancient tales about the serpent as enchanter and his power over the "natural instinct" of animals and even man's self-control are happily resurrected. For example, Newman repeats, with just a few variations, Cotton Mather's now familiar tale of the rattlesnake and the paralyzed squirrel and Joseph Breintnall's story about the man in Maryland who was "enchanted" by a snake. Fortified by such flimsy evidence, the Doctor convinces the Lady (who was not exactly skeptical to begin with) that "the power in man and the lower animals is essentially the same." Since the lower animals can fascinate man just as they seem to be able to fascinate each other, it is also self-evident that (1) man can fascinate the lower animals and that (2) man can fascinate man—a fact that, if well understood, can be put to good use.[70] The rather crudely drawn cover illustration of another popular treatise of the time, the Reverend Jacob Baker's *Human Magnetism* (ill. 40), represents the operator as a kindly father figure, facing a significantly smaller patient seated in a rocking chair and encouraging him to put his "trust in him" and "fear no evil."[71]

Still, esoteric or even sinister associations would persist and perhaps constituted part of the reason so many would become interested in mesmerism. Consider, for example, Edgar Allan Poe's 1844 tale "Mesmeric Revelation," in which the patient, glimpsing the ultimate life during his trance, imperceptibly crosses over into the domain of death, his corpse assuming the "stern rigidity of stone."[72] Rubens Peale's New York Parthenon featured shows in which mesmerism was hailed as a new scientific wonder; these were soon parodied by P. T. Barnum, with his characteristic sixth sense for popular entertainment. "Of course, all my 'passes' would not put a man in the mesmeric state," wrote Barnum, "At the end of three minutes he was wide awake as ever."[73]

As indicated in his unpublished notebooks from the time he attended medical lectures in Paris, Holmes knew about mesmerism and its possible uses and limitations in medical therapy.[74] Holmes's later medical writings establish him as a skeptical critic of the

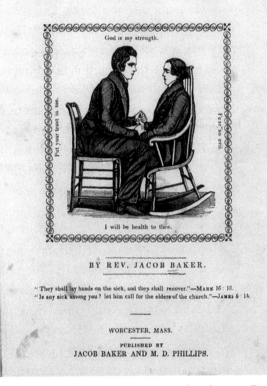

40. Jacob Baker, *Human Magnetism* (1843), cover. By permission of Houghton Library, Harvard University.

medical profession: they included an attack on the delusions of homeopathy as well as a groundbreaking study of the contagiousness of puerperal fever, which identified doctors not as deliverers from disease but as the unwitting carriers of the "black death of childbed."[75] It is more than a little strange, then, that this highly qualified Harvard physician, instead of continuing Braid's demystification of mesmeric trances, would compose a novel that in effect reinstated the "fascinating power" in all its occult unpredictability.[76]

Holmes's novel, to be sure, raises more questions than it answers. What it does suggest, though, is that "Master" Bernard—who arrived at Rattlesnake Ledge as "an active, muscular, courageous, adventurous young fellow" (*EV* 1: 253) and who returned, weakened and beaten, a coward saved by a woman—found himself, in a sense, emasculated by Elsie's reptilian powers: "If Master Bernard felt a natural gratitude to his young pupil for saving him from an imminent peril, he was in a state of infinite perplexity to know why he should have needed such aid" (*EV* 1: 253). Significantly, the plot of Holmes's novel requires Bernard to turn from a hypnotized victim into the hypnotizer himself. In order to "save" Letty Forrester, the wholesome Boston girl whom he will later marry, Bernard must stare down Elsie at a party, just as—the reader is invited to think—Elsie had earlier stared down the rattlesnake for Bernard. But the differences between Elsie's "counter-fascination" and Bernard's "counter-fascination" are obvious. Elsie wanted to save Bernard, while Bernard now wants to save himself and simultaneously keep a nubile girl's interest in him alive. The situation calls for immediate action. Having turned away from Bernard, although he is "making himself very agreeable" to her, Letty suddenly feels herself under an "unusual influence." This mysterious power the nervous Bernard quickly attributes to Elsie's fixed stare, which he realizes he now needs to counteract:

> He turned toward Elsie and looked at her in such a way as to draw her eyes upon him. Then he looked steadily and calmly into them. It was a great effort, for some perfectly inexplicable reason. . . . There was something intolerable in the light that came from them. But he was determined to look her down, and he believed he could do it, for he had seen her countenance change more than once when he had caught her gaze steadily fixed on him. All this took not minutes, but seconds. Presently she changed color slightly,—lifted her head, which was inclined a little to one side,—shut and opened her eyes two or three times, as if they had been pained or wearied,—and turned away baffled, and shamed, as it would seem, and shorn for the time of her singular and formidable or at least evil-natured power of swaying the impulses of those around her. (*EV* 2: 98–99)

Bernard's role in the novel is considerably complicated by the fact that he not only stares Elsie down but also, if unwittingly, finally kills her. Rejecting her love, he rather lamely offers her his "friendship." Elsie takes to her bed, and when the other students at Mr. Peckham's school collect flowers to cheer up the ailing girl, Bernard adds a sprig of white ash, by popular belief a powerful antidote against rattlesnakes.[77] Looking at the olive-purple leaves of the flower, Elsie, "as if paralyzed," shrinks up into herself and begins to fade away. The Ugly Things "never go where the white ash grows," explains

Sophie (*EV* 2: 260). Obligingly, Elsie dies. Some time later, the Mountain is shaken by an earthquake, which kills Elsie's black confidante Sophie and permanently closes up Rattlesnake Ledge.

It is as if the plot centering around Elsie's "fascination," as forced as it was from the beginning, had been brought to an even more artificial conclusion. In a world bent on maintaining the "broad sea-level of average" (*EV* 2: 52) Elsie has no place. For all the narrator's ominous talk about her "dark, cold, unmentionable instinct" (*EV* 2: 252), it is also clear that beautiful Elsie was not an Ugly Thing. And whatever her "bad habits"[78] might have been, they definitely did not spawn evil actions, apart from some vaguely described "mischief" done to a former governess (and unless one considers biting and perhaps attempting to kill Dick Venner, the most widely disliked character in the novel and a potential murderer himself, to be a disgraceful and not a beneficial act). While Elsie saves Bernard, Bernard kills Elsie. Coincidentally or not, the scientist in Doctor Holmes's novel is a pretty insufferable character, calculating, stiff, superior, and snooty, bereft of intuition and imagination.

By a curious logic, the narrator's words of farewell to the reader merge with Elsie's parting gesture. On the last page of *Elsie Venner,* the reader is given a brief glimpse of Letty Forrester enjoying a play in the company of Bernard, who has now become a "Professor" himself. Around her wrist there glitters a chain—Elsie Venner's serpentine golden bracelet, left for Bernard to pass on to someone worthier of his love than herself. The narrator's joy is mixed with tears, and the reader cannot help thinking that the bracelet should have glittered on somebody else's arm.

Holmes early in his life rebelled against his Calvinist upbringing, and his writings show a continuing interest in theological problems such as the natural depravity of man. In the second preface to *Elsie Venner,* Holmes explains that he wrote his novel in order "to test the doctrine of 'original sin' and human responsibility," and it has usually been read as an illustration of the terrible force of heredity, exempting Elsie from any kind of accountability for her actions. But what exactly are her unaccountable actions? According to her minister, Elsie tore the frontispiece, a picture of Eve's temptation, out of a Sunday-school book he had lent her, declaring that Eve was "a good woman" (*EV* 2: 33). A misogynistic text, *Elsie Venner* does not convince the reader that women can be snakes; there is in fact every indication that Holmes, as appalled and haunted as he was by the "disproved, or at best . . . unproved story" of the Fall, would have had a lot of sympathy for Elsie's iconoclastic gesture, even if the rashness of the act would have worried him. "The change of opinion" in religious matters, he wrote in his essay on Jonathan Edwards, would come "quite rapidly enough: we should hardly dare to print our doubts and questions if we did not know they will . . . have no apparent immediate effect on the great mass of beliefs."[79]

In one of his lectures, Holmes once identified science as "the topography of igno-rance": "We cast the lead, and draw up a little sand from abysses we may never reach with our dredges. . . . Nothing more clearly separates a vulgar from a superior mind, than the confusion in the first between the little it truly knows, on the one hand, and what it half knows and what it thinks it knows on the other."[80] It seems odd but perfectly in accordance with the odd topic that we have been exploring here that a scientist's first novel would deal with (and perpetuate) exactly this confusion between the little-known and the allegedly known. But even if it is true that, as R.W.B. Lewis said, Holmes often disappoints his readers, their disappointment here has a revela-tory function. Holmes's moralizing romance about "fascinating" *Elsie Venner* fails to convince the reader of the significance of the moral principles it flaunts (Elsie's "evil" nature), while, as a scientific or "medicated" text, it also fails to promote the science it pits against superstition.[81] Written just before the Civil War tore up the American na-tion, Oliver Wendell Holmes's novel about the power of fascination became, in its own way, symptomatic of the widening gap, whether real or imagined, between science and literature, reflecting only too accurately a predicament from which both of them would take a long time to recover.

Forever stripped of its "scientific" underpinnings, the notion of the rattlesnake's sinister "power of fascination" lived on. An entry in the diary of Arthur Inman, writ-ten around 1926, not only shows how easily the process of "fascination," imagined by humans, could turn into something that takes place, exclusively, between humans. It also demonstrates that a notion that had once served as a description of a supposedly "natural" event effortlessly lent itself to the expression and perpetuation of racial and, in this case, unabashedly anti-Semitic prejudice. In 1918, Arthur Inman's mother, wor-rying about the undiagnosable ailments of her twenty-three-year-old son, called in the Russian-born psychotherapist and hypnotist Boris Sidis. Inman's text, written with the stylistic panache that has made his strange diary famous, describes his encounter with a "compact Lucifer figure of a Jew," whose "malign influence" began, he felt, to assert itself the moment Sidis entered the room:

> The man pulled a chair close to mine, sat down, turned his black eyes full into mine. I cannot in mere words express the sensation of horror that overwhelmed me. Here was evil incarnate. The man talked as though I were insane, childish, to be impressed and persuaded all at the same time. He desired me to place my-self under his control. He wished me to come to his sanitarium. He promised to give me health if I would render myself completely and unreservedly over to him.[82]

Arthur shrinks back from Sidis's advances: "Such a stream of concentrated, malevolent will power directed against me I have never before or since felt." Sidis tries to lure Arthur with his "collection of fairy tales," the finest in the world. Arthur begins to perspire: "My lips and tongue grew dry. My hands trembled. I experienced the agonies of a human mouse faced by an omnipotent human snake." But, as Arthur would proudly remember later, as weak and debilitated as he was, he held fast to his determination not to yield to Sidis's alien influence: "I told him that I would not consider going to his sanitarium." Sidis, unimpressed, places his hand on Arthur's knee "as though to complete the circuit of contact. Then he regarded me with his penetrating black eyes. 'You will come,' he stated." Inman, the "human mouse," is afraid, but still he manages to stare down his opponent, the "human snake." Under the hypnotist's unflinching gaze, Arthur cringes, yet remains steadfast.

In his retelling of the scene, written eight years after the event, the sentence in which Arthur finally squeaks out his refusal to yield is dramatically severed by the intervening clause: "I summoned up the final reserves of my strength, faced the man, eye to eye. I was, melodramatic as it may sound, fighting to keep my soul my own. 'I will not,' I said, and the utterance of the words seemed to wrench my whole being, 'go with you!'" Triumphantly, Arthur witnesses the sinister enchanter's disenchantment: "Angry, insane, Jewish bafflement looked at me for an instant out of those hypnotic eyes, then was withdrawn of a sudden as though a glaring light had been switched off."

One of the central aims of my study has been to show that the claims of natural history continually mesh with, bend to, and reflect the designs and purposes of human history—and to point out how impossible it is, in the many stories that have been told (and made up) about nature, to separate the teller from the tale. In this context it is particularly instructive (and horrifying) to watch the anti-Semitic Arthur Inman's busy pen make the "Cain of the brotherhood of serpents"—as the rattlesnake is called in Holmes's *Elsie Venner* (*EV* I: 259)—take on the imagined features of a kind of Jewish anti-Christ. Sidis the human rattlesnake is defeated, yet—the ultimate proof of how strong Arthur wants us to believe his powers are—never loses his self-control. He merely, as it were, glides out of the room: "With no more notice of me, he got up, walked to the door, went out." Arthur remains behind—victorious but, the perhaps unintended moral of his story, no less a mouse for that: "For three days thereafter, I was in bed, or mostly so, with exhaustion."

Inman's diary entry brings to a preliminary (and particularly poignant) conclusion my sketchy history of rattlesnake "fascination," a collection of tales that tell of the mixture of elation, excitement, and horror with which several generations of Americans looked upon what were in fact different species of snakes but what, for all practical purposes, had merged in their imaginations into one composite, often larger-than-life animal. The next chapter, conceived as a kind of companion piece, tells of the fascination of one man with an entire class of animals, or at least with those myriad, infinitely beautiful and exciting representatives of this class that he discovered in the "gorgeous but melancholy" wildernesses of North America, "amid the tall grass of the far-extended prairies of the West, in the solemn forests of the North, on the heights of the midland mountains, by the shores of the boundless ocean, and on the bosom of the vast lakes and magnificent rivers" (*BA* 4: 79).

CHAPTER 5

Audubon at Large

Ocular Demonstration

Imagine, John James Audubon asked his readers, in a chapter on the Ivory-billed Wood-pecker from the long textual commentary he wrote for *The Birds of America*: imagine the woodpecker's abode. Imagine deep, swampy morasses, "overshadowed by millions of gigantic dark cypresses, spreading their sturdy moss-covered branches"; imagine the "massy trunks" of fallen, decaying trees, lakes filled with black muddy water, imagine walking over dangerous ground, "oozing, spongy, and miry," covered with mosses and water lilies, which yields instantly under the pressure of the anxious traveler's foot. Bet-ter still, imagine the bird itself: the "broad extent of its dark, glossy body and tail, the large and well-defined markings of its wing, neck, and bill, relieved by the rich carmine of the pendent crest of the male, and the brilliant yellow of its eye." Imagine its flight, "graceful in the extreme," when, for example, it crosses a large river, "which it does in deep undulations, opening its wings at first to their full extent, and nearly closing them to renew the propelling impulse." Then imagine it flying from tree to tree, in a single "elegantly curved line" (*BA* 4: 214–216).

An impossible task, it seems. Or so one thinks. "Nothing short of ocular demonstra-tion," writes Audubon, "can impress any adequate idea" of either bird or abode on the "mind's eye" (*BA* 4: 215). The lines quoted above are darkly evocative; they conjure up landscapes where ordinary people fear to tread and heighten the reader's respect for the selfless courage of the naturalist who has ventured there on her behalf. However, with remarkable poetic precision, they also detail colors (glossy black, rich carmine, bril-liant yellow) and movements (the wings at first open and close in "deep undulations"), as well as tactile impressions (the feel of the ground under the traveler's feet), thus goading on the reader's perhaps reluctant imagination. Some of the imagining that is required obviously has already taken place: "I have always imagined," proclaims Audu-

No. 52.   Pl. 256.

Ivory-billed Woodpecker.

1 Male. 2 & 3. Female

41. "Ivory-billed Woodpecker." Lithograph by J. T. Bowen after John James Audubon, for *The Birds of America* (Royal Octavo), vol. 4, pl. 256. Courtesy, The Lilly Library, Indiana University, Bloomington.

bon himself, "that in the plumage of the beautiful Ivory-billed Woodpecker, there is something very closely allied to the style of colouring of the great VAN-DYKE" (*BA* 4: 214).

The visual appeal of Audubon's carefully crafted prose matches the compositional force of the accompanying plate, which represents three Ivory-billed Woodpeckers,[1] one male and two females, dynamically arranged on the rugged, half-stripped branches of one of those mossy trees mentioned earlier (ill. 41). The male conspicuously displays the rich carmine crest so admired by Audubon, and while we see his back and his head in half profile, the female in the lower right half of the composition is shown in full profile. The third bird, partially hidden by the branch on the right that serves as the (slightly displaced) central axis, shows the front of her head and some of the underside of her body. Audubon's representation of these striking birds is ornithologically exact, since it demonstrates their ability, commented on elsewhere in the text, to balance themselves against trees or even grapevines by using their strong claws. But this amazing ballet of birds is also effective artistically, as an image of carefully controlled dynamism. The direction of the birds' beaks serves to focus the viewer's attention on the beetle about to be seized by the male or the female in the upper half of the composition and on the large piece of bark skillfully pierced by the bird in its lower half. Audubon might not be van Dyck, but with so much expert help as he provides, visual and verbal alike, the "mind's eye" can certainly travel a long way. Inattentive naturalists such as Buffon might have regarded woodpeckers as "miserable beings"; for Audubon, they are the kings of their domain.[2]

Audubon's tribute to the Ivory-billed Woodpecker first appeared in his *Ornithological Biography*, a five-volume series of essays on the habits and habitats of American birds, printed in Edinburgh between 1831 and 1839 in an edition of 750 sets.[3] *Ornithological Biography* was intended to complement *The Birds of America*, a sequence of 435 life-sized bird portraits engraved into beautiful, double-elephant aquatint plates between 1827 and 1838 initially by William Lizars in Edinburgh and then by Robert Havell and his son in London. None other than the great Baron Cuvier himself hailed Audubon's *Birds of America* as "the most magnificent monument which has yet been raised to ornithology,"[4] For the more affordable "miniature" edition of *The Birds of America*, the "Royal Octavo," published between 1840 and 1844, J. T. Bowen of Philadelphia produced colored lithographs from *camera lucida* copies of the original aquatints, to which Audubon then added the (slightly revised) texts from his *Ornithological Biography*, thus creating what some historians regard as "the single most influential work of natural history and art in the nineteenth century."[5]

Monuments, it is true, are rarely built by their architects alone, without the help of others. While Audubon was supported in his endeavors by his family as well as fellow ornithologists, engravers, colorists, and other collectors who freely shared their observations with him, he also relied readily on the labors of those who had preceded him. Nevertheless, it is the central thesis of this chapter that the images and the text that make up Audubon's *Birds of America* reflect a fairly consistent artistic vision, a coherent visual poetics unparalleled in either eighteenth- or nineteenth-century natural history.

Audubon's Ivory-billed Woodpecker was first drawn and described, under the name of *Picus maximus rostra albo* ("Largest White Bill'd Woodpecker"), by Mark Catesby, one of the earliest collectors of American birds. Catesby, to be sure, did not think of his work as artistic representation; rather, he was convinced that the plates included in the two volumes of his *Natural History of Carolina, Florida, and the Bahama Islands* (1731; 1743), plates that he had laboriously engraved and hand-colored himself, presented American nature as it really was: "I was not bred a Painter," the Essex-born gentleman wrote in the preface to the stunning work he had dedicated to the queen of England. Because he had received no specific training, Catesby hoped, "some faults in Perspective, and other Niceties, may be more readily excused, for I humbly conceive Plants, and other Things done in a Flat, tho' exact manner, may serve the Purpose of Natural History, better in some Measure than in a more bold and Painter like Way" (*NHC* 1: xi).

Sent to Her Majesty's Dominions in the New World to collect plants for his wealthy donors at home, Catesby soon found himself in a quandary. The birds he encountered in Carolina, Georgia, Florida, and the Bahamas were so beautiful, so much more "excellent," especially in terms of their coloring, than all other American animals, that

42. "Largest White Bill'd Woodpecker." Hand-colored engraving by Mark Catesby. From Catesby's *Natural History of Carolina . . .* vol. 1, 2nd edition (1754). Courtesy, The Lilly Library, Indiana University, Bloomington.

Catesby, overwhelmed by the "inimitable perfection" of Nature herself, soon lagged behind in his plant-gathering. He began to draw American birds, as it were, al fresco: "In designing the Plants, I always did them while fresh and just gather'd: And the Animals, particularly the Birds, I painted them while alive (except a very few) and gave them their Gestures peculiar to every kind of Bird, and when it would admit of, I have adapted the Birds to those Plants on which they fed, or have any Relation to" (*NHC* 1: xi–xii).

However much he wanted to avoid "painter-like" representations, Catesby's plates went far beyond the requirements of conventional natural history illustrations, presenting images that ranged from the playful (plate 73 in the first volume shows a Flamingo against an ornamental background provided by an underwater gorgonian coral) to the formal and to compositions almost abstract in their stylish beauty.[6] Plate 16, for example, one of a group of eight featuring woodpeckers, portrays Audubon's favorite bird clinging to the trunk of a willow oak (ill. 42). The branch extending somewhat inconsequentially from the oak's trunk just below the ivorybill's feet seems to have been included mainly for the sake of ornamental variety. The shape of the oak leaves ("long, narrow and smooth-edged," as Catesby himself describes them) reflects the form of the woodpecker's characteristic long, "channelled" beak, whose value as a collector's item for native tribes ("the *Canada* Indians") Catesby emphasizes in the accompanying text. The partially visible thick, curved trunk of the willow oak corresponds to the convex shape of this crow-sized bird, the largest woodpecker in North America (a fact pointed out in Catesby's concise text, which begins: "Weighs twenty ounces"). The long, knotty, gnarled feet of the bird, which look like tree bark, as if they were a growth on the trunk rather than an extension of the bird's body, provide the appropriate link between the massive body of the oak and the bulky body of the woodpecker. Small wonder, in view of such compositional ingenuity, that a discerning observer like John Bartram praised Catesby's book as "an excelant performance & an ornament of the finest Lybrary in the world" (*CJB* 153).

Several decades later, John Bartram's son William inspired another European-born nature enthusiast, Alexander Wilson, to take a closer look at American birds. Wilson was a weaver, poet, and social activist from Paisley, Scotland. One of his poems, "The Shark; or, Lang Mills Detected," written while a labor dispute was brewing in Paisley, had brought him such trouble that the author thought it advisable to seek his fortune elsewhere. After landing in New Castle, Delaware, on 14 July 1794, Wilson could not help but notice that the birds here were different from those he had known at home, as his first biographer and editor, George Ord, tells us. During a hike from Delaware to Pennsylvania, Wilson shot a red-headed woodpecker, the "most beautiful bird he had ever seen."[7] In Philadelphia, Wilson first tried his hand at drawing "landscapes and sketches of the human figure." The results were not impressive. At the suggestion of William Bartram, he made "a second attempt, upon birds and other objects of natural history, and in this he succeeded beyond his anticipations."[8] Wilson spent the next years either studying the collections in Charles Willson Peale's Philadelphia Museum or, shotgun in hand, traveling through the United States, seized by the overwhelming desire to make, as he said, "a collection of all the American birds," "from the shores of the St, Lawrence, to the mouth of the Mississippi, and from the Atlantic ocean to the interior of Louisiana,"[9]

It was Wilson's devout wish that in the pages of his *American Ornithology* "the figures and descriptions may mutually corroborate each other" (*AO* 1: 2). While Catesby had given his readers little more than a few notes (it had not been his intention, Catesby declared, "to tire the Reader with describing every Feather"; *NHC* 1: xii), Wilson saw himself as "the faithful biographer" of his birds, of their manners and dispositions. His descriptions, all of them the "result of personal observation" (*AO* 1: 2, 8), were extensive, prolix, and lively—an easy mix of the taxonomic with the anecdotal. On occasion, Wilson inserted poems he had written himself. He expressed respect for Catesby's "labours" (*AO* 1: 8) but insisted that his own "transcript from living Nature" (*AO* 1: 6) was different because his bird portraits were meant to "speak not to the eye alone" but also to the heart (*AO* 4: vi).

Wilson's chapter on the ivorybill is a case in point. It starts with an emotionally charged description of the woodpecker that evidently also inspired Audubon. In Wilson's eyes, the ivorybill was "majestic" and "formidable," the king of his tribe. "Nature seems to have designed him a distinguished characteristic in the superb carmine crest and bill of polished ivory with which she has ornamented him. His eye is brilliant and daring; and his whole frame so admirably adapted for his mode of life, and method of procuring subsistence, as to impress on the mind of the examiner the most reverential ideas of the Creator" (*AO* 4: 20).

In accordance with his "majestic" appearance, Wilson's woodpecker displays distinct class prejudice, in the form of a "dignity . . . superior to the common herd of woodpeckers." This bird shuns the shrubbery, fence posts, orchards, rails, and prostrate logs where his poor relations congregate, preferring instead the "most towering trees of the forest" or the "solitary savage wilds" of the cypress swamps, where, amidst "piles of impending timber," he remains the "sole lord and inhabitant" (*AO* 4: 20). Wilson goes on to stress the usefulness of the woodpecker: in an instant, he kills the insects that, in a matter of a season or two, can lay waste to "thousands of acres" of healthy pine trees. Then the author tells an anecdote—the story of how he actually found the woodpecker portrayed in the accompanying plate. Wandering through the woods, twelve miles north of Wilmington, North Carolina, he had picked up the "slightly wounded" bird and taken it with him, "under cover," to Wilmington. Unfortunately, the bird kept uttering the most pitiful cries, "exactly resembling the violent crying of a young child," which promptly attracted the attention of everyone out on the streets. Enjoying the looks of "alarm and anxiety" given him especially by the women of Wilmington, Wilson marched into a hotel, asking for "accommodations for myself and my baby," before revealing, amid general laughter, the actual source of the distressing cries (*AO* 4: 22). In the hotel room, the bird continued to show "unconquerable" (and very un-baby-like) resolve, plotting routes of escape, wrecking a mahogany table to which it was tied, and biting Wilson while he attempted to paint its portrait. "He lived with me nearly three days, but refused all sustenance, and I witnessed his death with regret" (*AO* 4: 22).

The differences between Wilson's and Audubon's "biographies" of the ivory-billed woodpecker are obvious enough. Wilson pontificates, admonishes, and generalizes, teaching lessons and telling stories, while in Audubon's biography meaning is conveyed in the *process* of a communication between the author, who has seen everything, and the reader, who will see more if she only listens. Wilson's illustration, plate 29 in the fourth volume of *American Ornithology*, shows the differences even more succinctly (ill. 43). Wilson crowds different species of woodpeckers, all collected by himself, into the space of just one, albeit beautifully colored, plate. The lower half of the picture consists of reduced images of the ivory-billed, the pileated, and the red-headed woodpecker perching on tree trunks (all of which are more sketchily suggested than fully drawn out), while in the upper half, the life-sized heads of the pileated and ivory-billed woodpecker appear as if floating in empty space, dissociated from their bodies. Seventy-five of the seventy-six plates in Wilson's *American Ornithology* feature more than one bird species, sometimes with strange-looking results (plate 75, for example, crams together a turkey vulture, a black vulture, and a raven). In this plate, Wilson's obvious intention was not only to stress the magnificence of his bird by including its life-sized head but also to show his readers that the ivory-billed woodpecker is larger than the pileated woodpecker, with which it is so often confused: "So little attention do the people of the countries where these birds inhabit, pay to the minutia of natural history, that, generally speaking, they make no distinction between the Ivory-billed and Pileated Woodpecker . . . and it was not till I shewed them the two birds together, that they knew of any difference" (*AO* 4: 24). While economic reasons had guided Wilson's hand in crowding several species into one plate, his intention had also been to provide the viewer with a basis for comparison.

John James Audubon's career as an ornithological artist began with drawings that adhered to the conventions of natural history illustrations. Like Wilson, he would show his birds motionless and in profile. But even then Audubon shied away from arrangements that would have required him to portray more than one species in a single image. A pastel sketch of an ivory-billed woodpecker, dated 30 July 1810 and now in the collection of the Houghton Library at Harvard, presents a standard profile view of the bird but already departs from convention in presenting the woodpecker as an individual, its single eye focused more or less directly on the viewer.[10] Two years later, in a pastel and pencil drawing dated 28 November 1812, Audubon portrayed a pair of ivorybills on opposite sides of a tree. This composition heralds the talent for creating dynamic arrangements through clever compositional shifts that would mark Audubon's mature work (ill. 44). The solid trunk of the tree almost completely hides the bill of the female woodpecker on the left, while the bill of the male on the right, partly opened, remains fully visible. This seemingly trivial device not only lends spatial depth to the composition but also individualizes the two birds and their activities. The tree itself seems to tilt to the right, thus creating a diagonal that neatly, effortlessly, parallels the diagonal formed by the back of the female bird. By

43. "Woodpeckers." Hand-colored engraving by Alexander Lawson after Alexander Wilson, for Wilson, *American Ornithology* (1808–1914), vol. 4, pl. 29. Courtesy, The Lilly Library, Indiana University, Bloomington.

44. *Ivory Bill Woodpeckers*, by John James Audubon, 1812. Ms Am 21 (31). By permission of the Houghton Library, Harvard University.

contrast, the male bird, his wing pointing straight down while his tail feathers veer to the left, has a much harder time than his mate in steadying himself against the skewed tree trunk—and, sure enough, Audubon depicts the male's firmly anchored, knotted claw, an image of strength and concentrated effort, in the center of the composition. Miraculously, it is exactly at this point that the trunk suddenly appears to straighten itself out again, if only for a moment, before it resumes its rightward ascent. While the female bird merges with her environment, the more colorful male appears to dominate it.

Over the next years, Audubon began to explore further ways in which he could introduce drama and movement into his compositions. The plate from *The Birds of America* showing three ivorybills (based on a watercolor completed in Louisiana in the early 1820s) proves that Audubon had come a long way since he drew his first woodpecker in 1810. As visitors to Audubon's public exhibitions in 1826 realized, these ornithological illustrations did not simply illustrate. Rather, they were highly charged works of art, vortices of violent activity, visual fields of force that made the viewer participate in an experience rather than, as had been the tradition, contemplate from a safe distance a scientific fact.[11] What Audubon successfully challenged, according to Ann Shelby Blum, was "the conceptual separation" between the wild bird and the viewer that had been central to conventional ornithological illustration.[12]

In 1810, Alexander Wilson's rambles in search of new members of the "feathered tribes of the United States" (*AO* 4: vi) also led him to spend a few days in Louisville, Kentucky, where Audubon, dreaming about a collection of drawings of all the birds of America, "all of Natural *Size*," and indulging his fondness for fine horses, was still attempting to run a store,[13] Audubon did not subscribe to Wilson's *American Ornithology* ("even at that time, my collection was greater than his")[14] and had reason to regret this decision. In 1820 and 1821, he scoured New Orleans in search of an affordable edition of

Wilson's work and had to borrow other people's copies to check on species identifica-
tions (*SJ* 94, 80). Some of the objections Audubon subsequently raised against Wilson's
work were dictated by professional jealousy rather than factual disagreements and are
not entirely free of condescension: "The great Many Errors I found in the Work of
Willson astonished Me," Audubon wrote in his diary on 20 October 1821. "I tried to
speak of them With Care and as seldom as Possible; Knowing the good Wish of that
Man . . . and the Hurry he was in and the Vast Many hear say he depended on" (*SJ* 133).
However, as Audubon himself probably knew only too well, his *Birds of America* would
have been unthinkable without Catesby or Wilson. Still, as the different representations of woodpeckers discussed here show, the concentrated force of the plates in *Birds
of America* was entirely Audubon's own invention, the source of a power that was still
evident more than a century later, when Lincoln Kirstein, Morton Gould, and George
Balanchine contemplated a (never realized) ballet titled *The Birds of America*.[15]

Audubon's cherished ivory-billed woodpeckers are practically extinct today. In 1820,
when Audubon was traveling down the Mississippi, the woods had still resounded with
their loud, wild cries, their insistent *"Pait Pait Pait"* (*SJ* 48). But, as Audubon noted
later in his *Ornithological Biography*, hunters already then liked to adorn their pouches
with the ivorybill's "rich scalp" and shot the bird "merely for that purpose." Travelers "of
all nations" were "fond of possessing the upper part of the head and bill of the male"
and would pay good money for it. "I have frequently remarked, that on a steamboat's
reaching what we call a *wooding-place*, the *strangers* were very apt to pay a quarter of a
dollar for two or three heads of this Woodpecker" (*BA* 4: 216). Such a dissevered bird
head, incidentally, is what we see in Wilson's drawing. While his plate, in a sense, seems
like a collection of trophies, rigid in their silent beauty, Audubon's picture shows not
birds that were hunted but rather birds that are, *themselves,* engaged in hunting. To
put it more generally, compared with Alexander Wilson's birds, all of which, to quote
the naturalist John Burroughs, reflect the author's "cautious, undemonstrative Scotch
nature," most of Audubon's birds, collected over a lifetime spent in search of specimens,
appear as if continually on the move, engaged in frenetic, "exaggerated activity."[16] The
remainder of this chapter—an essay on the indefatigable bird collector Audubon's extraordinary visual poetics—suggests why this might be so.

Inventing John James Audubon

"John James Audubon" was born in 1785 as Jean Rabine, the son of a French trader,
plantation owner, and slaveholder, Capitaine Jean Audubon, and his mistress, Jeanne

Rabine, in Les Cayes on what later came to be called Haiti.[17] His mother, a chambermaid from Les Touches in France, died shortly after his birth. In 1788, old Jean Audubon, terrified by the first signs of black insurrection on Saint-Domingue, took his young son back to France, perhaps not expecting that the shock waves of revolutionary upheaval would soon reach him there, too. Whatever the circumstances of his own life, the good captain, now "Citoyen Audubon," was determined that at least his son be spared. In 1795, he and his elderly wife legally adopted the boy, who early on showed less interest in formal studies than in roaming the woods around the family's country residence on the Loire, where he collected birds' nests and eggs, flowers, and curious pebbles. In 1803, Audubon pere sent his son, who now called himself John James Audubon, back to the New World, so that he could avoid being drafted into Napoleon's army and manage the family's estate, Mill Grove, near Philadelphia.

In 1808, John James Audubon married a neighbor's daughter, Lucy Bakewell, and then embarked on a brief entrepreneurial career at the end of which came bankruptcy and a stint in jail. While his flamboyancy had left him without credit, money, or a reputation to lose, it also gave Audubon the freedom to pursue the love of his youth, ornithology. "Ever since a Boy," Audubon writes in his diary on 28 November 1820, in typically idiosyncratic syntax, "I have had an astonishing desire to see Much of the World & particularly to Acquire a true Knowledge of the Birds of North America, consequently, I hunted when Ever I had an Opportunity, and Drew every New Speciman as I could, or dared *steel time* from my Business and having a tolerably Large Number of Drawings that have been generally admired, I Concluded that perhaps I Could Not do better than to Travel, and finish My Collection." Perhaps, hoped the destitute Audubon, such a collection would prove to be "a Valuable Acquisition" (*SJ* 54–55).

The Birds of America, first advertised to the public in 1827 and completed eleven years later, has its roots in Audubon's obsessive desire to shoot and draw every bird he could, literally, lay his hands on. And a "valuable acquisition" it became indeed. Full sets of the Double Elephant Folio of *The Birds of America* cost $1,000, and while they did not make Audubon rich, they earned him, after a "Life . . . strewed with Many thorns" (*SJ* 103), the career he had always desired: election as a Fellow of the distinguished Royal Society in London, membership in the American Philosophical Society, and, finally, the year before his death, inclusion in daguerreotypist Mathew Brady's *Gallery of Illustrious Americans* (1850). After a visit to the Académie des Sciences in Paris, to which he had been admitted at Cuvier's recommendation, he felt justified in calling himself "the first ornithological painter . . . of America" (*AJ* 1: 313). Audubon became "the best-known American naturalist of his period," and he remains the best-known today."[18] This is a surprising turn of events considering that, just like Catesby, Audubon was completely self-taught and had indeed not been "bred a Painter." His dubious claims that he had studied painting with the famous French artist Jacques-Louis David[19] should be taken as part of the same penchant for role-playing that made him feel "much tickled" when, on his return to France, a fanciful French customs officer mistook him for an American Indian and expected his face to be "copper-red" simply because of his passport (*AJ* 1: 304).

It has generally been assumed that if Audubon enjoyed any connection with "great literature," this must have been through one of his crazy business schemes: the purchase of a Mississippi riverboat, which entailed the financial involvement and the eventual bankruptcy of George Keats, brother of the English poet (no wonder that Keats thought that "Mr Audubon" was a "dishonest man").[20] However, although John Chancellor, one of his biographers, regards Audubon as only "marginally literate,"[21] and in spite of the fact that Audubon himself, with uncharacteristic modesty, claimed that he was only "a poor writer" (*AJ* 1: 63), his prose has a sparkle of its own. Often ungrammatical and sometimes incoherent, it possesses a strange, immediate beauty and has survived even the merciless bowdlerizing of his granddaughter Maria, who straightened out his syntax, corrected the wayward spelling in his journals and letters, and ruthlessly excised what she deemed improper.[22]

In composing the texts for his *Ornithological Biography*, Audubon received significant help from the young Scottish ornithologist William MacGillivray (1796–1852), who, as Audubon himself acknowledged in his "Introductory Address," smoothed "the asperities" of his style (*OB* 1: xix), toned down some of his poetic excesses, supplied the taxonomic details, and, beginning with the third volume, also contributed anatomical descriptions of the birds' respiratory and digestive tracts, an addition that reflected the professionalization of a discipline Audubon himself had entered as an inexperienced, if enthusiastic, autodidact. A comparison of Audubon's manuscript draft for the chapter on the Ruby-throated Hummingbird[23] with the version that was later published in *Ornithological Biography* reveals that MacGillivray, while tampering with some phrases and images, usually left the tone, the structure, and the central images of Audubon's text intact.[24] More often than not, MacGillivray's interventions made Audubon's prose more readable (for Audubon's clumsy construction "the generality of other birds" he substitutes the trimmer phrase "as birds generally do," and so on).

Admittedly, MacGillivray's scrubbing sometimes also dulled the original luster of Audubon's wild prose, sacrificing beauty to clarity. In his draft, Audubon first attempts to set the stage for the appearance of the little hummingbird in Louisiana during the first weeks of spring and, perhaps inadvertently, lapses into a rhythm resembling blank verse ("No sooner has the vivifying orb began [*sic*] to warm of spring once more the season"). Then he presents a startling image in which the bird, retrieving harmful insects from tender flower-cups, appears as an "anxious florist." In the published version, after the intervention of MacGillivray's busy pen, the "vivifying orb" has become "the returning sun," and Audubon's imaginative, lightly alliterative verbal play ("warms of spring the season") has given way to a construction consisting of a drab verb and an even duller object. In its new and more conventional form, the sun "introduces" the "vernal season" (*BA* 4: 190)—a needlessly stilted, trite substitution, to which Audubon's flights of verbal fancy in the manuscript version are preferable. Where Audubon calls the hummingbird a "lovely feathered miniature," the published text speaks of a "lovely little creature"; where he addresses it as a "richlly [*sic*] clad diminutive Seraph of the feathery tribe," it is now merely a "glittering fragment of the rainbow." But in spite of these and other minor changes, Audubon's central image of the bird as a kind of fairy "florist" has been preserved, and so have the many references to the activity of seeing ("on seeing this lovely little creature"; "the little Humming-bird is seen"; "it is observed peeping cautiously"; "Could you, kind reader, cast a momentary glance"; "Then how pleasing is it . . . to see"), which subtly redirect the emphasis from the things observed to the process of observation itself, allowing the eyes of the author and his readers to meet in mutual recognition.

"I wish," exclaims Audubon, in mock despair, "it were in my power at this moment to impart to you, kind reader, the pleasures which I have felt whilst watching the movements, and viewing the manifestation of feelings displayed by a single pair of these most favourite little creatures" (*BA* 4: 190–191). "See, as I have seen," he exhorts his readers. A few lines down, in the manuscript as well as the published text, a subtle shift in pronouns achieves the desired effect: the erasure not only of the boundary between the extratextual author and the on-the-spot experiencer but also of the boundary between the authorial persona and the reader's consciousness. Suddenly the reader finds herself paying a visit, however "unwelcome," to the hummingbird's nest, where she inspects a "newly-hatched pair of young, little larger than humble-bees" while the worried parents, "full of anxiety and fear," are "passing and repassing within a few inches of *your* face, alighting on a twig not more than a yard from *your* body" (*BA* 4: 191; my emphasis). The second-person pronoun, the rhetorical mode of address so often employed by Audubon, creates the usual sense of intimacy, and the measurements included in the sentence ("a few inches"; "not more than a yard"), with their suggestion of spatial proximity, further decrease the distance between the actual and potential observers in the text and the things they see. The fusion of the participant-observer, Audubon, and his heretofore nonparticipating readers is complete as both watch the sunlight pass through the outstretched diaphanous wing of the little hummingbird in the process of pluming its feathers: hummingbirds are, writes Audubon, "particularly fond of spreading one wing at a time, and passing each of the quill-feathers through their bill in its whole length, when, if the sun is shining, the wing thus plumed is rendered extremely transparent and light" (*BA* 4: 192).

Throughout the texts collected in *Ornithological Biography*, Audubon favors *seeing* at the expense of *writing*, which for him, he wants his readers to believe, usually consisted of nothing more than the casual collation of scattershot field notes. Audubon's finished text is often presented as more the fortunate result of happenstance than the product of literary premeditation (although the evidence of the many emendations and substitutions in Audubon's own manuscripts suggests otherwise). "I was ready to put my pen aside, kind reader," announces Audubon, "when, on consulting my journals, all of which are now at hand, I happened to read, that I have seen instances of this bird's plunging into the sea after small fry, at Powles Hook" (*BA* 4: 207). Or he confesses that, on looking over his notes, he suddenly discovers he has neglected to inform his readers of the extraordinary strength of the Whooping Crane (*BA* 5: 194). All these suggestions of immediacy are, in fact, part and parcel of a carefully planned performance, the main purpose of which is to render credible and convincing the main premise of the illustrations and the texts gathered in *Ornithological Biography*—which is that this author, a "practical ornithologist" if ever there was one (*BA* 6: 186), has seen for himself what other "closet-naturalists" (to use Audubon's favorite term for the enemy camp) have only conjectured about.

Because his own life is so often intimately connected with the lives of the birds he describes, this observer also has more authority than others with better literary skills to report on what he has seen and to instruct his readers in the proper study of nature: "No true student of nature ought ever to be satisfied without personal observation when it can be obtained. It is the 'American Woodsman' that tells you so, anxious as he is that you should enjoy the pleasure of studying and admiring the beautiful works of Nature" (*BA* 7: 274). If other naturalists were superior to him "in education and literary acquirements," Audubon told Dr. Thomas Stewart Traill, they could not equal him "in the actual course of observations of Nature at her best—in her wilds!—as I positively have done."[25] For him, personal, firsthand knowledge was more important than the often spurious authority of science: "Much has been said respecting the difference existing between the *Whippoor-will* and the *Night Hawk*, for the purpose of shewing them to be distinct species. On this subject I shall only say, that I have known both birds from my early youth, and I have seldom seen a farmer or even a boy in the United States, who did not know the difference between them" (*BA* 1: 157).

Not surprisingly, elements of Audubon's own biography, real or fabricated, frequently enter into his descriptions of the habits of birds. Connected with the biography of the Pewee Flycatcher (the eastern phoebe), claims Audubon, "are so many incidents relative to my own, that could I with propriety deviate from my proposed method, the present number would contain less of the habits of birds than of those of the youthful days of an American woodsman" (*BA* 1: 223). And deviate he does, slipping extraneous personal information into otherwise factual accounts, thus establishing, beyond doubt, his Americanness—his authority to speak as someone reared in and nurtured by the New World. Although Audubon did not arrive in the United States until he was eighteen years old and formally requested American citizenship only at the age of twenty-one, he is at pains to suggest to his readers a life that almost from the beginning was devoted to the study of the incomparable natural landscape of America: "While young, I had a plantation that lay on the sloping declivities of the Perkiomen creek. I was extremely fond of rambling along its rocky banks . . . observing the watchful Kingfisher perched on some projecting stone over the clear water of the stream" (*BA* 1: 223–224).

The phrase "my native land"—referring, of course, to the United States and not to Saint-Domingue—is a leitmotif in Audubon's texts, a weapon he brandishes in his many attacks on all the "closet-naturalists," those anemic compilers and systematizers of academic natural history who have not, like him, spent the course of a life "studying the birds of my native land, where I have had abundant opportunities of contemplating their manners" (*BA* 4: 78). Scornfully, Audubon, the "practical ornithologist" and "American woodsman," takes a stab at the European naturalists reluctant to pay a fair price for the birdskins American collectors have procured for them at great personal risk:

> How often, kind reader, have I thought of the difference of the tasks imposed on different minds, when, travelling in countries far distant from those where birds of this species and others as difficult to be procured are now and then offered for sale in the form of dried skins, I have heard the amateur or closet-naturalist express his astonishment that half-a-crown was asked by the person who had perhaps followed the bird when alive over miles of such swamps, and after procuring it, had prepared its skin in the best manner, and carried it to a market thousands of miles distant from the spot where he had obtained it. (*BA* 4: 215)

When Audubon arrived in the States in 1803, the year of the Louisiana Purchase, his forged passport that Anglicized his first names from Jean Jacques to John James also gave the new territory of Louisiana as his place of birth. As if to perpetuate this fiction, "my native Louisiana" (which Audubon himself never visited until 1820) figures prominently in *The Birds of America*, too: "Would that I were once more extended on some green grassy couch, in my native Louisiana," exclaims Audubon in his biography of the Purple Gallinule, that "I lay concealed under some beautiful tree, overhanging the dark bayou, on whose waters the bird of beauty is wont to display its graceful movements, and the rich hues of its glossy plumage!" Almost unnoticeably, the reader finds herself transported into a southern setting, lying among "the tall rushes that border the lake," where a "thick mass of withered leaves" covers the ground (*BA* 5: 128–129). Louisiana, with its "numerous lakes, creeks and lagoons, overshadowed by large trees" (*BA* 2: 89), is ideally suited for birds during the "love season" (Audubon's preferred term for the breeding season: *BA* 1: 92; 4: 12, 59; 6: 275; 7: 136). After all, as Audubon reminds his "dear reader," it is in Louisiana, "my favourite portion of the Union" (*BA* 6: 122), that the "bounties of nature are in the greatest perfection" (*BA* 2: 187). It is here that herons, egrets, and other magnificent water birds nest in large numbers; here the wonderful mockingbird makes its home for the summer. Here, and nowhere else, one "should listen to the love-song of the Mocking-bird, as I at this moment do" (*BA* 2: 187).

In many ways, then, the birds of America, elusive as they are, provide Audubon with a vast stage for his autobiographical self-construction.[26] As a rule, however, Audubon introduces and justifies his own subjective viewpoint in *Ornithological Biography* subtly and unobtrusively, with a cleverness that is an indication of his considerable literary talents: "Reader, imagine yourself standing motionless on some of the sandy shores between South Carolina and the extremity of Florida," begins his chapter on Wilson's Plover, which then continues: "or, if you dislike the idea, imagine me there. The air is warm and pleasant, the smooth sea reflects the feeble glimmerings of the fading stars" (*BA* 5: 214–215).

Audubon's blend of ornithology and autobiography needs to be seen against the background of the marked preoccupation with birds in earlier English and American natural history. Although birds were in their superficial anatomy the animals most different from humans, naturalists like Gilbert White believed that their social habits often had something to teach us.[27] In the preface to the fourth volume of *American Ornithology*, Alexander Wilson expressed his confidence that by "entering minutely into the manners of this beautiful portion of the animate creation . . . sentiments of esteem, humanity and admiration will necessarily result" (*AO* 4: vi). The marvelous intricacy of birds' nests, the monogamous faithfulness some birds show to their mates, the devoted care they take of their young, the vast distances they uncomplainingly traverse during their annual migrations—all of these bore witness to the wonderful order of nature of which man was a part, too.[28] In less uplifting form, this is a lesson Audubon had first learned from the favorite author of his youth, Jean de La Fontaine, and the talkative birds, the caring larks, solicitous doves, and cautious swallows, that populated the French poet's works. If La Fontaine's fables—written, in the writer's own words, not for herons but for the benefit of humans—also featured hungry kites, rapacious hawks, deceitful magpies, and murderous vultures, this was supposed to make their often cynical lessons even more useful to his readers: "C'est n'est pas aux hérons / Que je parle: écoutez, humains, un autre conte; / Vous verrez que chez vous j'ai puisé ces leçons."[29]

Audubon's texts, however, considerably complicate such anthropocentric readings. In *Ornithological Biography*, birds fascinate and repel, charm and elude Audubon. Many passages, it is true, evoke their human-like behavior, the tenderness and care they invest in their family life. Ivorybills, for example, generally stay together in pairs and continue their "mutual attachment" monogamously through life (*BA* 4: 218). While courting his "lovely mate," full of ecstasy because "his caresses are kindly received," the Ruby-throated Humming-bird, with loaded bill, offers insects and honey to his future spouse and fearlessly gives chase to the bluebird and the martin (*BA* 4: 191). But in this anthropomorphic avian world shaped by familial love and loyalty, it is ironically the human intruder whose behavior frequently strikes us as less than admirable. Some birds in fact seem more human than humans. The parental affection of a Canada Goose, for example, is so great that it attacks Audubon twice as he approaches the nest, striking him with its wings on the arm, which he then, for a moment, thinks is broken (*BA* 6: 184–185). And the little Arctic Puffins of Labrador put their pursuers to shame, pitting instinctive tenderness and selfless sympathy against human cruelty and callous greed, as Audubon realizes (he also admits that he would rather leave it to the reader to guess how many birds he killed during his first visit to the island on which the puffins nested). "I observed with concern the extraordinary affection manifested by these birds towards each other," wrote Audubon, "for whenever one fell dead or wounded on the water, its mate or a stranger immediately alighted by its side, swam around it, pushed it with its bill as if to urge it to fly or dive, and seldom would leave it until an oar was raised to knock it on the head" (*BA* 7: 240).

While these passages emphasize the strange familiarity of birds, other sections in *Ornithological Biography* describe their unfamiliar strangeness. They evoke the spectacular beauty especially of water birds: Great White Herons in Louisiana stand perched like "so many newly finished statues of the purest alabaster, forming a fine contrast to the deep blue sky" (*BA* 6: 114), admiring their own reflections "dipping as it were into the smooth water" (*BA* 6: 122); elegant Trumpeter Swans, "giddy with delight," incline their necks in graceful curves as they glide through the smooth waters of a secluded pond and make Audubon feel happier and more "void of care" than he can describe (*BA* 6: 222); wary Mallards elevate their heads, "glittering with emerald-green," and shine their beautiful "amber eyes" at the human intruder (*BA* 6: 237); flocks of Blue-winged Teals take flight in clear sunny weather, their vivid colors resembling "the dancing light of a piece of glass suddenly reflected on a distant object" (*BA* 6: 290).

Some of these amazing birds, it is true, suffer Audubon to keep them as pets: a little owl, for example, traveled in Audubon's pocket from Philadelphia to New York; it remained "generally quiet," ate out of his hand, and apparently never even attempted to escape (*BA* 1: 148). Audubon's pet Sparrow-Hawk (American kestrel) "Nero," adopted after he had fallen out of his nest, remained during his brief life in civilization unfailingly kind to his master, returning home every night to his favorite roost behind a shutter of Audubon's house (*BA* 1: 93–94). More usually, wherever Audubon goes, the birds do not follow him but are already there, ahead of him, having traveled their own inscrutable routes "with the quickness of thought," to use one of his favorite phrases.[30] In New York, for example, pigeons have been killed with their crops still full of rice from the fields of Georgia and Carolina, which means that they must have traveled "between three and four hundred miles in six hours." Such birds, muses Audubon, were they "so inclined," could visit Europe in less than three days (*BA* 5: 26).

Their America is vastly different from his. It knows no political and hardly any geographical boundaries, and whenever the habitations of humans appear in the backgrounds of Audubon's plates, they are miniaturized into insignificance. "Frantic with delight" (*BA* 7: 141), Audubon follows the movements of his birds around from "the mouth of the Mississippi . . . to Labrador" (*BA* 7: 146).[31] Thus, his *Birds of America*, plates and text alike, turns into an extended description of a continent still new to him, an often lyrical narrative dictated not by the needs and greeds of humans but by the undulating motions of speedy birds. Audubon's birds enter and leave America at will, defying, as Audubon recognizes not without satisfaction, the attempts of science to determine their typical habitats and thus to establish reliable classifications:

> It is difficult, for me at least, to understand how we should now have in the United States so many birds which, not more than twenty years ago, were nowhere to be found in our country. . . . That birds should thus suddenly make their appearance, and at once diffuse themselves over almost the whole of the country, is indeed a very curious fact; and were similar changes to take place in the other tribes of animals, and in other countries, the arrangements of systematic writers would have to undergo corresponding revolutions, a circumstance which would tend to add to the confusion arising from the continual shiftings, combinations, disseverings, abrasions of names, and alterations of method, which the interpreters of nature are pleased to dignify with the name of science. (*BA* 1: 212)

But then, Audubon somewhat distrusts the new tags naturalists give to birds anyway. Only "closet-naturalists" hope to found their fame on "the invention of useless names"

(*BA* 4: 75). Many birds, he says, are "strangely named," no less "in *pure* Latin, than in English, French and Dutch; and very many are every year receiving names still stranger than those they bore." Audubon professes to be "a kind of conservative" in the naming business, adhering to those terms that have been around for a long time (*BA* 4: 282). Whenever there is a chance, he takes care to mention the popular names of his birds, mostly those that are current in Louisiana: the Golden-winged Woodpecker (the northern flicker), he says, is "usually called *Pique-bois jaune* by the French settlers in Louisiana" (*BA* 4: 282); the Creoles of lower Louisiana refer to the Yellow-crowned Night Heron as "Cap-cap" (*BA* 6: 89) and call the Gadwall Duck, "on account of the whistling sound of its wings," "Violon" (*BA* 6: 254), while their name for the Buffel-headed Duck is "the Marionette" ("I think the name a pretty one"; *BA* 6: 370). These folksy designations are, of course, attempts at description rather than definition, but at least they come without the pretensions of permanence.

Some of Audubon's birds are in fact legitimate residents of the United States; others, however, like the Forked-tailed Flycatcher, appear here "as if they had lost themselves" (*BA* 1: 196). Audubon's birds are fiercely independent, and their fragile beauty is contingent on their liberty, an attitude with which Audubon, son of the New World that gave him "birth and liberty" (*BA* 1: 55), fully sympathizes.[32] In captivity, Audubon's birds refuse food (*BA* 7: 221); their plumage rapidly loses its brilliance, and they die. Alternatively, like the domestic ducks, they become pathetic slaves and lose their "native spirit": "Look at that Mallard as he floats on the lake. . . . How brisk are all his motions compared with those of his brethren that waddle across your poultry-yard! how much more graceful in form and neat in apparel!" (*BA* 6: 237).

In an autobiographical sketch entitled "Myself," written in 1835 but published posthumously, Audubon recounts an episode that he claims conditioned his later obsession with the beauty of birds: "One incident which is as perfect in my memory as if it had occurred this very day, I have thought of thousands of times since, and will now put on paper as one of the curious things which perhaps did lead me in after times to love birds, and to finally study them with pleasure infinite" (*AJ* 1: 8). His stepmother, Audubon explains, possessed a "monkey," "a full grown male of a very large species," as well as several "beautiful parrots," one of which was wont to ask for his breakfast in French. The details of the following scene sound like Poe's "The Murders in the Rue Morgue," though here we know right from the beginning who the murderer is:

One morning, while the servants were engaged in arranging the room I was in, "Pretty Polly" asking for her breakfast as usual, *"Du pain au lait pour le perroquet Mignonne,"* the man of the woods probably thought the bird presuming upon his rights in the scale of nature; be this as it may, he certainly showed his supremacy in strength over the denizen of the air, for, walking deliberately and uprightly toward the poor bird, he at once killed it, with unnatural composure. The sensations of my infant heart at this cruel sight were agony to me. I prayed the servant to beat the monkey, but he, who for some reason preferred the monkey to the parrot, refused. I uttered long and piercing cries, my mother rushed into the room, I was tranquillized, the monkey was forever afterward chained, and Mignonne buried with all the pomp of a cherished lost one. (*AJ* 1: 8)

But *do* we know who the murderer is? "American woodsman" was one of Audubon's preferred designations for himself; on the other hand, "man of the woods," the phrase used in our text, was a common eighteenth- and nineteenth-century translation of the Latin name for the "orangutan," *Homo sylvestris*. And this is indeed how orangutans (or the other anthropoid apes that went by that name) appear in popular natural history books of the time; for example, in a woodcut from Thomas Bewick's *General History of Quadrupeds* (1790), the "Wild-Man of the Woods" is represented as sitting on a bench and clutching a walking stick.[33] Even if his behavior is markedly less civilized, Audubon's "man of the woods" likewise acts as if he were human, since his murderous act is clearly intentional (he "thought the bird presuming upon his rights in the scale of nature"). But this is exactly what makes him so "unnatural" and ultimately despicable, too. Audubon's Polly is as vulnerable as she is pretty, a "denizen of the air" and, what is more, aristocratic by nature, as is shown by her habit of asking the servants for her breakfast and by her final "pompous" burial. Audubon's text suggests that nobility prefers birds who themselves prefer to be noble, whereas servants have an affinity for monkeys.

If this richly emblematic but ultimately self-contradictory scene illustrates a desire to identify with the "poor birds" that can be found in Audubon's water-colors as well as in the texts gathered in *Ornithological Biography*, it also reflects the violence used by the "woodsman" and hunter Audubon in his professional dealings with birds. In *Ornithological Biography*, Audubon cheerfully walks his readers through a landscape strewn with the feathered corpses. His "biographies" are studies as much of the last agonies of birds as of their habits in life. All too often, he assesses the abundance of a species in the United States by the number of individuals he has been able to shoot: seven Short-eared Owls in the course of one morning (*BA* 1: 140); "more than sixty" Lapland Lark-Buntings (Lapland longspurs) in just a few minutes (*BA* 3: 50); hundreds of Red-winged Starlings (red-winged blackbirds) "in the course of an afternoon, killing from ten to fifteen at every discharge" (*BA* 4: 33). Few birds escape "the clutch of the collector" (*BA* 3: 215). The word "draw" in "drawn from nature," the seal of quality that

Audubon put on all the watercolors from which Robert Havell in London produced his plates, is not just accidentally ambiguous: to draw means "to sketch," of course, but it can also, in the sense of "draw from" or "withdraw," signify the act of removing something from its original, in this case natural, context. Art is what nature is not—or, rather, is no longer. Almost all the birds represented in his books, bragged Audubon in his introduction to *Ornithological Biography,* had been killed by himself (*OB* 1: xii).

When Robert Penn Warren was asked what had inspired him to write his cycle of poems *Audubon: A Vision* (1969), he said that "Audubon was the greatest slayer of birds that ever lived: he destroyed beauty to create beauty."[34] A similar view of Audubon is proposed in a story Robert Penn Warren knew and wrote about,[35] Eudora Welty's "A Still Moment" (1942), in which the fictional character Audubon kills the bird he loves because this is the only way he can paint it: "In memory the heron was all its solitude, its total beauty. All its whiteness could be seen from all sides at once, its pure feathers were as if counted and known and their array one upon the other would never be lost. But it was not from that memory that he could paint."[36] Among the three characters of the story—Audubon, the Methodist minister Lorenzo Dow, and the outlaw James Murrell—the naturalist is the only one who can truly appreciate the beauty of the snowy heron, and yet he can see the bird "most purely" only "at its moment of death." Art (or science), it seems, vindicates the ruthlessness it requires.

Some commentators have found such and similar arguments more than just a little troubling[37] and a recent critic of Welty's fiction, from whom I have borrowed the previous sentence, hastens to assure us that Welty herself identifies not with Audubon's procedures but with the "living being of the heron."[38] The ideal observer of nature, said naturalist John Burroughs in 1908, "turns the enthusiasm of the sportsman into the channels of natural history, and brings home finer game than ever fell to shot or bullet." He, too, "has an eye for the fox and the rabbit and the migrating water-fowl," but he sees them "with loving and not with murderous eyes."[39] Audubon's defenders would, of course, argue, that he simply "had to kill in order to paint accurately," as Ben Forkner put it.[40] And they could with some justification quote a statement made earlier in John Burroughs's career—namely that a "bird in the hand is worth half a dozen in the bush" and that, for the sake of "sure and rapid progress" in science, lives need to be taken and specimens procured.[41]

Granted, a scientist needs specimens. But did Audubon really have to obtain birds by, literally, the basketful? A few shots, he brags in his chapter on the Carolina Parrot (the Carolina parakeet), procure plenty of parakeets; we can only guess at how many birds had to die so that only four of them could be represented accurately in Audubon's plate (*BA* 4: 307). Sometimes all that killing seems to be in vain, too. In his chapter on the American Golden-crested Kinglet (the golden-crowned kinglet), Audubon remarks, casually: "We killed a great number of them in hopes of finding among them some individuals of the species known under the name of *Regulus ignicapillus,* but in this we did not succeed" (*BA* 2: 165). The shotgun, to be sure, is a rough collector's tool. And as tidy as taxonomic categories appear on the page, scientific fieldwork isn't neat, Audubon seems to say. Real naturalists, as opposed to those hiding in the closet, have blood on their hands. To do his work well, Audubon had to become a mass murderer: bird plumages "within any given species vary widely, individual to individual, sex to sex, age to age, and season to season."[42]

Several passages in *Ornithological Biography* propose that the *quality* of the ornithologist's insights depends on the *quantity* of specimens he is able to obtain. For example, Audubon recommends that, if we want to find out whether or not the age of birds is related to the number of broods they produce in a season, we should shoot as many of the breeding birds as possible and should then determine their age through "either bending or breaking their bones, or tearing asunder their pectoral muscles, which will be found harder or tougher in proportion to their age" (*BA* 3: 19). In his biography of the Zenaida Dove, Audubon points out that the nineteen birds he shot gave him an opportunity for a thorough "internal and external examination . . . which enabled me to understand something of their structure" (*BA* 5: 12).[43] Science might seem like a blanket excuse for such acts as depriving little sandpipers of their precious eggs, but it is the best one Audubon can offer: "I was truly sorry to rob them of their eggs, although impelled to do so by *the love of science,* which affords a convenient excuse for even worse acts" (*BA* 5: 282; my emphasis).

But not only the love of science impels Audubon; he likes an appreciative audience, too. In his introduction to *Ornithological Biography* he informs his "kind reader" that these pages were written "with no other wish than that of procuring one favourable thought from you" (*OB* 1: v). And, alas, the only way to procure the reader's sympathy, as it were, is by procuring specimens. Writing about the Florida Cormorant (the double-crested cormorant), Audubon assures us that he killed as many birds as he did only so that we, the readers, could better appreciate the bird's true beauty in the illustration and read a description of its habits that we know is scrupulously exact. In order to wield his pen, so the argument runs, Audubon had to wield the gun first. "You must try to excuse these murders, which in truth might not have been nearly so numerous, had I not thought of you quite as often while on the Florida Keys, with a burning sun over my head, and my body oozing at every pore, as I do now while peaceably scratching my paper with an iron-pen, in one of the comfortable and quite cool houses of the most beautiful of all the cities of old Scotland" (*BA* 6: 432).

However, this chapter is not really concerned with an evaluation of Audubon's hunting practices. I am more interested in the image of himself that Audubon projects in his work and in the significant part this persona plays in his poetics. And this image is not as simple and conclusive as Audubon's apologists would have us believe. Rather, it is informed by a strange mixture of release and remorse that makes Audubon, as a character in his own text, almost as memorable as the birds he describes. For starters, this character knows that killing birds is wrong: "Who can approach a sitting Dove," he asks, "hear its notes of remonstrance, or feel the feeble strokes of its wings, without being sensible that he is committing a wrong act?" (*BA* 5: 9). Guiltily, he doesn't want to share with his reader the number of Canada Geese he kills in one day—it "would seem to you so very large that I shall not specify it" (*BA* 6: 192). And he admits that in his case the sheer temptation to kill has sometimes overwhelmed all scientific considerations. Hunting ibises in a bayou, Audubon is disappointed to find them "too wild to be approached." His longing gaze meets a flock of small swallows, and Audubon is suddenly seized by the desire to get a hold of these more readily available creatures: "How it happened I cannot now recollect, but I thought of shooting some of them, perhaps to see how expert I might prove on other occasions. Off went a shot, and down came one of the birds, which my dog brought to me between his lips. Another, a third, a fourth, and at last a fifth were procured" (*BA* 1: 193–194).

Audubon intended his *Ornithological Biography* as a guide for gourmets and "sportsmen" as well as for "students of nature" (*BA* 5: 146; 6: 281). He knew that in his quest for dead birds he was not alone: "Thousands of persons besides you and myself are fond of Woodcock shooting," he reminds us, and "once you have acquired the necessary . . . dexterity, you may fire, charge and fire again from morning till night" (*BA* 6: 19–20). Since birds, especially those that are breeding, are forever on the alert, what is crucial to the hunter's success is his ability to hide himself. Many of the narratives in *Ornithological Biography* take the characteristic form of a competition for the best hiding place, with the hunter ambushing his victims from a thicket or while lying in a ditch, whereas the birds, for their part, are trying to conceal themselves or their young as best as they can. Some birds even derive entertainment from eluding the hunter's sight, indulging in an exasperating avian form of hide-and-seek with their pursuer Audubon. The Pileated Woodpecker, for instance, frequently

found delight in leading you a wild-goose chase in pursuit of it. When followed it always alights on the tallest branches or trunks of trees, removes to the side farthest off, from which it every moment peeps, as it watches your progress in silence; and so well does it seem to know the distance at which a shot can reach it, that it seldom permits so near an approach. Often when you think the next step will take you near enough to fire with certainty, the wary bird flies off before you can reach it. . . . For miles have I chased it from one cabbage-tree to another, without ever getting within shooting distance, until at last I was forced to resort to stratagem, and . . . concealed myself. (*BA* 4: 227)

Canada Geese, too, are cunning, notes Audubon, who has seen them evade the hunter by silently moving "into the tall grasses by the margin of the water," where they lower their heads and lie "perfectly quiet" until the boat has passed by (*BA* 6: 188). The Loon prefers diving to flying off and on occasion even sinks backwards into the water, "like a grebe or frog," until he reaches "some concealed spot among the rushes," where he remains "until your eyes ache with searching" (*BA* 7: 288).

Most of these narratives of concealment in *Ornithological Biography*, however, inevitably turn into narratives of (conspicuous) consumption. Audubon is not partial to the flesh of the Pileated Woodpecker (tough and smells of worms) or the Loon (tough, rank, and dark), but the Canada Goose certainly affords "excellent eating" (*BA* 6: 190). In fact, many birds are tasty, and Audubon's biographies are filled with references to the availability and the current market value of birds: in the South, robins, because there is an ample supply of them, are very cheap (*BA* 3: 15); all over the United States, the markets are well supplied with bobolinks ("and the epicures have a glorious time of it"; *BA* 4: 13), while it is, except in New York and Philadelphia, more difficult to obtain Golden-winged Woodpeckers (whose flesh Audubon finds impalatable anyway; *BA* 4: 285). Long-billed Curlews are sold throughout the year in the markets of Charleston for "about twenty-five cents the pair" (*BA* 6: 38); in eastern markets, a pair of Velvet Ducks (white-winged scoters) costs between fifty cents and a dollar (*BA* 6: 334), while a pair of Pinnated Grouse (prairie-chickens), because they are rare, might sell for as much as ten dollars in Philadelphia, New York, and Boston (*BA* 5: 95), and so on.

However, in his chapter on the the Saltwater Marsh Hen (clapper rail), Audubon paints a grim portrait of a "sportsman" at work in the wetlands of South Carolina shooting edible birds in quantities that clearly exceed his need. While the killings benefit the gunner himself and his family, they inflict pain on other families, namely those of the birds. Here, in a deliberate anthropomorphizing move, Audubon uses the term "relatives" with reference to birds, establishing a disingenuous parallel between the joyous feelings of the hunter standing before the heaps of dead birds and the joyful feelings of those who have survived the carnage:

It is a sorrowful sight, after all: see that poor thing gasping hard in the agonies of death, its legs quivering with convulsive twitches, its bright eyes fading into glazed obscurity. In a few hours, hundreds have ceased to breathe the breath of life . . . , The cruel sportsman, covered with mud and mire, drenched to the skin by the splashing of the paddles, his face and hands besmeared with powder, stands amid the wreck which he has made, exultingly surveys his slaughtered heaps, and with joyous feelings returns home with a cargo of game more than enough for a family thrice as numerous as his own. How joyful must be the congratulations of those which have escaped, without injury to themselves or their relatives! With what pleasure, perhaps, have some of them observed the gun of one of their murderers, or the powder-flask of another, fall overboard! (*BA* 5: 169)

Killing birds, in Audubon's descriptions, often appears as the basest of acts. Yet, as many other passages suggest, murder one must, if not for the gratification of one's palate or for the sake of amassing scientific knowledge, then simply so that one can allay the mysterious "nervous anxiety" that, from time to time, besets human beings. However, as if prodded by his bad conscience, Audubon occasionally ceases to be merely the "two-legged monster, armed with a gun" (*BA* 6: 19) and casts himself in the role of the savior of birds in their time of need. For example, there is the female woodcock, harassed by "a pack of naughty boys," who would have been killed had it not been for Audubon's timely intervention on her behalf (*BA* 6: 15–16). And in Louisiana, Audubon successfully saves a wounded ibis from the relentless attacks of an alligator ("the Ibis, as if in gratitude, walked to our very feet, and there lying down, surrendered itself to us"; *BA* 6: 59).

The image of Audubon the inexorable procurer of birds is further complicated by the fact that birds in their natural state are often quite violent, too, or, to use one of Audubon's favorite terms in *Ornithological Biography,* "pugnacious" (*BA* 3: 65, 94, 117, 123, 199). Notice how Audubon tries to enlist the reader's sympathies in the following description of a Red-tailed Hawk's depredations:

The lively squirrel is seen gaily leaping from one branch to another, or busily employed in searching for the fallen nuts on the ground. It has found one. Its bushy tail is beautifully curved along its back, the end of it falling off with a semicircular bend; its nimble feet are seen turning the nut quickly round, and its teeth are already engaged in perforating the hard shell; when, quick as thought, the Red-tailed Hawk, which has been watching it in all its motions, falls upon it, seizes it near the head, transfixes and strangles it, devours it on the spot, or ascends exultingly to a branch with the yet palpitating victim in his talons, and there feasts at leisure. (*BA* 1: 34)

Hawks not only exterminate nimble rodents; they also steal, "on all occasions," food from each other and attack their mates with "merciless fury" (*BA* 1: 36). Worse still, they never know when to stop, as Audubon notes with disgust in the biography of the Red-shouldered Buzzard (the red-shouldered hawk): "The eating of a whole squirrel, which this bird often devours at one meal, so gorges it, that I have seen it . . . with such an extraordinary protuberance on its breast as seemed very unnatural, and very injurious to the beauty of form which the bird usually displays" (*BA* 1: 41).

Eat and be eaten—in order to be consumed, birds have to be good consumers themselves. A fastidious eater himself,[44] Audubon prefers those birds that are choosy about their fare; the rule of thumb is that birds that feed on fish or reptiles have a "fishy taste" themselves and are, like the Roseate Spoonbill, "oily and poor eating" (*BA* 6: 85, 76), while birds that feast on berries and fruits, like the Cedar Waxwing, are so "tender and juicy as to be sought by every epicure for the table" (*BA* 4: 169). But more often than not, birds are simply, sad to say, not "nice" in their eating habits.[45] Some birds eat so much that, gorged with food, they are incapable of flying away when humans approach (*BA* 6: 396). And not only do birds overeat—they eat what often defies description: vultures eagerly follow carts loaded with offal or the carcasses of dead horses around the city and lurk in the vicinity of the slaughterhouses (*BA* 1: 17–18); chickadees break the skulls of other birds and "feed upon their flesh" (*BA* 2: 148); oystercatchers devour "the hard particles of shells, pebbles, and other matters" along with their regular food, oysters, crabs, and sea-worms (*BA* 5: 239, 237); mallards are so omnivorous and greedy that they swallow "any kind of offals, and feed on all sorts of garbage, even putrid fish, as well as on snakes and small quadrupeds" (*BA* 6: 242); pintails feed on tadpoles in spring and on leeches in autumn and, should the opportunity arise do not spurn a dead mouse either (*BA* 6: 267); black-backed gulls, gluttons through and through, satisfy their "ever-craving appetite" with the "helpless young" of other birds (*BA* 7: 173).

With horror, Audubon recalls an episode in which two mallards were fighting for "the skin of an eel, which was already half swallowed by the one, while the other was engaged at the opposite end" (*BA* 6: 242). One of Audubon's friends shot and wounded a grouse, pursued it into the bushes, and found two blue jays already engaged in "picking out its eyes" (*BA* 4: 110). And there was the pair of herons roosting in John Bachman's garden: at first they merely devoured chickens and ducks, but then they graduated to pursuing the younger children in the house, whereupon they had to be killed (*BA* 6: 117).

Perhaps the most memorable example of avian brutality, however, is presented in the biography of the White-headed Eagle (the bald eagle), which Audubon thinks should never have been selected as "the Emblem of my Country" (*BA* 1: 63):[46] Audubon reports an incident that took place in Natchez, Mississippi, where several vultures were observed dismantling the corpse of a horse. When a White-headed Eagle approached, all vultures immediately took to wing, but one of them had managed to ingest only part of the horse's entrails and flew off with the remaining part, "about a yard in length, dangling in the air." Helplessly waving its sorry flag, the vulture inevitably attracted the eagle's attention. In Audubon's story, the eagle now commenced to prey on predator and prey alike. Struggling to divest itself of the treacherous leftovers, "the poor Vulture tried in vain to disgorge, when the Eagle, coming up, seized the loose end of the gut, and dragged the bird along for twenty or thirty yards, much against its will, until both fell to the ground, when the Eagle struck the Vulture, and in a few moments killed it, after which he swallowed the delicious morsel" (*BA* 1: 59). One thinks of the hawks and vultures in one of La Fontaine's fables who, engaged in bitter warfare over the rotting carcass of a dog, mercilessly attack each other and, when pigeons come along to intervene, kill and eat those, too. Let the bad ones fight their own wars, is La Fontaine's cynical lesson, and the rest of the world will remain safe.[47] In Audubon's sinister tale, however, even such a pessimistic moral is not immediately apparent.

Audubon's descriptions of the natural habitats of his "voracious" birds are filled with references to the foul odors emanating from the wasted food and excrements to be found there. He recoils from the memory of a trip to the breeding grounds abandoned by the Night Heron: "The stench emitted by the excrements with which the abandoned nests, the branches and leaves of the trees and bushes, and the ground, are covered, the dead young, the rotten and broken eggs, together with putrid fish and other matters, renders a visit to these places far from pleasant" (*BA* 6: 85). On Sandy Key, Florida, the breeding ground of the White Ibis, where the vegetation consists of nothing but thousands of wild plum trees, cacti the size of a man's body, and "the rankest nettles I ever saw," all of them tangled up to make the passage for humans difficult, the odor of the putrid eggs and of the ibis's "natural effluvia" was simply overwhelming (*BA* 6: 55). But the worst assault on Audubon's olfactory sense came during a visit to the Guillemots on Labrador. Never mind that part of the carnage was Audubon's own doing—his men who had invaded the island to pilfer eggs were assisted in their efforts by other birds, "rapacious Gulls": "Eggs, green and white, and almost of every colour, are lying thick over the whole rock; the ordure of the birds mingled with feathers, with the refuse of half-hatched eggs partially sucked by rapacious Gulls, and with putrid or dried carcasses of Guillemots, produces an intolerable stench; and no sooner are all your baskets filled with eggs, than you are glad to abandon the isle to its proper owners" (*BA* 7: 269).

Other, similar experiences could be added. In one of his most haunting chapters, Audubon and his friend Bachman, having fought their way through dark thickets of undergrowth, "tangled with vines and briars," finally arrive at the roosting place of the Carrion Crows of Charleston (note the funny, if unintended, visual alliteration), a swamp about two miles from the city. They are promptly overcome with nausea. Audubon's text evokes a truly surreal scene of universal slaughter, in which the lethal intentions of the human hunters mix with the desire of the birds to feed on the corpses of their companions. This swampy wasteland, inhabited by thousands of birds of prey perched on dead trees, is hardly the pristine American wilderness so often associated with Audubon's travels:[48]

> We found the ground destitute of vegetation, and covered with ordure and feathers, mixed with the broken branches of the trees. The stench was horrible. The trees were completely covered with birds, from the trunk to the very tips of the branches. They were quite unconcerned; but, having determined to send them the contents of our guns, and firing at the same instant, we saw most of them fly off, hissing, grunting, disgorging, and looking down on their dead companions as if desirous of devouring them. . . . The piece of ground was about two acres in extent, and the number of Vultures we estimated at several thousands. (*BA* 1: 18)

The fact that these uncouth birds, weather permitting, rise early in order to leave their hellish habitats and commute to the nearby city serves as an uncomfortable reminder of how closely the lives of men and the lives of even the most despicable birds are intertwined: "During very wet weather, they not unfrequently remain the whole day on the roost; but when it is fine, they reach the city every morning by the first glimpse of day" (*BA* 1: 18–19).

Considering the evidence of frequent murder in his work, with men killing birds by the thousands and birds feeding freely on the bodies of other birds, it is hard to believe that Audubon has often been hailed, even recently, as affirming that "man has more grace to gain than to lose by recognizing his eternal kinship with the birds and beasts."[49] A closer look at the (surprisingly coherent) narrative that binds the various volumes of *Ornithological Biography* together shows that Audubon's attitude is more conflicted, more tortured, more contradictory than such cheerful readings allow: he appears alternately as the murderer and the protector of birds, as their ardent admirer as well as their inveterate destroyer, "panting with heat and anxiety" (*BA* 6: 26), a kind of lethal father figure. In a particularly memorable scene from *Ornithological Biography*, we see Audubon hurling freshly killed birds into the air so that his "voracious" pet kestrel Nero can catch them before they fall to the ground (*BA* 1: 93). Audubon's pervasive paradoxes make him an unlikely identification figure but, arguably, a more interesting writer, as the following example confirms.

Apologists for Audubon's industrial-style killings have pointed out that he "was a man of his time," ignoring the fact that even in those days "in which marksmanship with living targets was an admired sport in itself"[50] there were some dissenting voices. The first laws for the protection of game were introduced at the time that Audubon was working on the octavo edition of *The Birds of America*. An early New York law of 1838, later repealed, prohibited the use of multiple guns in the killing of waterfowl; in 1844, the New York Association for the Protection of Game was established; a few years later Rhode Island outlawed the spring shooting of wood duck, black duck, woodcock, and snipe.[51] As early as 1788, Gilbert White, once an avid sportsman himself and not loath to cut up a cuckoo so that he could examine its stomach ("hard like a pin-cushion with food"), had spoken out against killing, "wantonly and cruelly," birds with young.[52] Closer to home, in James Fenimore Cooper's novel *The Pioneers*, published in 1823, Leather-stocking forcefully distanced himself from the indiscriminate havoc wrought by the inhabitants of Templeton on the millions of migrating passenger pigeons flying through their valley in the course of their northward migration. Cooper's novel was immensely popular after its publication in 1823, and there is some evidence that Audubon modeled his own appearance during his European subscription tour (long flowing hair, loose-fitting frontier garb, and leather stockings) after the image of Cooper's Natty Bumppo.[53] In the relevant scene from chapter 22 of *The Pioneers*, the whole village is assembled, and shots, arrows, and "missiles of every kind" are fired at the birds, which are so numerous that they darken the sky "like a cloud."[54] Even an old cannon is used in the concerted assault on the "immense masses" of pigeons that continue to arrive in the valley. In these horrific proceedings, considered a "princely sport" by the sheriff, Leather-stocking is an "uneasy spectator." He disapproves of "these wasty ways that you are all practysing." His sober motto is "Use, but don't waste," and he proves his own thrifty regard for nature's resources by demonstrating that a single carefully aimed shot is sufficient to bring down one bird, "without touching the feather of another" (the bird is brought to Leather-stocking "still alive").[55]

Audubon's own version of the slaughter of passenger pigeons in *Ornithological Biography* seems like an intentional revision of the scene from Cooper's novel, but it is a rewriting riddled with paradox. His account of a similar "scene of uproar and confusion" (*BA* 5: 29) that took place at one of the roosting places of the passenger pigeon, at the Green River in Kentucky, is more sinister, more apocalyptic, than Cooper's rendering, which for all its dark overtones ends with the anthropomorphizing assertion that on this day nearly as many pigeons had been killed as Frenchmen on the occasion of General Rodney's victory over Admiral de Grasse.

Audubon arrives at the scene when the first round of killings is already over and a "great number of people" have assembled. Sitting amidst "large piles" of dead birds, they are waiting for more flocks to come. If Cooper's attention was on the actions of the people, Audubon evokes the dismal landscape in which the killings take place. The ground is covered with dung, "several inches deep." Some of the largest trees, "two feet in diameter," have been broken off, "as if the forest had been swept by a tornado" (presumably to make it easier for the gunners to take aim). When the birds finally arrive, they come in a great rush of air and with a noise that reminds Audubon of gales he experienced at sea. (In an earlier description of the flight of pigeons from the same chapter, Audubon writes that "the light of noon-day was obscured as by an eclipse; the dung fell in spots, not unlike melting flakes of snow" [*BA* 5: 27].)

While the birds continue to pour in, settling down on the denuded, mutilated tree branches, which snap under their weight and kill the other birds perched below, a "magnificent" and "terrifying" sight presents itself to the observer. In the general din and confusion, heroic acts of individuals—such as Leather-stocking's "wonderful exploit," the shooting of a pigeon on the wing—have become quite impossible.

> The Pigeons, arriving by thousands, alighted everywhere, one above another, until solid masses were formed on the branches all round. Here and there the perches gave way under the weight with a crash, and, falling to the ground, destroyed hundreds of the birds beneath, forcing down the dense groups with which every stick was loaded. . . . I found it quite useless to speak, or even to shout to those persons who were nearest to me. Even the reports of the guns were seldom heard, and I was made aware of the firing only by seeing the shooters reloading. (*BA* 5: 29)

Cooper's scene ends with the industrious gathering of the dead pigeons, which are loaded on horses and carted away, but Audubon lets the "uproar" continue through the night, with the sounds of the slaughter to be heard miles away: "I sent off a man, accustomed to perambulate the forest, who, returning two hours afterwards, informed me he had heard it distinctly when three miles distant from the spot" (*BA* 5: 30). When the steady stream of prospective victims finally subsides, wolves, foxes, lynxes, cougars, bears, raccoons, opossums, polecats, eagles, hawks, and buzzards start helping themselves to their share of the booty. Only afterwards, the gunners enter among the "dead, the dying, and the mangled," piling them up in heaps, "until each had as many as he could possibly dispose of, when the hogs were let loose to feed on the remainder." Again, there is a clear indication of gratuitous waste: the number of birds killed clearly exceeds the capability of the killers to carry the corpses. But Audubon does not use this realization as a starting point for a critique of human destructiveness. Pigeons, he assures his readers, "not un-frequently quadruple their numbers yearly," and just recently he has seen them "so abundant in the markets of New York, that piles of them met the eye in every direction" (*BA* 5: 30).

Audubon's account of the Green River massacre reappears in naturalist John Muir's autobiography, *The Story of My Boyhood and Youth* (1913). Muir, born in 1838, recalls reading Audubon's essay in grammar school while still in Scotland and describes how he himself witnessed the arrival of these "beautiful wanderers" in Wisconsin in the 1850s: "I have seen flocks streaming south in the fall so large that they were flowing over from horizon to horizon in an almost continuous stream all day long, at the rate of forty or fifty miles an hour, like a mighty river in the sky, widening, contracting, descending like falls and cataracts, and rising suddenly here and there in huge ragged masses like high-plashing spray." Decades later, the passenger pigeon, killed like so many other birds by "preaching, praying men and women," is no more than a memory for Muir: "Think of the passenger pigeons that fifty or sixty years ago filled the woods and sky over half the continent."[56]

For Audubon and his friends, America was still the land of plenty, at least as far as its birds were concerned. In 1833, for example, John Bachman wrote Audubon from South Carolina that he had just seen so many tree swallows that "the air was positively darkened. As far as the eye could reach, there were Swallows crowded thickly together. . . . There must have been many millions" (*BA* 1: 176). About the bobolink, Audubon himself remarked that, though millions of these birds were destroyed each year, "yet millions remain" (*BA* 4: 13). Audubon was well aware that some birds, such as the grouse, were "decreasing at a rapid rate" and were threatened with extinction, like the original inhabitants of the American continent.[57] "When I first removed to Kentucky, the Pinnated Grouse were so abundant, that they were held in no higher estimation as food than the most common flesh, and no 'hunter of Kentucky' deigned to shoot them" (*BA* 5: 94). In those days, grouse would walk in the very streets of the villages, enter the farmyards, and mingle with the poultry. Now they "have abandoned the State of Kentucky, and removed (like the Indians) every season farther to the westward, to escape from the murderous white man" (*BA* 5: 95). Carolina parakeets, likewise, were vanishing rapidly: "scarcely any are now to be seen. . . . I should think that along the Mississippi there is not now half the number that existed fifteen years ago" (*BA* 4: 309).[58] Sadly, even pelicans, in spite of their limited culinary value, were being driven, like the more coveted wild turkey, farther and farther away from their accustomed habitats, "until to meet with them the student of nature will have to sail round Terra del Fuego, while he may be obliged to travel to the Rocky Mountains before he find the other bird" (*BA* 7: 38). But if the "progress of civilization" is so harmful to the parakeet, the grouse, the turkey, the pelican, and the ivorybill, it does let the chimney swift and the cardinal thrive, as Audubon feels compelled to point out in other chapters.[59]

In spite of his frequent invocations of the Creator's benevolence and of the wonderful order He has instituted among the birds and the beasts in the New World, Audubon's universe in *The Birds of America* is, in the final analysis, dominated by waste. Man, "the tyrant of the creation" (*BA* 5: 31), stumbles around in it, often "knee-deep in the mire" (*BA* 6: 125), his clothes "smeared with the nauseous excrements of hundreds" of birds (*BA* 7: 45). Clutching his double-barreled shotgun, he seems like a helpless god of destruction—aware of his own baseness yet incapable of keeping his own lethal impulses at bay. But there are mitigating factors. Granted that Audubon's birds never kill their prey or each other on the same scale that Audubon kills them, the overall effect of Audubon's literary strategy is to make himself part of what he describes. John James Audubon the ornithologist, painter, and writer becomes "Audubon" the protector and slayer of birds, irascible, passionate and pugnacious as well as tender, considerate and caring, impressive as well as flawed, attractive as well as repulsive, approachable and then again elusive, like the birds he writes about and to which he would, only half in jest, occasionally compare himself.[60] If Audubon later became a popular literary figure,[61] this happened partly because, in his own writing, he had already provided an excellent blueprint. Thus, as a character, he lived on, not only as "Audubon" in Eudora Welty's fiction but also in the restless young hunter from Sarah Orne Jewett's story "A White Heron" (1886). He has been working on his "collection of birds" since he was a boy and therefore ruthlessly kills "the very birds he seemed to like so much," to the confusion of the nine-year-old girl Sylvy, who has a special affinity to "all sorts o' birds" and eventually refuses to help him. Not for all the money in the world can she "tell the heron's secret and give its life away."[62]

"As if Alive"

Having killed his heron, Eudora Welty's Audubon despairs that, "apart from his hands," the bird would always remain dead—"a dead thing and not a live thing, never the essence, only a sum of parts."[63] The fictional Audubon's despair grew out of the fears that haunted his historical counterpart. In his "Introductory Address" to *Ornithological Biography*, Audubon looked back on his first attempts at procuring birds: "The moment a bird was dead, however beautiful it had been when in life, the pleasure arising from the possession of it became blunted" (*OB* 1: vii). Audubon wished "to possess all the productions of nature," but he "wished life with them" (*OB* 1: vii). This paradoxical desire ultimately led him to develop new methods of representation. "My style is new and different from what has preceded me," Audubon noted in his diary in Edinburgh on 3 December 1826.[64] In an essay titled "My Style of Drawing Birds" (1831), Audubon reflected in more detail on the genesis of his own unique process. Having started with "some pretty fair signs for poulterers" and driven by a desire to represent nature as if it were alive, he would make "hundreds of outlines" of the birds he had caught: "I continued for months together, simply outlining birds as I observed them, either alighted or on the wing, but could finish none of my sketches. I procured many individuals of different species, and laying them on the table or on the ground, tried to place them in such attitudes as I had sketched. But, alas! they were *dead*, to all intents and purposes" (*AJ* 2: 523). Audubon's pencil had again given "birth to a family of cripples" (*OB* 1: viii).

After many detours he discovered a method by which he would mount freshly killed "specimens" on wooden boards and use wires to breathe "real" life into their limp corpses again and arrange them "according to my notions." The kind of nature that is represented here is actually a wire construction, and the drawings based on it consequently have to be seen as so many representations of what is itself already a representation:

> I was off to the creek, and shot the first Kingfisher I met. I picked the bird up, carried it home by the bill, sent for the miller, and bade him bring me a piece of board of soft wood. When he returned he found me filing sharp points to some pieces of wire. . . . I pierced the body of the fishing bird, and fixed it on the board; another wire passed above his upper mandible held the head in a pretty fair attitude, smaller ones fixed the feet according to my notions, and even common pins came to my assistance. The last wire proved a delightful elevator to the bird's tail, and at last—there stood before me the *real* Kingfisher. . . . This was what I shall call my first drawing actually from nature, for even the eye of the Kingfisher was as if full of life whenever I pressed the lids aside with my finger, (*AJ* 2: 524–525)

Audubon's method, which would shun the stuffed specimen yet rely on the wired model, usually required him to finish his drawings "at one sitting," often, as he adds, "of fourteen hours." In fact, Audubon's journals are punctuated by references to the "disgusting" smell emanating from the decomposing birds he had killed and the relief he felt when he had accomplished his task,[65] In his work, Audubon dealt daily with death, hoping to achieve life in his drawings. In retrospect, it does not seem to be a coincidence that, as Audubon also reports in "Myself," at the outset of his career, as a penniless artist in Kentucky, he would paint deathbed portraits so well that his fame spread throughout the entire county. A clergyman at Louisville, for example, "had his dead child disinterred, to procure a fac-simile of his face," which Audubon "gave . . . to the parents as if still alive, to

45. *Green-blue, or White-bellied Swallow,* by John James Audubon. 1824. Watercolor, gouache, and graphite on paper, 47.9 x 29.5 cm. Purchased for the Society by public subscription from Mrs. John J. Audubon. New-York Historical Society, 1863.17.100.

their intense satisfaction" (*AJ* 1: 36). Audubon's watercolors for the Double Elephant Folio Edition of *The Birds of America*,[66] as well as the plates Robert Havell made from them, actually reflect, right into the very details of their composition, their own ambivalent genesis, thus offering more than a glimpse into the artist's own workshop. In the words of Audubon's favorite author, Jean de La Fontaine, as quoted by Audubon himself in his *Mississippi River Journal: "A l'oeuvre on connoit L'Artizan"*—the artist by his work is known (*SJ* 137).[67]

Audubon himself believed that even the "most learned of this country" considered him "unrivaled in the art of Drawing," as he told his son Victor in 1826.[68] This judgment seems justified when we look at such an effective image as the one captured in the watercolor for DEF 98, painted in Philadelphia on 17 May 1824 (ill. 45). Two White-bellied Swallows (tree swallows) are attacking each other: frozen in midair, with their beaks barely but—and this is the point—*not really* touching in the very center of the picture, a compositional trick that lets the viewer experience the full force of the drawing. The visual correspondences between the birds' bodies, with their elegantly curved wings and tails, make them look like mirror images of each other, two aspects of the same body (belly and back) revolving around an unseen axis that cuts diagonally through the center of the picture. For all its lifelikeness, Audubon's arrangement also serves the purpose of "ocular demonstration" (*BA* 4: 215). It satisfies the scientist's curiosity in that it effortlessly catches in one flat image different aspects of birds of the same species, thus presenting us with an almost three-dimensional view of the specimen. At the same time, Audubon's watercolor also pleases, by its sheer economy of presentation, the viewer's aesthetic sensibilities.

The same purpose is achieved by another unseen diagonal, which reaches from the solid, life-sized egg in the lower left corner of the painting—which the viewer perceives as the composition's "point of departure," in Rudolf Arnheim's words[69]—to the light feather that can be seen drifting toward the middle of the picture's right margin. Indeed, the fighting of the swallows often has its roots in their nesting habits, which are metonymically represented by the egg and the feather that will line the nest. Both egg and feather might also, in a more symbolic way, represent the earth-bound—or, rather, nestbound—start of a bird's life and the ultimate goal of its maturity, the freedom to fly wherever it wants. Again, the constructedness of the image and the desire for naturalness seem to meet, but what the viewer should not forget is that their meeting point is exactly between the beaks of the two birds. The center of life in the painting is also the focal point of its potential violence.

Audubon never budged an inch from his decision to reproduce his birds life-sized, in monumental plates printed on sheets of paper over three feet high and over two feet wide. Some of Audubon's good friends were "much against" his plates "being the size of life," he wrote in his diary on 24 November 1826. "I must acknowledge it renders it rather bulky, but my heart was always bent on it, and I cannot refrain from attempting it so."[70] But even Audubon's generous format is too small for the tallest American bird, the Whooping Crane. In his painting of a Whooping Crane, done in late 1821 in New Orleans (ill. 46), Audubon represents the bird bending over, apparently so that he could keep it in the frame. In April 1822, he added two young alligators, as if to justify the bird's posture as an attempt to kill its prey.[71] But what looks like a reduction of size paradoxically leads to a reaffirmation of the sheer largeness of the bird; the tiny alligators seem like a visual joke added to underscore Audubon's own technical prowess. The crane's feet, by the way, are of approximately the same dimension as the entire body of the second young alligator, already lying prostrate on its back.

As is evident from the examples already cited, many of Audubon's compositions, in spite of their insistent commitment to the re-creation of life out of death, are very often dominated by the experience of death, or at least impending death. In his "Introductory Address" to *Ornithological Biography*, Audubon tried to defend himself against objections that he had not arranged his birds in systematic fashion:

I can scarcely believe that yourself, good-natured reader, could wish that I should do so; for although you and I, and all the world besides, are well aware that

46. *Whooping Crane*, by John James Audubon. 1821–1822; 1829–1833. Watercolor, oil paint, graphite, gouache, white lead pigment, black ink & pastel, with glazing on paper, laid on Japanese paper. 94.7 x 65.2 cm. Purchased for the Society by public subscription from Mrs. John J. Audubon. New-York Historical Society, 1863.17.226.

a grand connected chain does exist in the Creator's sublime system, the subjects of it have been left at liberty to disperse in quest of the food best adapted for them, or the comforts that have been so abundantly scattered for each of them over the globe, and are not in the habit of following each other, as if marching in regular procession to a funeral or a merry-making (*OB* 1: xix).

But despite the fact that Audubon's plates in the Double Elephant Folio come to us, as Adam Gopnik has said, in "democratic disorder,"[72] there are certain thematic and structural continuities that allow us to regard this "enormously gigantic Work" as the kind of "book" Audubon wanted it to be.[73] One of these continuities is violence: very frequently, we see birds hunting for or already mangling their prey, songbirds whose beaks close around small insects, water birds mauling fish or small reptiles. DEF 81, based on a drawing made in New Jersey in June 1829, depicts a Fish Hawk (osprey) flying with a large weakfish in its talons: the open mouth of the fish uncannily corresponds with the open beak of the bird, and the body of the fish forms a line that is roughly parallel to the feathers of the tail and the bird's left wing (ill. 47). In DEF 166 we see a Rough-legged Falcon (a rough-legged hawk) feeding on a smaller dead songbird in a dead tree (ill. 48). The neck of the small bird is bent downward and thus responds, in a kind of miniature visual

47. "Fish Hawk, or Osprey." Aquatint engraving by Robert Havell Jr. after John James Audubon. Ca. 1829. *The Birds of America* (Double Elephant Folio), pl. 81. Sheet: 97 x 65 cm. Courtesy, The Lilly Library, Indiana University, Bloomington.

48. "Rough-legged Falcon." Aquatint engraving by Robert Havell Jr. after John James Audubon. 1832. *The Birds of America* (Double Elephant Folio), pl. 166. Sheet: 97 x 65 cm. Courtesy, The Lilly Library, Indiana University, Bloomington.

echo, to the movement indicated by the curved neck of the hawk descending on the exposed breast of his victim.

DEF 171, one of Audubon's most effective images, represents two Barn Owls against the background of an uncanny nightscape, which was later added in the plate (ill. 49). One of the owls is grasping in its talons a dead ground squirrel, which the other one is trying to catch, as if in jest, with its wide-open beak, its masklike face contorted into a devilish grin. The parallelism of the upper bird's left wing and the lower bird's right is striking and gives the picture a powerful element of dynamism, which is offset, yet also enhanced, by the pendant body of the lifeless squirrel. The squirrel is not only the center of attention for both owls but also marks the slightly displaced central vertical axis of the composition of the picture which is as off-bal-ance, as "out of joint," as the artistically represented world of the night owls seems to be. The absence of foliage on the branches underscores the sinister effect of the plate.[74]

An even more unsettling watercolor, produced in 1829, features two black vultures about to feed on the head of a deer (ill. 50). The power of the picture derives at least in part from the fact that the two vultures and the head of their victim fuse into a gruesome triangle in the center of the plate, a heap of bird body and deer body, so to speak. The left bird's head is presented in profile, its eye fully visible, whereas the second bird's head (which Audubon later pasted onto the picture, along with the feet, after he had painted in the body) is slightly tilted and appears as if seen from above. Together with the deer's right eye, wide open in death, the eyes of the two birds form a second, smaller equilateral triangle, which is all the more powerful because it is so obviously off-center. This second triangle, the base and the right side of which

are indicated by the birds' beaks, effectively dramatizes the relations between the three animals and anticipates the violation of the deer's body that will soon take place. Notice also the bizarre correspondences between the right vulture's left talons and the deer's left antler which constitutes the bottom right angle of the central large triangle.

An interesting narrative dimension emerges from the changed sequence of Audubon's plates in the Royal Octavo edition of Audubon's work, his "people's *Birds of America*." Here, for the first time, the plates are arranged so that the families of birds (from the "Vulturine Birds" to the "Divers and Grebes") appear together. But occasionally Audubon seems to want us to see different forms of coherence between plates. The sequence of plates 6 and 7 constitutes a visual pun, which could be intentional on Audubon's part. Plate 6 (plate 376 in

49. "Barn Owl." Aquatint engraving by Robert Havell Jr. after John James Audubon. 1832. *The Birds of America* (Double Elephant Folio), pl. 171. Sheet: 97 x 65 cm. Courtesy, The Lilly Library, Indiana University, Bloomington.

50. *Black Vulture*, by John James Audubon. 1829. Watercolor, pastel, collage, blank ink, graphite, and charcoal with selective glazing on paper, laid on thin board. Purchased for the Society by public subscription form Mrs. John J. Audubon. New-York Historical Society, 1863.17.106.

51. "Common Buzzard." Lithograph by J. T. Bowen after John James Audubon. *The Birds of America* (Royal Octavo), vol. 1 (1840), pl. 6. Courtesy, The Lilly Library, Indiana University, Bloomington.

52. "Red-tailed Buzzard." Lithograph by J. T. Bowen after John James Audubon. *The Birds of America* (Royal Octavo), vol. 1 (1840), pl. 7. Courtesy, The Lilly Library, Indiana University, Bloomington.

DEF) shows a Swainson's hawk (called Common Buzzard by Audubon) descending on a frightened hare (ill. 51). In the Royal Octavo edition it is followed by plate 7, a representation of two Red-tailed Buzzards (red-tailed hawks) fighting over the corpse of a hare, which the female hawk is clutching in her left foot as she uses her right to fend off her male attacker (ill. 52). Again, Audubon is using compositional devices to depict a bird species from several possible angles, but what is more important in our context is that, to an untrained eye, the birds featured in the two plates appear to be similar. Through the sequence of these plates, a narrative continuity other than that established by the accompanying text is achieved, and we experience visually the shift between the killer and the victim, the hunter and the prey: the bird, which in one plate figures as the perpetrator, reappears in the next plate in the role of the potential victim.

The original watercolor for one of the most spectacular plates in the Double Elephant Folio includes what is generally believed to be Audubon's self-portrait (ill. 53). Audubon's engraver Havell probably eliminated it when he produced the plate simply for reasons of consistency: in *The Birds of America,* humans are only represented metonymically through buildings in the backgrounds of plates featuring, as a rule, waterfowl or shore-birds.[75] Audubon's painting is dominated by the central and, of course, life-sized representation of a Golden Eagle taking off with a snowshoe hare in his left talons. Below is a vista of massive snow-covered mountains. In the left corner of the painting we see a fallen tree stretching from one rock to another, across which a man is slowly and laboriously making his way with the corpse of a bird (most probably another Golden Eagle) as well as a gun strapped to his back. The dead bird's body has fused in a rather strange way with the man's body: its still extended wings and the barrel of the rifle, together with the man's spread-out legs, form a curious, almost starlike

53. *Golden Eagle,* by John James Audubon. 1833. Watercolor, pastel, graphite, black ink, and black chalk, with touches of gouache and selective glazing on paper, laid on card. 96.8 x 64.8 cm (paper). Purchased for the Society by public subscription form Mrs. John J. Audubon. New-York Historical Society, 1863.17.181.

geometric pattern, which seems to demonstrate the precariousness of the hunter's balancing act.

In the corresponding essay in *Ornithological Biography*, Audubon has a different story to tell about how he came to possess the specimen depicted in this water-color. In 1833 he bought a living Golden Eagle, caught in a trap set for foxes in the White Mountains, from Nathan E. Greenwood, one of the subscribers to *The Birds of America*.[76] Audubon takes his new acquisition home, protected with a blanket, in order to save the noble bird "in his adversity, from the gaze of the people" (*OB* 2: 464). Evidently, the gaze of the people does not include Audubon's own, because as soon as the naturalist is home he places the cage "so as to afford me a good view of the captive." As he was watching the eagle's eye, observing "his looks of proud disdain," Audubon suddenly felt toward him "not so generously as I ought to have done" (*OB* 2: 464). Notice how important the imagery of seeing and being seen now becomes in Audubon's account. Irritated but obviously also impressed by the bird's attitude of condescension, Audubon first toys with the idea of setting his Golden Eagle free again: "I several times thought how pleasing it would be to see him spread out his broad wings and sail away towards the rocks of his wild haunts" (which is *not* what we see in the picture). But finally the desire to paint the bird supersedes all these doubts: "Some one seemed to whisper that I ought to take the portrait of the magnificent bird, and I abandoned the more generous design of setting him at liberty, for the express purpose of shewing you his semblance" (*OB* 3: 464–465). The "semblance," the artistic transformation of the living bird into the dead material of art, obviously no longer needs to be guarded from the "gaze of the people."

While on the first day of his strange encounter with the bird Audubon still patiently observes the eagle's movements, the second day is devoted to a more focused activity: the determination of the best and most effective position in which the bird is to be portrayed. On the third day Audubon attempts to kill the bird so that it may become a better model for his art. What follows is a nightmarish scene that sends us again to the work of Edgar Allan Poe, most notably "The Tell-Tale Heart," first published a decade later. If in Poe's story the pale blue "vulture eye" of the old man is the reason the narrator wants to kill him,[77] the eagle's unflinching gaze, his "looks of proud disdain," prove equally unsettling to Audubon. The bird's cage is again covered with a blanket, but this time not in order to shield him from "the gaze of the people." Rather, the aim is to smother the eagle gradually to death in a carefully sealed small room into which Audubon introduces a pan of burning charcoal. Audubon's attempts to kill the Golden Eagle are all the more gruesome for their obvious clumsiness, an amateurism that is effectively dramatized when after several hours the naturalist "peeps" into the cage and meets the steadfast gaze of the bird, still alive after having been exposed for hours to the deadly charcoal fumes: "I waited, expecting every moment to hear him fall down from his perch; but after listening for *hours*, I opened the door, raised the blankets, and peeped under them amidst a mass of suffocating fumes. There stood the Eagle on his perch, with his bright unflinching eye turned towards me, and as lively and vigorous as ever!" (*OB* 2: 465).

The room is sealed again, and when around midnight Audubon takes, as he says, another "peep" at his victim, the Golden Eagle is still breathing, while Audubon and his son ironically feel as if *they* were going to die from the fumes in the room: "He was still uninjured, although the air of the closet was insupportable to my son and myself" (*OB* 2: 465). When the next day fails to bring the desired change, even though sulphur has now been added to the charcoal, Audubon resolves to kill the eagle by piercing his heart with a long, sharp piece of steel:

> Early next morning I tried the charcoal anew, adding to it a quantity of sulphur, but we were nearly driven from our home in a few hours by the stifling vapours, while the noble bird continued to stand erect, and to look defiance at us whenever we approached his post of martyrdom. His fierce demeanour precluded all internal application, and at last I was compelled to resort to a method always used as the last expedient, and a most effectual one. I thrust a long pointed piece of steel through his heart, when my proud prisoner instantly fell dead, without even ruffling a feather. (*OB* 2: 465)

Significantly, the process of drawing the bird proves to be nearly fatal to Audubon as well and lasts, quite uncharacteristically for a fast worker like him, a fortnight: "I sat up nearly the whole of another night to outline him and worked so constantly at the drawing, that it nearly cost me my life" (*OB* 3: 465–466). If we look at the finished drawing that nearly killed Audubon just as he had killed the bird, the analogy between the human and the animal is visible enough (both carry their prey), and we might even, as has been suggested recently, read the majestic bird in the foreground as an elaborate metaphor for the artist's struggle with this composition.[78] Interestingly, although Audubon's own inscription on the painting reveals the specimen depicted as an "adult female," the corresponding text from *Ornithological Biography* insistently masculinizes the bird, casting the encounter in the terms of a heroic struggle between males.

But the differences between the bird and the man in the picture are obvious as well: while the bird is ascending, with its head and beak pointing upward, the hunter on his log is clearly descending, with his head bent downward so that he will not fall. The miniature axe the hunter is using to stabilize himself against the tree further underscores the contrast between the bird's flight and the man's crawling motion. Then there is, of course, the contrast between the imposing life-sized body of the bird in the center of the plate and the tininess of the man who has been relegated to the margins of the composition, an ironic counterpoint to Charles Willson Peale's life-sized self-representation at the very center of his famous painting *The Artist in His Museum* (1822), discussed in the second chapter of this book.

The texts in *Ornithological Biography* are filled with references to birds peering at human intruders and following the motions of hunters, whose guns they usually recognize for what they are. In his room in Boston, the "fine eyes" of a captive Snowy Owl never lose sight of Audubon: "If I attempted to walk round him, the instant his head had turned as far as he could still see me, he would open his wings, and with large hops get to a corner of the room, when he would turn towards me, and again watch my approach" (*BA* 1: 115). Most birds are "shy," "vigilant," and "suspicious" (*BA* 6: 188; 7: 300), keen of sight (*BA* 7: 288), fast in action, and cunning in the choice of different modes of concealment from the man who enters, unbidden, into the carefully guarded circle of their private lives. Even when Audubon is not using his gun, his voyeurism makes him equally unwelcome, as in the following visit to the nests of the Common Cormorant (great cormorant) on the rocks in the Gulf of St, Lawrence:

> I lie flat on the edge of the precipice some hundred feet above the turbulent waters, and now crawling along with all care, I find myself only a few yards above the spot on which the parent bird and her young are fondling each other. . . . At this moment the mother accidentally looks upward, her keen eye has met mine, she utters a croak, spreads her sable wings, and in terror launches into the air. . . . Meanwhile the little ones, in their great alarm, have crawled into a recess, and there they are huddled together. I have witnessed their pleasures and their terrors, and now, crawling backwards, I leave them. (*BA* 6: 412).

In the case of the Great Blue Heron, such bizarre eye-to-eye encounters take an extreme form, because this bird, which is "ever on the look-out" and has a sight "as acute as that of any Falcon," when threatened, actually aims its powerful bill at the eye of the trespasser (*BA* 6: 123).

One of the strangest eye-to-eye confrontations between man and bird in *Ornithological Biography*, so extraordinary that Audubon himself thinks it sounds like a "traveller's tale," involves a Broad-winged Hawk. Audubon's essay begins in a deceptively peaceful manner, with an evocation of gently rocking dewdrops, but the ending of the sentence sounds a jarring note: "One fine May morning, when nature seemed to be enchanted at

the sight of her own great works, when
the pearly dew-drops were yet hanging
at the point of each leaf, or lay nursed in
the blossoms, gently rocked, as it were,
by the soft breeze of early summer, I
took my gun" (*BA* 1: 43). In the course of
their excursion, Audubon and his hunting companion, William Bakewell, capture a female hawk ("I looked at it with
indescribable pleasure, as I saw it was
new to me"). Not only does the hawk
not resist when she is taken away from
her nest (Audubon, in fact, feels "vexed"
that this bird is "not of a more spirited
nature"), it later on even willingly poses
for Audubon's portrait as if it were already mounted on one of the naturalist's
wooden boards and supported by one of
his inevitable wire constructions (ill. 54).
Directing a "sorrowful" look toward her
potential killer, the bird is, in fact, mim-
icking, *playing her own birdness,* at least
the kind of birdness Audubon wants for
his lifelike, life-sized paintings. Therefore she survives her own representation
and is finally allowed to sail off out of
the draftsman's sight: "Its eye, directed
towards mine, appeared truly sorrowful.
I measured the length of its bill with the

54. "Broad-winged Hawk." Lithograph by Robert
Havell Jr. after John James Audubon, *The Birds of
America* (Double Elephant Folio), pl. 91. Sheet: 96.8 x
64.8 cm. Courtesy, The Lilly Library, Indiana University, Bloomington.

compass, began my outlines, continued measuring part after part as I went on, and finished the drawing, without the bird ever moving once. . . . The drawing being finished,
I raised the window, laid hold of the poor bird, and launched it into the air, where it
sailed off until out of my sight" (*BA* 1: 44).

John Berger has argued that the look exchanged between animal and man has
played a crucial role in the development of human society.[79] If we accept this view,
these episodes from Audubon's *Ornithological Biography* present an extremely reduced
form of such encounters, leaving us with the indifferent, hopelessly knowing look of

the animal at the man that brings home the essential loneliness of both of them. Using "perspective" and all the "other niceties" Mark Catesby had spurned, Audubon's bird drawings represent, rather than present, images "drawn," withdrawn, "from nature." Unlike Catesby's and Wilson's birds, Audubon's birds seem alive, but what ultimately makes them so attractive as well as interesting is that they signal an awareness that this is exactly what they do: they *seem*. This awareness manifests itself in the carefully applied tricks of composition in Audubon's plates (which almost make us see the wires he used to prop up his freshly killed birds) as well as in the frequent references to death and killing, displaced reminders that an act of killing had to precede the representation of life we see here. Every representation of nature is a construction; but what Mark Catesby's and Alexander Wilson's pictures still, in the full sense of the word, flatly deny is the central point of Audubon's compositions. How self-conscious an artist the allegedly half-literate Audubon must have been is, finally, underscored by the passages on the genesis of his paintings which I have quoted from his written work—texts in which we meet the Golden Eagle's "bright unflinching eye" and in which we also encounter and then lose sight of the female Broad-winged Hawk who has, while still alive, turned herself into an Audubon bird, a work of art, so that she might stay alive.

A few decades after its publication, Audubon's octavo edition of *The Birds of America* became a major point of reference in Darwin's attempt to offer a theory of the origin of man, for which Darwin had been collecting notes since 1837, but which appeared in book form, titled *The Descent of Man, and Selection in Relation to Sex,* only in 1871. As an old man, Darwin still remembered Audubon's appearance in 1826 in Edinburgh and his "interesting discourses on the habits of N. American birds."[80] In the anecdote-laden, heavily footnoted *Descent of Man,* belligerent *and* beautiful birds (to which Darwin devoted four of the twenty-one chapters of his book), next to quarrelsome mammals, bore the brunt of Darwin's argument that the laws of sexual selection have determined the differences between the sexes in the animal kingdom as well as in the process of human evolution. Among birds, male competition as well as female choice influences the pairing of members of the species. As Darwin shows, relying on Audubon's description of the behavior of the wild turkey and the grouse, the "season of love" (he adopts Audubon's favorite term here and elsewhere in the book) is also the season of "battle," but it is also more than probable that "females are excited, either before or after the conflict, by certain males, and thus unconsciously prefer them."[81] A female bird is invariably pursued by different males but reserves the right to choose from among those suitors whichever pleases her the most. As Darwin writes: "Audubon—and we must remember that he spent a long life in prowling about the forests of the United States and observing the birds—does not doubt that the female deliberately chooses her mate." And then he goes on to quote from Audubon's biography of the Golden-winged Woodpecker: "He says the hen is followed by half-a-dozen gay suitors, who continue performing strange antics, 'until a marked preference is shewn for one.'" And when the choice is made, these pairs continue their attachment: mated Canada geese, for example, "careful to keep in pairs," renew their courtship each January. Adds Darwin, "Many similar statements with respect to other birds could be cited from this same observer."[82] In short, birds have fine powers of discrimination, "and in some few instances it can be shewn that they have a taste for the beautiful." Hence the tails, the beautiful plumes, elongated feathers, topknots, and other ornaments exhibited by so many male birds—among them the "bright crimson" head of the ivory-billed woodpecker, colored, we remember from Audubon's description quoted at the beginning of the chapter, as if he had been painted by van Dyck himself.[83]

For Darwin, human beings, because of, or rather despite, their peculiarly large, "god-like" intelligence, merely constitute one particular problem in the general process of selection that has affected the "history of the organic world" (Darwin, it should be remembered, found the ant's brain a more amazing instrument than that of man). When it comes to marriage, writes Darwin, "man" is "impelled by nearly the same motives as the lower animals, when they are left to their own free choice, though he is so far superior to them that he highly values mental charms and virtues." And even if, during "the earlier periods of our long history," men generally seemed to have been the selectors, women had always been able to tempt the men they prefer and reject those they disliked.[84] In *The Origin of Species* and *The Descent of Man*, animal nature no longer functions *in analogy* to human nature: where the history of man has become part and parcel of a vast network of family relations, where "man" has become brother to the barnacles and the bears,[85] natural history has ceased to be a stage for autobiographical self-definition. Anxious to purge the language of natural history of indications of human intention and agency, Darwin created texts in which "man" is no longer the omnipresent *subject* of his own discourse about nature, the proud product "of a separate act of creation,"[86] but appears, along with the birds, the beasts, and the insects, as one of its *objects*. However, for all its talk about American birds and African tribes, *The Descent of Man* is essentially a book of "armchair adventures," not the record of a life lived in the thick of experience, not the lively chronicle of an engagement with the multifold energies of nature itself.

If autobiographical disclosures had thus begun to disappear from the discourse of natural history, the example of Walt Whitman shows that, in turn, the language of scientific objectivity was causing problems in literature, too. In 1883, Whitman published his "wayward" autobiography, *Specimen Days,* a gathering of notes "from that eternal tendency to perpetuate and preserve all that is behind Nature." (Obviously, the title of the work itself, along with that of its companion piece, *Collect,* is an allusion to the ambitions of natural history collectors).[87] Whitman included not only passages from his Civil War notebooks, literary tributes, and reflections on painting and music but also entries from his diary, written after 1876, during time spent at his beloved Timber Creek in Camden County, New Jersey, where Audubon had once gone to watch the migrations of warblers (*BA* 4: 149). In May 1881, under the title "Birds—and A Caution," Whitman mentions some birds he has noticed "lately" in the "Jersey woods." He comments, specifically, on the song of the "russet-back" ("I like to watch the working of his bill and throat"), on the sounds produced by the woodpecker, the whippoorwill, and the cat-bird, and on the "delicious" gurgle of the thrush.

Note that Whitman does not bother to distinguish between species of woodpeckers or thrushes and that he chooses a deliberately unscientific name ("russet-back") for what seems to be the bird now known as Swainson's thrush. Many birds, Whitman admits, "I cannot name; but I do not particularly seek information." Liberate the birds from the scientists, was the message that Whitman had imparted to the naturalist John Burroughs, who was Whitman's friend and biographer before, much later, he chronicled Audubon's life.[88] Thus, in a deliberately placed parenthesis, Whitman ends his sloppy little field note about the birds at Timber Creek with a stab at scientific exactitude and with a cheerful plea for the creative powers of ignorance.

(You must not know too much, or be too precise or scientific about birds and trees and flowers and water-craft; a certain free margin, and even vagueness—perhaps ignorance, credulity—helps your enjoyment of these things, and of the sentiment of feather'd, wooded, river, or marine Nature generally. I repeat it—don't want to know too exactly, or the reasons why. My own notes have been written off-hand in the latitude of middle New Jersey. Though they describe what I saw—what appear'd to me—I dare say the expert ornithologist, botanist or entomologist will detect more than one slip in them.)[89]

John James Audubon had already been dead four years when Whitman published his *Leaves of Grass* in 1855. Unlike Audubon, Whitman was born in the United States, "of parents born here from parents the same, and their parents the same," as he famously said in "Song of Myself." But Whitman's dream—to represent the natural and national whole within one and the same, eternally unfinished work—had been Audubon's, too. *The Birds of America* marked the pinnacle of the "poetics of natural history," the zenith of a tradition in which knowing "the reasons why" and enjoying the reality of "things as they are" had not seemed, as they now suddenly did for Whitman, mutually exclusive.

It is ironical but poetically appropriate that another naturalized American would finally intone what became the swan song of American natural history as it has been described in this book. In 1855, Whitman's annus mirabilis, the Harvard naturalist Louis Agassiz first announced his intention to write a book that would classify and exhaustively describe the animal life of the entire nation, a work "not . . . less honorable to the country than to its author," based on collections made by amateur naturalists from all regions of the United States and financed exclusively by public subscription.[90] Agassiz's work, jubilated the *Boston Daily Advertiser,* would "exalt the character of America"; finally, the "Science of the New World" was "raising her head and claiming to sit side by side with the Science of the Old." In the preface to the first volume of *Contributions to the Natural History of the United States of America,* which came out two years later, Agassiz stated proudly: "I expect to see my book read by operatives, by fisherman, by farmers, quite as extensively as by the students in our colleges, or by the learned professors; and it is but proper that I should make myself understood by all." He had planned a description of the "whole animal kingdom," of the "mode of life of all our animals," including their geographical distribution, natural affinities, internal structure, embryonic growth, and fossil remains. He never got past the turtles and the jellyfish.[91]

CHAPTER 6

Agassiz Agonistes

"The First Naturalist of His Time"

When Louis Agassiz, the professor of geology and zoology in the Lawrence Scientific School at Harvard University, launched his unprecedented subscription campaign for *Contributions to the Natural History of the United States of America*, it never occurred to him that he was courting disaster. The purported modesty of the work's title belied the scale on which the Swiss-born naturalist had conceived it. If he had been virtually silent for ten years, Agassiz stated in a letter to the British geologist Charles Lyell, now the scientific world would hear his voice again: "My subscription is marvellous."[1] And while not exactly announcing that there was "never any more inception than there is now"—as Walt Whitman boasted in "Song of Myself"—the Harvard naturalist at least made it plain that he would not rest until ten volumes had been printed, one, as it were, for each year of his past silence.

Neither personal gain nor profit was his goal; instead, the funds Agassiz wished to solicit from potential subscribers were wholly intended to provide these splendid volumes with even "more text, more illustrations, and more expensive lithography than originally planned."[2] In a letter to Senator Charles Sumner, Agassiz revealed that the wish to make his contribution to the natural history of the United States was not his sole motivation. "When my subscription list reaches Europe," he wrote to Sumner, "my friends will not credit their own eyes. I do not think Humboldt himself could obtain in all Europe put together such a subscription for so expensive a work."[3]

The terms in which Agassiz announced his project are not surprising, coming from a man who, while still an adolescent, had asked his sister, only half in jest: "Will it not seem strange when the largest and finest book in Papa's library is one written by his Louis?" (*LC* 1: 81). No slouch when it came to promising predictions of future glory for himself, Louis shortly thereafter informed his father: "I wish it may be said of Louis Agassiz that he was the first naturalist of his time" (*LC* 1: 98). Now, decades later, he compared himself to Alexander von Humboldt and implied that, through his *Contributions,* he would become to North America what Humboldt had been to Europe. Given time, propitious circumstances, and, most important, a lot of money, Louis Agassiz might soon out-Humboldt Humboldt himself. The first volume of Agassiz's *Contributions* began, appropriately, with a twenty-seven-page list of contributions made by others. Among the subscribers to Agassiz's grand work were Spencer Fullerton Baird, Ralph Waldo Emerson, John Edwards Holbrook, Oliver Wendell Holmes, Henry Wadsworth Longfellow, Theodore Parker, Titian R. Peale, and Jared Sparks, as well as local natural history associations, literary societies, booksellers, mechanics' institutes, the Library Company of Philadelphia, the Library of the U.S. Military Academy of West Point, the Smithsonian Institution, and His Majesty, the King of Prussia.

The beginnings of Agassiz's American career had been auspicious indeed. Born in Motier, Switzerland, on 28 May 1807, Agassiz was already a world-renowned geologist and ichthyologist when in the winter of 1846 he accepted an invitation to deliver a series of public lectures at the Lowell Institute in Boston. Thousands vied for seats in Tremont Temple to hear the European professor with the charming Continental accent lecture on "The Plan of Creation in the Animal Kingdom." This was the beginning of Agassiz's systematic conquest of the world of American science, which culminated in his appointment to the faculty at Harvard's Lawrence Scientific School in 1848. In the years that followed, Agassiz's lecturing inside and outside of Harvard made him a legendary figure.[4] He helped found the American Association for the Advancement of Science and impressed the likes of William James and Henry Adams (Adams later noted in his autobiography that Agassiz's courses had exerted "more influence on his curiosity than the rest of his college instruction altogether").[5] Among the students who later pursued successful careers in science were Frederick W. Putnam, the director of the Peabody Museum, the first anthropological museum in the United States; Samuel H. Scudder, the founder of insect paleontology in the United States; and Nathaniel Southgate Shaler, who became Dean of the Lawrence Scientific School.

But Agassiz's influence extended far beyond the classrooms of Harvard. As soon as he started work on a "Natural History of the Fishes of the United States," circulars were distributed instructing interested citizens in the proper modes for mailing fish to Cambridge. As Americans all across the country, fishermen, nature enthusiasts, and specialists alike, responded, specimens began pouring in. Even if most of us today would have more sympathy with Thoreau's injunction to think of the bream as a "contemporary and neighbor" and our "little fishy friend in the pond,"[6] this is not how the overwhelming number of Agassiz's contemporaries related to the inhabitants of the rivers and ponds of North America. A kind of benign Captain Ahab of American natural history, Agassiz established a vast network of natural history collecting that spanned the entire continent. Even Thoreau succumbed and sent some snapping turtles. With a mixture of personal charisma, chutzpah, and cunning, Agassiz accomplished the nearly impossible feat of simultaneously popularizing and professionalizing the world of American natural science. The crowning achievement of Agassiz's American career came in 1860, when his collections of natural history specimens found a permanent home. The Museum of Comparative Zoology at Harvard University was dedicated and opened to the public in Cambridge on 13 November. "Since Benjamin Franklin," said William James decades later, "we had never had among us a person of more popularly impressive type."[7] The widower Louis Agassiz entered the upper echelons of Boston society as early as 1850 by marrying Elizabeth Cabot Cary; in 1863, he completed the process of his Americanization by formally becoming an American citizen.

Enter the second major protagonist of the story to be told in the pages of this chapter. In the fall of 1861, William James, after abandoning his earlier plans to become a painter, enrolled as a student of chemistry at the Lawrence Scientific School.[8] In his brother Henry's eyes, William's interest in art had always been complemented by more scientific inclinations, in which he was actively supported by his philosopher-father, who wished that his son would one day provide the facts that would help bolster his Swedenborgian spiritualism. During the years the boys spent in Switzerland and France, his brother had been, wrote Henry, in ways "often appalling to a nature so incurious as mine in *that* direction," much "addicted to 'experiments' and the consumption of chemicals," cultivating "stained fingers" and wasting hours on "the maintenance of marine animals in splashy aquaria."[9] At Harvard, William eagerly attended the lectures of the anatomist Jeffries Wyman but derived even more satisfaction from his encounters with Agassiz, an "admirable, earnest lecturer, clear as day," as he informed his family in September 1861 (*CWJ* 4: 43). In November 1861, he followed his brother Wilkinson in submitting "a resume of his future history" to his family. He explained that he planned to study chemistry first and would then "spend one term at home, then 1 year with Wyman, then a medical education, then 5 or 6 years with Agassiz, then probably death, death, death with inflation and plethora of knowledge" (*CWJ* 4: 52). The following month, on Christmas Day, William declared that during a conversation with one of Agassiz's students it had dawned on him for the first time "how a naturalist could feel about his trade in the same way that an artist does about his." Agassiz did not allow his students to linger over their textbooks: he "makes *naturalists* of them, he does not merely cram them." After two years with Agassiz, James's friend now "felt ready to go any where in the world . . . with nothing but his note book and study out anything quite alone." Agassiz certainly was, concluded James, "a great teacher" (*CWJ* 4: 63). Even after James had decided to enter medical school because he felt "very much the importance of making soon a final choice of my business in life," he still seriously considered what he referred to as "the prospects of a naturalist": "If I can get into Agassiz' museum I think it not improbble I may receive a salary of $400 or 500 in a couple of years" (*CWJ* 4: 86).

James's dreams of a paid position in Agassiz's museum never materialized, and when two years later he accompanied Louis Agassiz, Agassiz's wife, Elizabeth, several professionally trained naturalists, and five other volunteers on an expedition to Brazil, he did so at his own or, rather, his family's expense. This chapter will continue to follow the paths of Agassiz's and James's lives, which, for a brief period, touched and then, in ways interesting for the further development of American science, diverged. Taking Louis and Elizabeth Agassiz's *Journey in Brazil* (1868) and William James's Brazilian letters and diaries as focal texts, this chapter argues that, in Louis Agassiz's hands, natural history officially ceased to be what it once was, an instrument for the discovery of what appeared "curious" (i.e., noteworthy, interesting, remarkable) in American nature. Instead, as William James's comments on Agassiz show, it became a "curiosity" itself, in the sense of the adjective "curious" that still persists today: something "odd" and "strange." Following Agassiz to a strange country, the strangeness of the demands of pre-Darwinian natural history, its desire to offer a comprehensive collection, classification, and description of things both nonhuman and human, no longer made sense to William James; *his* descriptions turned their attention to the describers themselves.

It is important to realize that at the time James began to study with him, the quality of Agassiz's scientific work had already fallen behind his pedagogical and entrepreneurial accomplishments. The man who had set out to "rebuild" the entire "system of zoology," the scientist who was the world's foremost authority on fossil fishes, the founder of one of the world's great natural history museums, and the most popular lecturer in the United States"[10] was forced to recognize that others were already busily replacing the foundations of the entire edifice. On 1 July 1858, a joint paper "On the Tendency of Species to form Varieties; and on the Perpetuation of Varieties and Species by Natural Means of Selection" by Charles Darwin and Alfred Russel Wallace was delivered to a remarkably indifferent audience at a meeting of the Linnean Society in London.[11] But while the new theory of natural selection at first took hold only slowly in British and American academic circles, Agassiz's carefully fortified status as North America's premier naturalist crumbled fast, based as it was on the advocacy of a theory that allowed change or "evolution" to become visible chiefly in the individual's phases of development from embryo to adult. High-flying plans, ambitious goals, and expensive projects increasingly took the place of actual accomplishments. For example, Agassiz did not, as he had promised, publish one volume per year of his projected masterwork, *Contributions to the Natural History of the United States of America*. The four volumes that he did complete were, upon closer inspection, really only two, if lengthy, monographs on, of all things, American turtles and jellyfish. The first two volumes, after a general introduction, dealt with the zoology and embryology of the turtle; and the third and fourth offered a detailed look at the class of Acalephs, "under which name naturalists now include the so-called jelly-fishes or sea-blubbers or sun-fishes, and the animals allied to them."[12]

The third volume of *Contributions* came out in 1860, only a few months after Charles Darwin had presented his Harvard colleague with a copy of *The Origin of Species*. Darwin begged Agassiz not to think that he was giving him the book "out of a spirit of defiance or bravado." His interest was not in stirring up controversy for controversy's sake, and however "erroneous" his illustrious colleague thought his conclusions were, Agassiz should give him credit "for having earnestly endeavoured to arrive at the truth."[13] The ideological differences between Darwin and Agassiz are too familiar to warrant detailed repetition here. While Agassiz held fast to his belief in the fixity of species, praised the immutability of the Creator's well-conceived plan for the universe, and cheerfully invoked an "ideal unity holding all parts . . . together,"[14] Darwin sketched out a view of nature in which unfit species perished and chance mutations thrived, in which benign design seemed to have been replaced by the vast machinery of inevitable change. In the famous concluding passage to *The Origin of Species*, Darwin asked his readers to contemplate an "entangled bank, clothed with many plants of many kinds, with birds singing on the bushes, with various insects flitting about, and with worms crawling through the damp earth," and he invited them to reflect that these complex, beautiful forms of life, so different from and yet so dependent on each other, were all the result of a "Struggle for Life," of "the war of nature," of "famine and death"; thus from death, from the extinction of forms less improved, sprang life. There was, said Darwin, "grandeur" in this view of life: "Whilst this planet has gone cycling on according to the fixed law of gravity," from simple beginnings "endless forms most beautiful and most wonderful have been, and are being, evolved,"[15] Agassiz's vision of natural grandeur was different: for him, God had come up with a complete plan for the history of life and then, in the course of geological time, created species in their appropriate sequence, fully adapted to their respective habitats. Ironically for someone who had himself strayed from his own original habitat, Agassiz insisted that species did not normally migrate but, rather, originated where they live. Environments were not endless challenges to the adaptability of organisms struggling to survive but had been shaped to fit God's precise schedule for the development of life. Nature was the work of God's thought; it was the scientist's duty to read and transcribe as accurately as possible "a system that is his and not ours." Pockets of individual development, i.e., the transformations of the embryo in which Agassiz showed a special interest, did not affect but in fact confirmed the static nature of God's plan for the world. The fetus, he claimed, was nothing but a recapitulation of the themes in God's creation, a summary of the chapters in the Book of Nature, of which Agassiz was convinced there was only one possible reading.[16]

Change, for Agassiz, was merely the revelation, the unfolding of an underlying essence. If nature as a whole does not really change, our *knowledge* of nature does, and it is here that Agassiz was prepared, at least in theory, to allow for growth, progress, and—surprise. God's Good Book required patient readers, as a beautifully written passage from the magisterial beginning of the third volume of Agassiz's abortive masterwork demonstrates. The history of nature, Agassiz says here, is the history of our attempts to grasp its perennially wonderful complexity. We look, we touch, feel, and destroy, yet we have still not learned a thing. As the jellyfish "melts away" on our naked, clumsily interfering hand, so does our knowledge of its "slight" existence, unless we learn to look properly:

> When we first observe a jelly-fish, it appears like a moving fleshy mass, seemingly destitute of organization; next, we may observe its motions, contracting and expanding, while it floats near the surface of the water. Upon touching it, we may feel the burning sensation it produces upon the naked hand, and perhaps perceive also that it has a central opening, a sort of mouth, through which it introduces its food into the interior. Again, we cannot but be struck with their slight consistency, and the rapidity with which they melt away when taken out of the water.

As we move from first innocent sense impressions through a gradual refinement of the appropriate "methods of investigation" to an awareness of how each perceived detail will assume its assigned place within a well-ordered taxonomic whole, our initial surprise unmasks itself as what it always was—the product of ignorance. The microscopic opens up a window onto the macroscopic. Where there was a jellyfish, we see the world. Now we know in part, but soon we shall know all:

> It is not until our methods of investigation are improved; and when, after repeated failures, we have learned how to handle and treat them, that we begin to

perceive how remarkable and complicated their internal structure is;—it is not until we have become acquainted with a large number of their different kinds, that we perceive how greatly diversified they are;—it is not until we have had an opportunity of tracing their development, that we perceive how wide the range of their class really is;—it is not until we have extended our comparisons to almost every type of the animal kingdom, that we can be prepared to determine their general affinity, the natural limits of the type to which they belong, and the peculiarities that may distinguish their families, their genera, and their species.[17]

If anything, Louis Agassiz, intimate of the likes of Longfellow, Lowell, and Emerson, worked hard to earn his place as archfiend in the cosmology of all good Darwinians.[18]

From Lake Superior to the Amazon River

In Agassiz's antievolutionary view, nature did not travel. In the 1860s, however, it dawned on him that, in order to prove this, a modern naturalist might have to travel himself. Agassiz's first American expedition, to be sure, had been a success. In 1848, a portable blackboard tucked under his arm, he led a group of nine Harvard students, two New York doctors, and two European naturalists on a specimen-collecting trip to Lake Superior. The hefty volume that introduced a wider audience to the results of the party's expedition also contained a "narrative" written by James Elliot Cabot. If, in William Bartram's *Travels*, storytelling had easily merged with scientific description, and description in turn had frequently gained narrative momentum as well as poetic brilliance, the different genres were kept strictly separate in *Lake Superior*. The literary and the scientific inhabited, at least in principle, different parts of the book.

Cabot's lively text conjures up memorable images of the Harvard professor and his team in a wilderness to which, with their more academic inclinations, they were not too well adapted, as other, more experienced travelers who encountered the group, also recognized.[19] Soaking wet in his canoe but "enraptured by the variety of the scaly tribe," Agassiz tirelessly extolled the beauties of his "favorite science," ichthyology, to an audience of students safely sheathed in waterproof coats (*LS* 23–24). The professor would gently coax the local natives into providing him with a complete set of fish wherever his group pitched their tents, and he would then spend the afternoons, scalpel in hand, standing before an improvised table on which "fishes little and big" were heaped—pointing, describing, and slicing until nothing was left intact (*LS* 24). At one point, at the professor's request, Cabot sketched a green pickerel ("a new species") while the local Indians crouched around the group of naturalists—out of curiosity, as Cabot first believed. But what he and his companions had mistaken for rapt wonder was in fact hunger: "The whole family had been without food all day, and were waiting to eat the fish as soon as we were done with it" (*LS* 72). The faces of these starving natives were amazingly "round, full and rather flat," their lips "thickish" as those of "negroes," noted Cabot (*LS* 54); their habitat, the extensive woods, in which one could walk for hours without hearing a sound, seemed often less picturesque than desolate (*LS* 109).

Agassiz's men were endlessly pursued by mosquitoes and flies until their faces were "speckled with blood" and their eyes surrounded by red rings (*LS* 42, 61). They suffered under the scorching sun, which blistered their skin. Nevertheless, they persisted, the indefatigable harbingers of science and culture in a world that looked as if it had just emerged from the primeval miasma, from an archaic time when mosses and pines had just begun to cover the rocks, and the animals "as yet ventured timidly forth into the new world" (*LS* 124). Significantly, a frying pan served as the figurehead of Agassiz's canoe, a metonymic reminder of the advantages of civilization in an area where, as Cabot also noted, "nature provides nothing that can be eaten raw except blueberries" (*LS* 38).

His mind teeming with details, Agassiz returned to Boston. His collection alone occupied four barrels and twelve boxes, most of them large (*LS* 125). Not only did the geological evidence confirm his theories of glaciation, but he had also accumulated plenty of evidence to support his general belief that animals and plants were not randomly scattered over the surface of the globe. Rather, he proclaimed, they had "remained almost precisely within the same limits ever since they were created," except in those cases where, under the active influence of man, these limits had been "extended over large areas" (*LS* 248). Under the stern eyes of the professor and disemboweled by the sharp knives of his eager helpers, the fish of Lake Superior had little choice but to fit neatly into the categories provided by the Boston naturalists. "The fishes and all other freshwater animals of the region of the great lakes," judged Agassiz, "are, and were from the beginning, best suited for the country where they are now found" (*LS* 376). And, as if anticipating the main point of contention in his great controversy with Darwin, he described the territory through which his party had traveled, mosquitoes and all, as a marvelous exemplification of the insight contained in the divine master plan. Here, God's creation was as fresh as on the first day; such wisdom as Agassiz had found in the Great Lakes could be found anywhere else, an observation that should keep anyone from ascribing the order of the natural world "to the creatures themselves" or, worse still, "to physical influences or mere chance" (*LS* 377).

It might appear a little strange at first that the preeminent naturalist of the United States should have selected Brazil as the destination of his second major expedition.

Upon closer inspection, however, this seems a more than logical choice. Agassiz's first book, written in Latin when he was a twenty-two-year-old student in Munich, was a monumental study of the fish indigenous to Brazil, consisting of the drawing, classification, and ordering of more than five hundred species found mostly along the Amazon River.[20] Agassiz's account, which won him the admiration of the greatest anatomist of the time, Baron Cuvier, was based not on firsthand observation but on a close analysis of the samples gathered, in the course of a three-year-long expedition to Brazil, by his teacher, the German naturalist Carl F. P. von Martius, and his colleague, Johann Spix.[21] As Agassiz explained in the preface to *Journey in Brazil,* ever since the completion of this first major scientific project, the desire to travel to Brazil himself had stayed with him.

In 1865, Agassiz felt worn out by the work that had gone into his museum, and it might really have been his intention, as he wrote to his mother, just to "loaf a little in Brazil" (*LC* 2: 626). But when the wealthy Nathaniel Thayer offered to pay all the scientific expenses of an expedition to South America, Agassiz immediately began to daydream about the large collections he could amass there for his Cambridge museum. Thousands of Brazilian fish began to drift by before his inner eye, all of them ready to contribute their share to the grand refutation of Darwinian theory that he had in mind.[22] He knew that the Amazonian river system was the largest freshwater basin in the entire world; silent armies of fish were waiting to be plucked from the rivers and then subsumed into the folds of Agassiz's taxonomic system. And freshwater fish, because they lived "within narrower limits than either terrestrial or marine types" (*LS* 247), did, after all, carry the burden of proof in Agassiz's patient interpretation of God's static master plan.

There was yet a third reason why the prospect of an expedition to Brazil would have tempted Agassiz in 1865. The American Civil War was drawing to a bloody close. It had marked the end of the conception of the United States as vast unencumbered spaces. But if battle lines and strategic moves and countermoves had now indelibly marred the once boundless North American landscape, vast unexplored and unclassified territories elsewhere still held a similar promise for enterprising naturalists. Early in the twentieth century, the amateur naturalist, big-game hunter, and ex-president Theodore Roosevelt would still describe Brazil as the last "frontier between civilization and savagery" and prophesy: "Decades will pass before it vanishes."[23]

Interestingly, *Journey in Brazil* contains very few references to the Civil War itself, none of which is more than incidental, with the exception perhaps of the first. While Louis Agassiz and his crew of naturalists and volunteers were trundling along the North American coast aboard the steamer *Colorado* (provided to the Harvard naturalist, free of charge, by the Pacific Steamship Company), their enjoyment of the "delicious weather" was slightly disturbed by "a singular cloud" on the horizon, which the captain identified as smoke coming from the direction of Petersburg, Virginia. "We think," commented Elizabeth Agassiz, "it may be the smoke of a great decisive engagement going on while we sail peacefully along" (*JB* 1–2). As they discovered after their arrival in Rio, it had been on this day that one of the most violent confrontations of the Civil War, trench warfare that anticipated World War I, had taken place: "The mass of smoke gathered above the opposing lines of the two armies" at Petersburg had produced the disagreeable cloud that blotted the sky as Agassiz and his party were passing by (*JB* 2). On 9 April, the day Elizabeth Agassiz opened her journal to praise the "large and commodious" staterooms on board the *Colorado* ("in short, nothing is left to be desired except a little more stable footing," *JB* 20), General Lee, with fewer than thirty-thousand soldiers left under his command, surrendered his army. While troops clashed in deadly battle under the "otherwise stainless" sky over Virginia, Agassiz and his men disappeared on their "peaceful" mission to collect palm leaves, fish, and evidence of glacial activity in Brazil. Natural history, to be sure, was beyond the fray and frazzle of politics.

However, Agassiz's expedition was not completely devoid of political importance. "Science is . . . bloodless in her conquests," Massachusetts Senator Charles Sumner rhapsodized to Agassiz on 20 March 1865, "You are a naturalist; but you are a patriot also. If you can take advantage of the opportunities which you will surely enjoy, and plead for our country . . . and the hardships it has been obliged to endure may be appreciated, you will render a service to the cause of international peace and goodwill" (*LC* 2: 634). Sumner hoped, of course, that Agassiz would favorably impress those slaveholding South Americans who still sympathized with the cause of the Confederacy.[24] But above all, he recognized the importance of Brazil, a country covering an area greater than that of the United States, to the balance of power on the two American continents. While the United States was being torn apart by civil strife, Brazil had quietly emerged as one of the world's suppliers not only of coffee but also of cotton. Brazil was, as Oliver Wendell Holmes told Louis Agassiz after his return, "a new world to most of us,"[25] and yet not everything in it seemed entirely unfamiliar. In many ways Brazil shared the problems faced by its neighbor the United States, and in many ways it was very different.

Like the United States, Brazil had cut its ties with a European colonial power and had passed from "colonial to national life" (*JB* 498). Unlike the United States, however, it had become independent more or less quietly, through relatively harmless evolution rather than bloody revolution.[26] For more than a decade, the Portuguese royal family, exiles from their own country because of the Napoleonic Wars, had resided in Rio de Janeiro. When it became evident in 1822 that Brazil would soon be reduced to its former colonial insignificance, Dom Pedro, the son of the king of Portugal who had been left behind to rule the country, exclaimed "Independence or death!" and forced the few remaining Portuguese troops to leave. He became Pedro I, emperor of an independent Brazil, a country "more populous, more wealthy, more energetic, and as wise as Portugal," as Thomas Jefferson, who predicted Brazilian independence several years before the fact, remarked to the marquis de Lafayette in 1817.[27]

Like the United States, Brazil also had to cope with a history of slavery. Here it had taken longer to ban the actual importation of slaves, and even after the end of slave trade in 1850, slavery was still widely tolerated.[28] Also like the United States, Brazil was composed of people from different cultures and traditions.[29] So far, however, it had survived without internal, fratricidal strife, although the war with Paraguay (1865–1870) created a feeling of national crisis that would have appeared uncannily familiar to traveling naturalists from the United States.

Louis and Elizabeth Agassiz, then, planned their expedition to Brazil at a time when a country that was becoming cautiously democratic was also discovering its identity as a nation and when the two giant countries in South and North America were, though not without difficulties, gradually approaching each other.[30] Such circumstances, as the Agassizs correctly anticipated, would benefit their own work, and it was with considerable satisfaction that during the trip Elizabeth noticed "a strong desire throughout Brazil to strengthen in every way her relations with the United States" (*JB* 252). In Elizabeth's diary, she envisioned a glorious future in which "the twin continents will shake hands and Americans of the North come to help Americans of the South in developing its resources" (*JB* 257).

Thus drawn to Brazil by a "lifelong" personal desire (*JB* [v]), the precarious political circumstances at home, and the propitious opportunities abroad, Louis and Elizabeth Agassiz were also mindful of the distinguished naturalists before them who had braved the massive rivers and forced their way through the forests of Brazil. The ranks of previous explorers in Brazil included Agassiz's mentor, Baron Alexander von Humboldt, who discovered the Cassiquiare River (and thus the connection between the Orinoco and Amazon river systems) and had been prevented from traveling further into Amazonia only because of the stubbornness of the Brazilian authorities who thought that the foreign naturalist was surveying lands "that did not belong to him."[31] More important, it was on the Brazilian coast that the *HMS Beagle* had dropped off the young British naturalist Charles Darwin in 1832, the first important member of the cast of characters populating the stage on which Louis Agassiz now also planned to play a significant role.

While Captain FitzRoy sailed up and down the coastline to check existing admiralty charts for the area, Darwin settled in Rio de Janeiro and enjoyed himself: "Delight itself," wrote Darwin in his diary, "is a weak term to express the feelings of a naturalist who, for the first time, has wandered by himself in a Brazilian forest."[32] On excursions north from Rio that brought him "deeper pleasure" than he felt he would ever experience again, Darwin tested the accuracy of Humboldt's "glorious descriptions" and concluded that even the German explorer's "rare union of poetry with science" fell far short of the truth.[33] The tropical forest, which reminded Darwin of some great opera house or theater,[34] presented a true *embarras de richesse* to the inquiring naturalist, who didn't know where to look first: "If the eye attempts to follow the flight of a gaudy butterfly, it is arrested by some strange tree or fruit; if watching an insect one forgets it in the strange flower it is crawling over.—if turning to admire the splendour of the scenery, the individual character of the foreground fixes the attention. The mind is a chaos of delight."[35] Brazilian scenery was, wrote Darwin, "nothing more nor less than a view in the Arabian nights, with the advantage of reality." Deeply disturbed by the insolence of the government officials with whom he had to haggle for passports, Darwin was sufficiently enchanted by the prospect of "wild forests tenanted by beautiful birds, Monkeys & Sloths" to concede that any naturalist would be prepared to "lick the dust even from the foot of a Brazilian."[36] Significantly, when Louis and Elizabeth Agassiz finally surprised the American public with their own five-hundred-page answer to Darwin's *Voyage of the Beagle,* their book showed the Brazilians licking the feet of the Agassizs, not vice versa.

The first traveler from the United States to voyage up the Amazon was William H. Edwards (1822–1909), who had grown up in the Catskills and had studied law in New York City, neither of which experiences had particularly equipped him for his Brazilian sojourn.[37] Nevertheless, he enjoyed the trip: surrounded by "ease-loving Indians," Agassiz's predecessor reveled in the "wild luxuriance" of the Brazilian forest, where monkeys were "frolicking through festooned bowers" and squirrels scampered "in ecstasy from limb to limb" (V3, 29). As he tells his readers in *A Voyage up the River Amazon, Including a Residence at Pará* (1847), Edwards delightedly stretched out in comfortably swaying hammocks, gingerly tasted the flesh of monkeys and sloths ("delicious, though not laid down in the cookery-books"), shot and cooked some of the birds he saw ("fricasseed parrot might rank favourably with most kinds of wild game"), and gratefully acknowledged the elegance of others ("The heron seems to me . . . one of the most beautiful, graceful beings in nature"; V 39, 69, 100). He collected shells and saw so many different kinds of fish he sorely regretted that his ignorance of ichthyology "rendered it impossible for us to distinguish them, and that our want of facilities made it equally impossible to preserve them" (V 69). Sampling the beauty of a species he said he could not find in Cuvier—Indian girls with "their long floating hair and merry laugh"—Edwards willingly contributed to the "popularity of the universal Yankee nation," freely passing out liquor, cigars, and, we may assume, other "favors" as well (V 113, 105).

Edwards felt wistful when he said farewell to the Amazon: "The months that we had passed upon its waters were bright spots in our lives. Familiarity with the vastness of its size, the majesty and the beauty of its borders, the loveliness of its islands, had not weakened our first impressions. He was always the king of rivers" (V 175). But, raptures or not, Edwards never forgot where his real allegiances were. On the Fourth of July, he contemplated "a celebration on our own account," if only to show "the benighted Amazonians how glorious a thing it is to call oneself free and independent." In a textbook case of pathetic fallacy, Brazilian birds trumpeted sympathy with Edwards's patriotic sentiments, making the air "vocal with a hundred different notes." Asked Edwards, "Was it a fancy that one red-coated fellow, as he tossed himself up, greeted us with a 'viva' to the independence of America?" (V 128).

Edwards's spirited narrative made no pretensions to scientific accuracy. Frequently we see him as the rapt but speechless admirer of nature's handiwork. For all practical purposes, he was content to praise the beauty of Brazilian nature without always knowing what exactly it was that he saw before him: "We . . . observed many rich flowers of which we know not the names" (V 45). When he discovered, for example, a new fruit, Edwards would describe it merely by saying that it resembled something he already knew from home, say, a strawberry, "except that its red skin was smooth" and that it was, of course, "most beautiful" (V 49–50).

Small wonder that this unassuming little book found avid readers among other aspiring naturalists who were mesmerized by the simple power of Edwards's descriptions, while remaining comfortably convinced that more serious work still needed to be done. In England, Henry Walter Bates and Alfred Russel Wallace, two young, penniless amateur naturalists from Leicester, felt inspired enough by Edwards's account to plan their own expedition to Belé do Pará (now known as Belém), hoping that the specimens they would collect there would help defray the cost of the trip.[38] They arrived in Pará, near the mouth of the Amazon River, on 28 May 1848. At the end of a period of two years. Bates and Wallace agreed that they should separate, Wallace to explore the northern parts and tributaries of the Amazon, and Bates the main stream and the Upper Amazon, also known as the Solimões. Racked by bouts of malaria and disenchanted with the tropics, Wallace left Brazil two years later and lost his collection in a fire on board the ship that was to take him back to England. Bates stayed on, for "eleven of the best years of my life," as he later said, and returned home with "no less than 8,000 species new . . . to science" (*NRA* I: 2, v). Wallace's account of his travels, *A Narrative of Travels on the Amazon and Rio Negro,* did not sell well, but Bates's *Naturalist on the River Amazons,* published ten years later, was an instant success: Darwin hailed it "the best work of Natural History Travels ever published in England," while Elizabeth Agassiz lauded it as "a very pleasant companion to us in our wanderings" (*JB* 243).[39]

Edwards, Bates, and Wallace more or less directly set the stage for the narrative offered a few years later to the American public by the Agassizs, especially since they all drew special attention to the multitude of fish they had found in Brazil: "The Amazon has most of its fishes peculiar to itself, and so have all its numerous tributaries . . . so that the number of distinct kinds inhabiting the whole basin of the Amazon must be immense" (*TA* 187). But no matter how plethoric the details, no matter how extensive the descriptions of Brazilian nature, both Bates's and Wallace's books were already informed by an insight that must have chagrined Louis Agassiz—namely the recognition that there had to be "some other principle regulating the infinitely varied forms of animal life" than the one which was so often evoked in natural history books, that is, "the marvellous adaptation of animals to their food, their habits, and the localities in which they are found" (*TA* 58). One feverish afternoon a few years later, while traveling in the Malay Archipelago, Wallace, independently of Darwin, hit upon the idea of "natural selection" and sent Darwin an abstract of his thoughts that made the latter shiver: "All my originality . . . will be smashed."[40] And if Wallace's thoughts, "admirably expressed and quite clear,"[41] precipitated the publication of the book that Louis Agassiz thought was "truly monstrous,"[42] Bates's work, notably his observations on the function of mimicry among the insects of the Amazon Valley, further fortified and substantiated Darwin's findings."[43]

Travel with Comfort: Elizabeth Agassiz

Journey in Brazil, by "Professor and Mrs. Agassiz," was well received upon its publication in early 1868, even though professional scientists remained largely silent about it.[44] The *New-York Daily Tribune* praised the "richness of its details" and the "graceful freedom and simplicity of its style." The *Boston Evening Transcript* felt that it was "impossible to give the reader an idea" of the book's "wealth," and the *Springfield Republican* exclaimed: "A more charming volume of travels we have seldom met with."[45] Oliver Wendell Holmes liked it and, in a note addressed to both the great ichthyologist and his wife, joked: "So exquisitely are your labors blended, that as with the mermaidens of ancient poets, it is hard to say where the woman leaves off and the fish begins."[46]

And Ralph Waldo Emerson declared his satisfaction that such a "very cheerful book" should have come out of an expedition sent "in such dark times." Somewhat ambiguously, Emerson went on to praise the "adequate results" of Agassiz's investigations and professed to be proud of "the good naturalist" who had been so "fully up to his own" and given "wholesome counsel."[47]

While *Journey in Brazil* was written for the general reader, its scientific pronouncements demanded to be taken seriously by layman and specialist alike. (Agassiz never wrote the exclusively scientific report that he had originally promised his colleagues.) In many ways, Louis and Elizabeth Agassiz's book is a conventional exploration narrative in that it faithfully charts, in straight diary form, the course of what became known as the Thayer Expedition, from New York to Rio de Janeiro and then up the coast to Pará, from Pará to Manáos, from Manáos to Teffé and Mauhes, and then back again to Manáos and Pará. Elizabeth's entries, composed on site, convey the usual impression of immediacy, allowing the reader to participate in the day-to-day occurrences of the expedition: "At this moment the laboratory rings with click of hammer, and nails, and iron hoops" (*JB* 242). The book ends where it began, in Rio, and then, instead of a conclusion, offers the reader some "General Impressions of Brazil" (chapter 16), followed by five appendices, ranging from comments on the Gulf Stream to observations on flying fishes to an outline of Louis Agassiz's own pet theories of race ("Permanence of Characteristics in Different Human Species").

However, *Journey in Brazil* also differs significantly from its precursors. Wallace had to patch his narrative together from the few notebooks he was able to save from submersion in the Atlantic. Henry Walter Bates composed his account ("an arduous task") only reluctantly and finished it mainly at the instigation of his mentor and model Darwin (*NRA* iv). Apparently, no such problems existed for Louis and Elizabeth Agassiz. *Journey in Brazil,* published less than two years after Agassiz had returned from the tropics, grew, as Louis Agassiz rejoiced in his preface, "naturally" into its present shape: the effortless result of circumstances rather than of "preconceived design" (*JB* ix).

Journey in Brazil is a cowritten book. Collaboration and multiple authorship are in themselves not unusual in travel narratives, but *Journey in Brazil* is different from its precursors, including Agassiz's own *Lake Superior,* written in collaboration with J. Elliot Cabot, in that here the narratives of science and personal adventure seemed to blend effortlessly. Agassiz's own observations, which he said he shared daily with his wife, "knowing that she would allow nothing to be lost which was worth preserving" (*JB* x), are framed by, and merge with, the entries Elizabeth made in her diary, "partly for the entertainment of her friends, partly with the idea that I might make some use of it in knitting together the scientific reports of my journey by a thread of narrative" (*JB* ix). In

Lake Superior bits and pieces from Agassiz's lectures and other scientific pronouncements had laced Cabot's text, but in *Journey in Brazil* the scientist and the writer, we are invited to think, firmly joined hands: "Our separate contributions have become so closely interwoven that we should hardly know how to disconnect them, and our common journal is therefore published . . . almost as it was originally written" (*JB* x).

The quaint, domestic metaphor of knitting used by Louis Agassiz indicates a mutuality of effort, the happy togetherness of husband and wife not just in life but also in work—a suggestion that, however unintentionally, is disavowed by Elizabeth herself when later in her text she distinguishes between the strict "progress" of her husband's science and the looser "thread" of her own narrative. Justifying the inclusion of one of her husband's letters in her text, she writes that it was meant to benefit "those of my readers who care to follow the scientific progrese of the expedition as well as the thread of personal adventure" (*JB* 157). In their collaboration on *Journey in Brazil*, the professor gave the cues while his wife performed the actual work of writing. While the professor pinned, pickled, plucked, and dissected his way through the Brazilian flora and fauna, it was his wife who sat, spun, knitted, and wove the narrative.

Edwards, Bates, and Wallace are mostly reticent about their personal circumstances, to the extent that they rarely ever mention their traveling partners. Striving to keep the autobiographical impulse at bay, they usually remain silent even about their loneliness. When Wallace complains about his "dull evenings" in the forest, the lack of human companionship and the absence of proper reading material characteristically appear within a single sentence (*TA* 175). And even though he must have suffered dreadfully from having to go without news from home for considerable lengths of time, Bates was careful to limit descriptions of metaphysical despair to those experiences he had in an early phase of his Brazilian sojourn, when months had elapsed "without letters or remittances" and he had read from beginning to end even the advertisements in the few copies of the *Athenaeum* that he owned: "Towards the end of this time my clothes had worn to rags: I was barefoot, a great inconvenience in tropical forests, notwithstanding statements to the contrary that have been published by travellers; my servant ran away, and I was robbed of nearly all my copper money" (*NRA* 2:187–88).

Compared with the experiences of these bachelor-naturalists in Amazonia, the Agassizs' *Journey in Brazil* is a very busy book—a book in which people are always together and in which there is little time for boredom, despondency, or loneliness. Since the narrative is mainly written in her own words, Elizabeth frequently succeeds in establishing her own perspective on life in Brazil, as is evident, for example, in her comments on the situation of the women she encounters in the cities: "There is not a Brazilian senhora . . . who is not aware that her life is one of repression and constraint"

(*JB* 479). It is on those occasions that Elizabeth's writing gains special momentum, driven by a holy anger that seems to come from the autobiographical experiences of someone who, with little formal schooling[48] had to develop her considerable talents entirely on her own: "It is sad to see these stifled existences; without any contact with the world outside, without any charm of domestic life, without books or culture of any kind, the Brazilian Senhora in this part of the country either sinks contentedly into a vapid, empty, aimless life, or frets against her chains, and is as discontented as she is useless" (*JB* 270). By contrast, Elizabeth admires the women she met on the Upper Amazon who had been "brought up in the country and the midst of the Indians" and were therefore "very energetic, bearing a hand at the oar or the fishing-net with the strength of a man" (*JB* 232). But even as she wielded her energetic pen to comment on her husband's investigations, Elizabeth herself always remained his unfailingly admiring and supportive partner in life and never aspired to becoming his partner in science. In *Journey in Brazil,* there is no reference to her own accomplishments as a science writer,[49] and when Elizabeth addresses her readers directly, she often assumes a position of graceful innocence: "If any of my readers are as ignorant as I was myself before making this voyage, a bit of geography may not be out of place here. As everybody knows . . ." (*JB* 303).

Many passages in the book extend and cultivate this carefully crafted persona. For instance, while watching her husband dragging his net in an Amazonian lake, Elizabeth Agassiz rejoices that, with every fish he finds, the confirmation of his long-cherished theories seems to come closer: "It is impossible not to be gratified when the experience of later years confirms the premonitions of youth, and shows them to have been not mere guesses, but founded upon an insight into the true relation of things" (*JB* 233–234). But ultimately the pleasures of the fishing net are not for her: "Wearied after a while with watching the fishing in the sun, I went back into the forest, where I found the coffee-pot already boiling over the fire. It was pleasant to sit down on a fallen, moss-grown trunk, and breakfast in the shade" (*JB* 233). The specialized names of botany are tiresome to her: "I have a due respect for nomenclature, but when I inquire the name of some graceful tree or some exquisite flower, I like to receive a manageable answer, something that may fitly be introduced into the privacy of domestic life, rather than the ponderous official Latin appellation" (*JB* 90). The beautifully textured net enclosing the stem of an unusual mushroom she discovers on one of her rambles reminds her of the fairy web of Queen Mab. It remains for her husband to supply, in a footnote on the same page, the mushroom's "ponderous . . . Latin appellation," which in this particular case indeed is unfit for polite drawing-room conversation: "This mushroom belongs to the genus Phallus, and seems to be an undescribed species. I preserved it in alcohol" (*JB* 144).

The general parameters of this story seldom change. For example, when Elizabeth experiences "delicious . . . rest and refreshment in Pará," her husband—invariably referred to as "Mr. Agassiz" or "the Professor"—does what he does everywhere: he busily collects, preserves, lectures and supervises. "The week, so peaceful for me, has been one . . . of intense interest for Mr. Agassiz. The very day of his arrival, by the kindness of our host, his working-rooms were so arranged as to make an admirable laboratory, and, from the hour he entered them, specimens have poured in upon him from all quarters" (*JB* 144). When Louis peers into baskets of Brazilian fish, Elizabeth reaches for a book, heads for the beach, and takes a nap. "The wind in the trees overhead, the water rippling softly around the montarias moored at my side, lulled me into that mood of mind when one may be lazy without remorse or ennui. The highest duty seems then to be to do nothing" (*JB* 259).

Such passages can easily be seen as a gentle subversion, on Elizabeth Agassiz's part, of her husband's all-too-serious scientific pursuits.[50] To contemporary readers, however, the more immediate effect of this technique was to enhance Louis Agassiz's reputation as the well-nigh heroic "man of pure science" (*JB* 271), whose Midas touch turns everything, if not into gold, then at least into valuable collectibles—that is, into natural history specimens to be housed in the ever-growing museum back home: "Mr. Agassiz passed the morning in packing and arranging his fishes," Elizabeth notes on 30 October, "having collected in those two-days more than seventy new species" (*JB* 273). While Elizabeth sleeps, rests, saunters through "fragrant lanes," takes notes, or does nothing at all, Louis and his assistants sweat in the laboratory.

Moments of introspection and humble repose, in which even Louis Agassiz feels compelled to admire "this wonderful tropical world" (*JB* 371), fade in view of the seriousness of Agassiz's historic mission, often referred to simply as "the work." After all, Agassiz had traveled to Brazil to prove, through the zealous accumulation of new data, that Darwin's "transmutation theory is wholly without foundation in facts" (*JB* 33). To that end, he apparently also turned an entire foreign country into a reservoir of unpaid research assistants. "As usual, wherever we go, everybody turns naturalist in his behalf," observed Elizabeth Agassiz (*JB* 315). In every town where they stopped on their way from Rio to Pará, "the ready, cordial desire of the people" to help Agassiz enabled "him to get together collections which it would otherwise have been impossible to make in so short a time" (*JB* 138). Elizabeth Agassiz records one significant example of such spontaneous support. Traveling in a canoe through a narrow boat-path, concealed on either side by "long reedy grasses and tall mallow-plants with large pink blossoms," Agassiz simply reaches out and indiscriminately grabs whatever his eager hands can hold: toads, snails, beetles, grasshoppers, and flowers—"in short, an endless variety of living things, most interesting to the naturalist" (*JB* 359–360).⁵¹ Suddenly his oarsmen pitch in, too: inspired by "Mr. Agassiz's enthusiasm," they become "almost as interested as he was; and he had soon a large jar filled with objects quite new to him" (*JB* 360). Small wonder that sometimes Agassiz simply dispatches others to do his collecting for him: "The Indians here are very skilful in fishing, and instead of going to collect, Mr. Agassiz, immediately on arriving at any station, sends off several fishermen of the place, remaining himself on board to superintend the drawing and putting up of the specimens as they arrive" (*JB* 160).

Agassiz's compulsive collecting not only benefited from the willing participation of the local population, it also depended on the unprecedented institutional support he received. Prior to his trip, he kept up an extensive correspondence with the Pedro II, praising the emperor's talents as an amateur naturalist and even offering to name after him an extraordinary specimen that had arrived in Cambridge from Brazil, a fish that carried its eggs in its mouth until they are hatched: "Je demanderois à Votre Majesté la permission d'attacher Son Nom à cette espèce je le décrirai."⁵² While Charles Darwin and Alfred Russel Wallace had to beg for passports that would allow them to go from one part of Brazil to another, and Wallace had to bamboozle reluctant Indians into steering him and his badly built canoe through dangerous rapids, Louis Agassiz traveled in a steamer provided by the Brazilian authorities. In no other part of the world, he felt, "could a private scientific undertaking be greeted with more cordiality or receive a more liberal hospitality than has been accorded to the present expedition" (*JB* 145–146). Some of the "uncommon lack of cultural arrogance"⁵³ that Linda Bergmann has sensed in Elizabeth Agassiz's descriptions of Brazil can be traced directly to the fact that Agassiz's party was exploring the country in a semiofficial capacity, an aspect acknowledged in the preface to *Journey in Brazil*. What goes around comes around: "The Emperor of Brazil was deeply interested in all scientific undertakings, and had expressed a warm sympathy with my efforts to establish a great zoological museum in this country, aiding me even by sending collections made expressly under his order for the purpose. . . . I knew that the head of the government would give me every facility for my investigations" (*JB* [v]).

The steamer *Icamiaba*, owned by the Amazon Steamboat Company, transported Agassiz and his crew from Pará to Tabatinga on the Brazilian border (*JB* 145). On their return to Manáos, the war vessel *Ibicuhy* awaited them, placed at Agassiz's disposal by order of the Brazilian minister of public works (*JB* 351). The hardships the Agassizs expected to encounter on their South American journey seemed to "retreat" at their approach (*JB* 153); it would hardly be possible, reflects Elizabeth, "to travel with greater comfort than surrounds us here." Her suite of rooms on the steamer alone consisted of a stateroom with adjoining bath- and dressing-room, "and if the others are not quite so luxuriously accommodated, they have space enough" (*JB* 153). In fact, people at home and even some Brazilians themselves had an "exaggerated idea of the . . . difficulty of a voyage on the Amazons." A breakfast in Manáos, for example, had "all the comforts, and almost all the luxuries, of a similar entertainment in any other part of the world" (*JB* 285–286).

In his wife's admiring vignettes, Louis Agassiz appears like a latter-day, steam-propelled Columbus, surrounded by eager natives proffering natural history specimens: "At dusk we returned to the steamer, where we found a crowd of little boys and some older members of the village population, with snakes, fishes, insects, monkeys, &c. The news had spread that the collecting of 'bixos' was the object of this visit to their settlement, and all were thronging in with their live wares of different kinds" (*JB* 155). Along the Amazon at least, the hair-raising perils that attended and complicated the recent voyages of Bates and Wallace had indeed disappeared, and the days of romantic adventure and hair-breadth escapes were over: "The wild beasts of the forest have disappeared before the puff of the engine; the canoe and the encampment on the beach at night have given place to the prosaic conveniences of the steamboat" (*JB* 443).

Traditionally, travel is not comfort but travail, torture of one's body and mind, a sense that is reflected in the word's etymology and was still evident to earlier travelers. Jean de Léry, for example, at the end of his *Histoire d'un voyage en terre de Brésil* (1599–1600) included a short poem in which he reckoned with his God, who had made him suffer "tant de travaux."[54] Travel usually implied the departure from home, a separation from what is familiar, followed by a head-on plunge into the unfamiliar—experiences recorded, in exemplary fashion, in Alexander von Humboldt's *Personal Narrative of Travels to the Equinoctial Regions of America*. Venturing into regions so little frequented that even the natives could not identify the objects by which they tried to set their compass, Humboldt and his companion Aimé Bonpland slept in the open air and desperately tried to compensate for the lack of adequate food by ingesting small portions of dry cacao. Both eventually felt, while not "positively ill," "in a state of languor and weakness, caused by the torment of insects, and food, and a long voyage in narrow and damp boats." Crocodiles and boas they found to be "masters of the river," while the jaguar, the peccary, and the monkey ruled, "as in an ancient inheritance," in the rainforests of the South American interior. "Here, in a fertile country, adorned with eternal verdure, we seek in vain the traces of the power of man; we seem to be transported into a world different from that which gave us birth."[55]

The differences between the experiences of Humboldt (whose presence is frequently invoked in *Journey in Brazil*)[56] and those of Louis and Elizabeth Agassiz

55. "Adventure with Curl-Crested Toucans." Frontispiece, Henry Walter Bates, *The Naturalist on the River Amazons* (1863), vol. 1.

are not only due to the fact that the nature of the country itself had changed. They also indicate a shift in the nature and aims of scientific exploration itself. What used to be the hazardous undertaking of an individual had now become a corporate, well-funded enterprise, in which sufficient provisions and superior equipment guaranteed the successful outcome of the trip as much as the scientific training, sharp eyes, and daring of the participants. This difference is captured emblematically in the different treatments accorded to what Wallace and Bates regarded as the birds most characteristic of the Amazon district. The frontispiece of Bates's *Naturalist on the River Amazons* features the bespectacled, somewhat confused naturalist himself, inadequately dressed in a checkered shirt and loose pants fastened with a rough cord (ill. 55). In the process of retrieving his "dinner" (a toucan he had just shot) from a nearby thicket, Bates apparently found himself being attacked by a multitude of other long-beaked birds: "In an instant, as if by magic, the shady nook seemed alive with these birds, although there was certainly none visible when I entered the jungle. They descended towards me, hopping from bough to bough, some of them swinging on the loops and cables of woody lianas, and all croaking and fluttering their wings like so many furies" (*NRA* 2: 344). Unlike Bates, Agassiz and his group did not have to penetrate dark thickets to salvage their meager fare. In Elizabeth Agassiz's narrative, toucans appear as the decorative ingredients of a dish lovingly served to herself and her husband by the inhabitants of a little village near the Lake of Hyanaury: "They brought an enormous cluster of game as an offering. What a mass of color it was!—more like a gorgeous bouquet of flowers than a bunch of birds. It was composed entirely of Toucans, with their red and yellow beaks, blue eyes, and soft white breasts bordered with crimson; and of parrots, or papagaios as they call them here, with their gorgeous plumage of green, blue, purple, and red" (*JB* 260).

Elizabeth Agassiz, to be sure, liked her Brazil neat and tidy. In *Journey in Brazil,* she deplores the carelessness with which the botanical garden in Rio is kept, although she realizes that the very idea of establishing a garden in the midst of a tropical paradise is a bit preposterous: "The very readiness with which plants respond to the least culture bestowed upon them here makes it very difficult to keep grounds in that trim order which we think so essential" (*JB* 61). But one of the garden's features, "as unique as it is beautiful," she admires: a long avenue of palms, some of them eighty feet in height. "I wish it were possible to give in words the faintest idea of the architectural beauty of this colonnade of palms, with their green crowns meeting to form the roof." The palms are "straight, firm, and smooth as stone columns," and to make visible what her words cannot express, Elizabeth includes a woodcut, based, like the majority of the illustrations in her book, on a photograph (ill. 56). The majestic trunks of the palms hold the encroaching tropical vegetation at bay and bring the dim background of mountains and forests into focus. Her husband's scientific footnote identifies the palms as the "beautiful *Oreodoxa oleracea*" (*JB* 61).

Occasionally, Brazil seems even a little tidier than home, and here it wins Elizabeth's unqualified praise. The Indians of Brazil, she notes gratefully, are so much cleaner than the poor in the United States, even though they share their mud houses with all sorts of animals, rats, bats, and birds: "The open character of the houses and the personal cleanliness of the Indians make the atmosphere fresher and purer in their houses than in those of our poor." However "untidy" these natives may be in other respects, they unfailingly take baths once or twice a day, "if not oftener, and wash their clothes frequently." Never had the travelers' tender nostrils been offended by disagreeable odors in these households: "We certainly could not say as much for many houses where we have lodged when travelling in the West, or even 'Down East,' where the suspicious look of the bedding and the close air of the room often make one doubtful about the night's rest" (*JB* 263).

For Elizabeth Agassiz, the rainfor-

56. "Vista down the Alley of Palms." Woodcut from a photograph by Stahl & Wahnschaffe, Rio de Janeiro. From L. and E. Agassiz, *A Journey in Brazil*, 1868.

est is not the menacing and merciless place that Wallace had known but a welcome opportunity to see, in their natural context, the same plants she had admired in the greenhouses at home. The famous *Victoria regia*, the water lily, is simply charming when seen with all the appropriate birds, insects, and fish above, beneath, and around it: "Wonderful as it is when seen in the tank of a greenhouse, . . . in its own home it has the charm of harmony with all that surrounds it,—with the dense mass of forest, with palm and parasite, with birds of glowing plumage, with insects of all bright and wonderful tints, and with fishes which, though hidden in the water beneath it, are not less brilliant and varied than the world of life above" (*JB* 356). On her nature walks, Elizabeth thoroughly enjoys the variety of fragrant flowers she discovers. They remind her of "hot-house plants": "There often comes a warm breath from the depths of the woods, laden with . . . perfume, like the air from the open door of a conservatory" (*JB* 352). Ultimately she prefers the hothouse to the heat itself: "Much as I enjoy the verdure here, I appreciate, more than ever before, the marked passage of the seasons in our Northern hemisphere." Drenched with perspiration whenever she ventures outside, even for a few minutes, she finds Brazil a "continual vapor-bath" (*JB* 343). And in a syntactically well-crafted passage that is characteristic of her talents as a writer, she complains: "In this unchanging, green world, which never alters from century to century, except by a little more or less moisture, a little more or less heat, I think with the deepest gratitude of winter and spring, summer and autumn. The circle of nature seems incomplete" (*JB* 343).

Like Elizabeth Agassiz, Alfred Russel Wallace hankered after the gardens of his home country ("In the whole Amazon, no such thing as neatness or cultivation has ever been tried"; *TA* 232). But then, he apparently had more reason to yearn: many of the situations described in *Travels on the Amazon and Rio Negro* conjure up the image of the wrong man in the wrong place at the wrong time. Reluctantly following his Indian guides into a morass, Wallace "floundered about" in mud and water while his feet got caught "in tangled roots of aquatic plants, feeling warm and slimy, as if tenanted by all sorts of creeping things" (*TA* 48). Slowly he limps through the rainforest, in "barefooted enjoyment" of the hard pebbles and rotten leaves that cover his path: "Some overhanging bough would knock the cap from my head or the gun from my hand; or the hooked spines of the climbing palms would catch in my shirt-sleeves, and oblige me either to halt and deliberately unhook myself, or leave a portion of my unlucky garment behind" (*TA* 149). Six feet two inches tall, Wallace had no stateroom at his disposal in which he was able stretch out and suffered enormously in canoes whose cabins had berths "just five feet long" (*TA* 60). The mosquitoes kept him in a "state of feverish irritation" (*TA* 111), and fleas burrowed into his foot, infecting his toe and rendering "wearing a shoe, or walking, exceedingly painful" (*TA* 259). Though not inclined, as he said, to "despond in sickness," Wallace did become alarmed when he fell ill with dysentery and realized he had "no medicines or even proper food of any kind" (*TA* 196).

Worst of all were the conditions under which he had to work. Louis Agassiz commanded a fully equipped laboratory, in which his troops drew, dissected, and preserved the fish they had collected. Presiding over this "abode of Science" like a foreign potentate, Agassiz received Brazilian visitors "curious to see the actual working process of a laboratory of Natural History" (*JB* 59–60). He employed a scientific illustrator and, with the help of one of his assistants, Walter Hunnewell, as well as of local photographers, also made ample use of the new medium of photography.[57] By contrast, Alfred Russel Wallace had to draw his specimens outdoors, exposed to millions of flies that bit him on the face, ears, hands, and feet. "Often have I been obliged to start up from my seat, dash down my pencil, and wave my hands about in the cool air to get a little relief" (*TA* 174). His hands "as red as a boiled lobster" (*TA* 174) and his feet "of a dark purplish-red colour, and much swelled" (*TA* 213), Wallace often was unable to prepare his specimens adequately.

Some of the difficulties he encountered on this trip up and down the waterways of the Amazon, the Rio Negro, and the Uaupés were, admittedly, of his own making, the result of personal clumsiness rather than the challenges of the environment, and it is interesting that Wallace should have made them such a prominent part of his narrative.[58] Wallace loses his balance while his companion is shooting a coot and nearly swamps the boat; he knocks off his spectacles when running away from wasps that attack him; the birds he shoots get stuck in trees; he accidentally discharges his own gun when reaching for it by its muzzle and nearly kills himself. After the last mishap, he has to walk around with his arm in a sling, "unable to do anything, not even pin an insect, and consequently rather miserable" (*TA* 57). The only comparable disaster Agassiz mentions in *Journey in Brazil* is, significantly, someone else's fault. Apparently Agassiz kept "a very extensive collection of fish brains" in an open barrel, an appealing treasure he hoped to bring home to his museum. "In an unguarded moment, however, while landing, one of our assistants capsized the whole into the Rio Negro. It is the only part of my collections which was completely lost" (*JB* 244).

This brings us to the last, and cruellest, irony in Wallace's *Travels on the Amazon*. While Agassiz's immense Amazonian collections, including 1,800 new species of fish, arrived in Cambridge in good condition, "suffering little loss or injury in the process of transportation" (*JB* 396), Wallace came back from Brazil empty-handed. A fire on board the *Helen*, the ship that was taking him back to England, forced him and the crew into badly leaking lifeboats from which they were rescued only after ten harrowing days at sea. Helplessly, Wallace had to watch his monkeys, macaws, and parrots go down with his instruments, books, and drawings; only one bird, wet and confused, survived.

But Wallace's narrative is not a simple cautionary tale of the terrible things that may befall a naturalist. It is, instead, a consciously crafted book whose many contradictions and incongruities are part and parcel of a general refusal to emulate the common policy of travel narratives and to synthesize into one coherent image what in reality has been a plurality of different impressions. Only in time, says Wallace, will the peculiarities of vegetation, animals, and people form a "connected and definite impression" in the traveler's mind (*TA* 3–4).[59] Perhaps they never will. At a crucial point in his narrative, Wallace recites a poem he wrote at the border between Brazil and Venezuela. There, in a state of general "indignation against civilised life," Wallace apparently went native, at least in his excitable imagination. He envisioned himself hunting and fishing like an Indian, "rich without wealth, and happy without gold." After quoting his long Indian fantasy, Wallace laconically reports, in a decidedly antiromantic vein, the desertion of the "noble savages" who had been traveling with him at the time: "The day before I had just bought a fresh basket, and the sight of that appears to have supplied the last stimulus necessary to decide the question, and make them fly from the strange land and still stranger white man, who spent all his time in catching insects, and wasting good caxaça by putting fish and snakes into it" (*TA* 180).

Wallace's "natives" had followed him from Brazil into the Spanish-speaking border-lands of Venezuela, where even they were suddenly not native anymore and felt like foreigners. However, what remained constant for them even in such a new and strange environment was the abiding strangeness of the "still stranger white man" they had accompanied—out of place wherever his travels lead him, in a continent that has little need for all his fish-preserving, insect-catching, and snake-pickling. Here, the white naturalist's traditional look at all the "strangeness" around him is exchanged for the look of one of these so-called "strangers" at the white naturalist himself; suddenly *he* appears as the only real outsider in a world not meant for him, wearing his inappropriate garments and clutching his useless equipment. Passages such as this one from Wallace's book point to a quality of ironical self-reflexiveness that is virtually absent from the Agassizs' *Journey in Brazil*.

Louis Agassiz had gone to Brazil in order to find what he already knew would be there. He came to the Amazon to contemplate, in his wife's words, "what every investigation demonstrates afresh, namely, the distinct localization of species in each different water basin" (*JB* 259). This perfect accord of anticipation and achievement eliminates the element of dramatic surprise from the Agassizs' account in which knowledge increases only quantitatively, never qualitatively.[60] Louis Agassiz travels, but the more that things around him change, the more they remain the same. If Darwin's work had been, in George Levine's apt phrasing, concerned with "the strangeness of what we have always thought we understood,"[61] Agassiz's investigations revealed the ultimate familiarity of everything we only thought we didn't know. For Agassiz, the task of the naturalist was not discovery; it was the uncovering and happy recognition of what had always been there. *Journey in Brazil* is less a description of a country, the record of a risky search for the unknown, than a tangle of chatty stories about everything ranging from "sand-dunes resembling those of Cape-Cod" (*JB* 147) to the "frank geniality" of the local Indians, "so different from our sombre, sullen Indians, who are so unwilling to talk with strangers" (*JB* 226). In this context, the discourse of natural history appears as little more than a quotation, framed by a narrative written mostly from the informed tourist's point of view. And even though Elizabeth's and Louis's contributions to *Journey in Brazil* differ in emphasis, tone, and perspective, they eventually both have the same effect, a tendency to stabilize a foreknown, always familiar world. *Journey in Brazil*, in a sense, anticipates the professionalization of travel ridiculed, almost a hundred years later, by another visitor to Brazil, Claude Levi-Strauss, who in *Tristes Tropiques* laments that the literature of exploration has deteriorated to the status of the laundry list, "a story or two about the misdemeanors of the ship's dog, and a few scraps of information." Where mere mileage has replaced discovery, platitudes all too often take the shape of revelations.[62]

At the end of *Journey in Brazil*, Professor and Mrs. Agassiz travel back to Cambridge, their "northern home," not with surprising new insights but with countless barrels of fish and "a store of pleasant memories and vivid pictures to enrich our life hereafter with tropical warmth and color" (*JB* 494). On the last page of the book's final chapter, Louis Agassiz reflects that what he missed most in Brazilians was a certain lack of strength and persistence. But even in this, he added, he merely recalled "a distinction which is as ancient as the tropical and temperate zones themselves" (*BA* 517). The reader closes a book that has shown her, often brilliantly, what she only expected—namely that, well, Brazil is Brazil and Massachusetts is Massachusetts. By contrast, the hapless Wallace, bereft of even a single specimen to "illustrate the unknown lands I trod,"[63] leaves his readers with an image that illustrates perfectly how in *Travels on the Amazon* the narrator's willingness *to let himself be surprised* survives even the loss of his collections. Drifting in a lifeboat into an uncertain future, Wallace's glance falls upon the dolphins swimming about the boat: "Their colours when seen in the water are superb, the most gorgeous metallic hues of green, blue, and gold." And, he adds, touchingly, "I was never tired of admiring them" (*TA* 276).

"I Hate Collecting": William James

An increasingly irreverent outsider's view of the Thayer Expedition is supplied in the diaries, letters, notes and sketches of Agassiz's student William James.[64] The reasons James decided to participate in the trip, as a volunteer who depended on the financial support of his family, are as complex as (and perhaps directly related to) the reasons for his nonparticipation in the American Civil War, which left him, like his "helplessly absent" brother Henry, hushed and embarrassed.[65] His younger siblings Wilky and Bob, who were mere youngsters when they enlisted, instilled in James lifelong feelings of personal inadequacy. (When Wilky returned from the assault on Fort Wagner on 17 July 1863 with a wound in his foot, Henry James Sr. was surprised at "so much manhood so suddenly achieved.")[66] Was James aware of the connection between his failure to serve his country in its time of need and his desire to prove his manhood and mettle abroad? Biographers have also speculated on whether William James's scientific interests and the wish to explore the possibility of a career as a naturalist with a teacher he still regarded as "fascinating" were also operative when he signed up for Agassiz's expedition or whether he still secretly wanted to be a painter and yearned for the lushness of the tropical scenery he had never seen.[67]

Some of James's more deeply personal motives become evident in a letter he addressed to his parents shortly after his arrival in Rio, on 21 April 1865: "I have felt more sympathy with Bob and Wilk than ever from the fact of my isolated circumstances being more like theirs than the life I have led hitherto. Please send them this letter. It is written as much for them as for any one" (*CWJ* 4: 103). The boat trip had been hellish, to be sure, accompanied not by spice-laden "gentle zephyrs" but "hideous moist gales." However, even from seasickness a useful lesson could be learned: "The awful slough of despond into which you are there plunged," explained James, stiffly, "furnishes too profound an experience not to be a fruitful one" (*CWJ* 4: 100). But James's doleful mood didn't last long. He found the weather in Rio to be quite "like Newport." The city itself seemed redolent with "fraternal love," and the streets and shops in town were pleasantly reminiscent of Europe. The cuisine in the restaurant they went to was perfect, and the bookstore he patronized was well stocked with "french scientific & philosophical works" (*CWJ* 4: 102–103).

If James had really wanted his trip to Brazil to be a kind of private descent into hell,[68] life soon—to modify a bitter line from a poem by Elizabeth Bishop—proved "nothing if not amenable." In May 1865, left behind in Rio to collect jellyfish with Walter Hunnewell, James soon succumbed to a variety of smallpox that left him jaundiced and temporarily blinded. Eighteen days later, his face still resembling an "immense ripe raspberry," he was released from the "hard straw bed" and the "eternal chicken & rice, &

extortionate prices" of the hospital in Rio (*CWJ* 4: 105). He lamented the money wasted on medical treatments and announced to his father that he wanted to return home sooner rather than later. Rationalizing his reasons for escape, he listed what a modicum of foresight could have told him before he embarked on Agassiz's collecting expedition: "I find that by staying I shall learn next to nothing of Natural History as I care about learning it. My whole work will be mechanical, finding objects & packing them, & working so hard at that & in travelling that no time at all will be found for studying their structure. The affair reduces itself thus to so many months spent in physical exercise" (*CWJ* 4: 106). He was no Humboldt, after all, and exploring trips such as Thayer's definitely weren't his "forte": "Good Heavens! When such men are provided to do the work of travelling, exploring, and observing for humanity, men who gravitate into their work as the air does into our lungs, what need, what *business* have *we* outsiders to pant after them and toilsomely try to serve as their substitutes?" (*CWJ* 4: 107). Before he had come here, the "romance of the thing" had seemed so great, but now, "here on the ground, the romance vanishes & the misgivings float up" (*CWJ* 4: 108).

But James stayed.[69] And it would be wrong to regard his expedition as an "errand into the wilderness" or, more simply, as another giant waste of time in a life that really did not take shape, as James's biographers would have us believe, before 1878, the year that James married Alice Howe Gibbens, when he "acted, finally, alone, in his own interest, to fulfill his own needs."[70] A photograph taken during the Thayer Expedition shows the members of Agassiz's motley crew assembled in a makeshift photographic studio (ill. 57). As if un-

57. Group portrait of the Thayer Expedition. Photographer unknown. Rio de Janeiro, 1865. Museum of Comparative Zoology, Harvard University.

sure about the camera's actual position, everybody is looking in a different direction, an unintended but apt commentary on the trip as a whole. A comparatively relaxed James is resting on the floor, next to Monsieur Bourget ("a Frenchman . . . whom the Prof, picked up in Rio"; *CWJ* 4: 111). In the foreground, from left to right, are Walter Hunnewell, who served as the expedition's photographer, the Swiss painter and illustrator Jakob (Jacques) Burkhardt, and Newton Dexter. Standing in the rear is the son of the expedition's scientific sponsor, Stephen Thayer, who has placed his hands, in proprietary fashion, on the shoulders of Hunnewell. Next to Thayer we see the Brazilian interpreter, who had joined the party in Rio and funnily enough seems to be the only one who is looking squarely at us.[71] James's serious, bearded face and slightly lowered eyebrows betray the same inner detachment that his outward position in the picture also suggests—as if he were still wondering why he had ended up here in the first place, along with all these strange people. Interestingly, while the hands of the other sitters are at rest, James's right hand is fidgeting with some invisible object.

William's letters, diary entries, and drawings reveal that the self-conscious pose he adopted for the photograph was more than just a passing fancy. In fact, James was an acute, often satirical observer of Agassiz's maneuvers, starting with a comical vignette that characterizes the self-celebratory mood on board the steamer that brought the group to Brazil, In Agassiz's daily lectures, the members of the party were indoctrinated into the main tenets of the Harvard professor's creationism, a program that was actively endorsed by another passenger, the ailing Bishop Alonzo Potter ("the Bish"), who used his Sunday sermons on the ship to preach obedience unto Agassiz: "Last Sunday he . . . told us we must try to imitate the simple child like devotion to truth of our great leader. We must give up our pet theories of transmutation, spontaneous generation &c, and seek in nature what God has put there rather than try to put there some system wh. our imagination has devised &c &c. (Vide Agassiz passim.) The good old Prof, was melted to tears, and wepped profusely" (21 April 1865; *CWJ* 4: 101). But Agassiz seldom wept; more frequently, apparently, he supervised, ordered, and commanded, which was all the more necessary since of his eleven assistants, as James unsparingly observed, "3 are absolute idiots; Tom Ward, Dexter & myself know nothing; of the 5 who know something, one is superannuated & one in such a feeble condition that the least exertion renders him unwell" (to Henry James, 3 May 1865; *CWJ* 1: 7).

However, as James's notes from the "Encampment of Savans" (*CWJ* 1: 5) suggest, for him the main problem was not lack of intelligence or experience but the absence of the same enthusiasm that fueled Agassiz's compulsive activities. In *Journey in Brazil,* Agassiz praises his own method of dividing up his party in order to ascertain whether the differences in the distribution of fish in Amazonia were permanent and God-ordained (the preferred result) or—horrible to think—had been the result of migration. The distribution of species, Agassiz decided, could only be adequately determined by means of a corresponding distribution of naturalists: "I therefore determined to distribute our forces in such a way as to keep collecting parties at distant points, and to repeat collections from the same localities at different seasons. I pursued this method of investigation during our whole stay in the Amazons, dividing the party for the first time at Santarém, where Messrs. Dexter, James, and Talisman separated from us to ascend the Tapajoz, while Mr. Bourget remained at Santarém, and I, with the rest of my companions, kept on to Obydos and Villa Bella" (*JB* 169n).

What Agassiz presented as the result of rational calculation and ingenious premeditation seemed like untoward haste to James. When Agassiz ordered James, Dexter, and a Brazilian American with the colorful name Talismão de Figueiredo de Vasconcelos ("Tal for short," *CWJ* 4: 114) to collect fish on the Tapajós River, Agassiz rejoiced at the golden opportunity he was offering his assistants: "A collection of the fishes of the Tapajos wd. be extremely interesting and he wished we wd. try it." But James and friends considered the trip, as he told his sister Alice on 31 August 1865, "a very foolish expedition because the Prof only gave us 8 days to do it in, and no collection worth a cent can be made in that time; but the Prof is very apt to do things in that rash manner" (*CWJ* 4: 114). One of the Indians who was supposed to join the fishing party absconded, so "Tal" simply kidnapped another, "a wild Indian boy" with "blue black hair," "a real willing young savage" (*CWJ* 4: 114). Not only did they catch hardly any fish ("a wild goose chase"), they also ended up in a rainstorm that perversely reminded William of home: "I never suffered more pain since Father used to spank me with a paper cutter in fourteenth street" (*CWJ* 4: 116). Grounded for a few days in a lovely little cove, complete with palm huts and palm trees, they discovered the pleasures of lounging in hammocks and sipped coffee while an old Indian tried to fish for them but caught nothing either. When it was time again to head downstream, James's party had "very little of a collection" to offer to the "Prof" (*CWJ* 4: 117).

Back in Santarém, however, natural history collecting at first took a back seat to attempts at intercultural understanding: "At Brazilian tables we grin and bow a great deal, try a Portuguese sentence, flounder about in it & then give it up, & end by conversing among ourselves without paying any more attention to our hosts" (*CWJ* 4: 118). In retrospect James especially regrets that he did not too effectively communicate with the lovely native maidens who threw a ball for them at Santarém. "Ah Jesuina, Jesuina, my forest queen, my tropic flower," he exclaims, "why could I not make myself intelligible to thee?" His Portuguese was unfit, alas, to express "all those shades of emotion" that penetrated his soul, and in conclusion he conjures up the image of lovely "Jesuina" hankering after the foreigner she couldn't understand: "She now walks upon the beach with her long hair floating free, pining for my loss" (*CWJ* 4: 120). Modern readers might decide that, in a letter directed to his "beloved Alice" whom her brother wanted to "kiss . . . to death" (*CWJ* 4: 103–104), this is more rhetorical posturing than a faithful report of actual experiences, and in view of what we know about James's inhibitedness, such an interpretation is indeed likely, if not compelling. It is revealing that the list of useful practice phrases in the *lingoa geral* that occupies one page in one of James's Brazilian diaries included not just the words for "comb" and the question "Do you find many mosquitoes on the river Solimões?" ("chá purá rá ere té carapaná Solimoes oropi") but also equivalents for "woman" ("cunhá"), "servant girl" ("cunhá mucú") and "sweetheart" ("miraira") as well as the sentence "Serendéra ére mendáre potáre sére seirúma?" In James's own translation: "My dear, will you marry me?"[72]

James's narrative of a collecting trip gone awry appropriately finishes with the ironic sentence "This evening we reach Manaos & see again the principal light of modern science" (*CWJ* 4: 120). But how different do the meager results of James's collecting appear in the light of "modern science"! "The collections," noted Elizabeth, with irrepressible good cheer, "are very large from both our stations" (*JB* 183).[73]

A page from one of William James's Brazilian notebooks, in spite of the random doodlings that clutter it, epitomizes his own satirical view of the expedition, focusing on the "triumphal return" of one of the "idiots" who had accompanied Agassiz to Brazil: Newton Dexter, described in one of James's letters from Brazil as "a sunburnt & big jawed devil . . . from Providence" (*CWJ* 4: 99).[74] In the foreground of the cartoon (ill. 58), beside the number 7, Dexter himself is visible in an advanced stage of intoxication, with a bottle of "Old Tom" next to him—an image that is in marked contrast with the endless pageant of successes sketched out in the upper half of the page. No. 1 is Dexter's own, horse-drawn carriage; no. 2, a "large diamond

58. "Triumphal Return of Mr. Dexter from Brazil," by William James. From James's Diary from the Brazilian Expedition, 1865, Houghton Library, Harvard University, bMS Am 1092.9 (4498). By permission of Houghton Library and Bay James, literary executor for the James family.

from the Emp[eror?]." Carriage no. 3 contains the poems written by "Mr. D[exter]," while carriage no. 4 holds the remains of the unlucky draftsman himself. The hearse is followed by "New and hitherto unknown genera of animals discovered and captured by Mr. D."—a gigantic ostrich, a tailless hybrid of crocodile and iguana, and a round-eyed combination of elephant and horse are led on leashes held by tiny human figures. A little sign announcing the discovery of forty-thousand new species of fish (a spoof directed at Agassiz himself) heralds the appearance of an entourage consisting of "Young and beautifull Ladies in Love with Mr. D." The last (unnumbered) part of the pageant is, significantly, a monstrous book on wheels.

Unlike Agassiz, James did not write a hefty book about the expedition in which he had more or less reluctantly taken part. In fact, when he finally borrowed a copy of *Journey in Brazil* from an American dentist in Dresden, "that unauthorized sort of U.S. ambassador who is to be found in all large European cities" (*CWJ* 4: 296), he was disappointed, although he acknowledged that the book could have been even "bulkier & duller." Elizabeth Agassiz had no ability to "describe landscape, or in fact anything, worth a damn," wrote William James to Thomas Ward, who had also participated in the Thayer Expedition, on 24 May 1868 (*CWJ* 4: 310). When James dutifully offered his compliments to Elizabeth herself on 15 June 1868, his praise was mealy-mouthed and his choice of metaphors ambiguous. Elizabeth and her husband had, James wrote, "*sandwiched* the narrative with science so artfully that one . . . finishes the book without knowing it" (my emphasis). He could, he said, think of hardly anyone who would start the book and then not care to finish it. What was absent from Elizabeth's descriptions, however, (and here James didn't mince words) was the monotony and vast sadness of the Brazilian landscape: "On those side rivers at dawn when the forests used to reveal themselves standing as if painted the soberness was particularly striking—whereas in our climates there has always seemed to me something exhilarating about the 'jocund morn'" (*CWJ* 4: 316). Here, James was not only quoting Byron's *Childe Harold*;[75] but also indulging in self-quotation, directly referring to a passage in one of his diaries, written in Manáos shortly before the departure for his second trip up the Solimões in November 1865: "Whilst we were chatting the solemn sad dawn began to break and to show the woods standing, standing, as if in a picture. Surely no such epithet as the 'jocund morn' could ever have suggested itself to a dweller in these regions. The mysterious stirring of the fresh cool perfumed air, while the sky begins to lighten & redden & and all the noises of the night to cease as the day birds begin their singing & crying, all make these early hours the most delicious of the whole day here."[76]

By his own standards, James's scattered notes were more successful than Elizabeth Agassiz's strained descriptions, in terms of atmospheric evocation as well as in their characterization of people. His portrait of Louis Agassiz is surprisingly complex, vacillating between embarrassed admiration and bemused detachment.[77] In spite of his accent ("On con'trāry, as Agassiz says"; *CWJ* 4:122), Agassiz was James's General Sherman—but a Sherman painfully out of touch with the reality of his situation. "The Professor has just been expatiating over the map of South America and making projects as if he had sherman's army at his disposal instead of the 10 novices he really has." Still, for all his pathos, it was difficult to resist this particular leader: "*Offering* your services to Agassiz is as absurd as it wd. be for a S. Carolinian to *invite* Gen. Sherman's soldiers to partake of some refreshment when they called at his house" (to Mary Robertson Walsh James, 31 March 1865; *CWJ* 4: 98–99). Other remarks by James show the same kind of ambivalence: Agassiz's "charlatanerie is almost as great as his solid worth," wrote William to his brother Henry (3 May 1865; *CWJ* 1: 6), and when he admitted to his parents that he had profited from hearing Agassiz talk, he added that this had not been so much on account of what the professor actually said ("never did a man utter a greater amount of humbug"; *CWJ* 4: 122) but because of what he represented. Louis Agassiz was "a vast practical engine," a veritable fact-machine. "You have a greater feeling of weight & solidity about the movement of agassiz's mind, owing to the continual presence of this great background of special facts, than about the mind of any other man I know" (to Henry James Sr., 12 September 1865; *CWJ* 4:122). Warts and all, Agassiz, for James, had "a greater personal fascination than any one I know" (to Henry James Sr., 3 June 1865; *CWJ* 4: 108).

As much as the professor's proverbial energy impressed him ("I never saw a man work so hard"; *CWJ* 4: 122), the process by which Agassiz accumulated his facts wearied and appalled him. While James, his legs ulcerated with countless mosquito bites, slept on beaches and squatted in rickety canoes to catch fish, the professor usually advised, counseled, and instructed from a safe distance. One of James's notebooks from Brazil contains not only lists of expenses and debts, as well as remedies against snakebite, but also detailed instructions for the use of preserving fluids from the professor's lectures: "Preserve in alcohol all reptiles fishes and the very large insects. The small insects are packed dry in boxes in layers with paper moistened with alcohol & corrosive sublimate or alcohol & camphor . . . Of mollusks, make drawings while alive. Paint fishes fm. life."[78] One of Agassiz's last Brazilian letters to James shows how seriously the "Prof" took these hints: "Ascertain the condition of all the collections," he ordered James, "and give them a little help if needs be. Take also care that all the barrells that may come in . . . are distinctly labelled that I make no mistake when examining them."[79] James genuinely dreaded each time their ship was forced to stop, because wherever this happened, as he lamented in a letter to his parents, "the Prof, is sure to come around and say how very desirable it wd. be to get a large number of fishes from this place, and willy nilly you must trudge. . . . If there is any thing I hate it is collecting" (21 October 1865; *CWJ* 4: 127–128). In December 1865, he repeated: "I thoroughly *hate* collecting, and long to be back to books" (*CWJ* 4: 131), a wish that was granted to him soon after Christmas, when he left the Amazon for New York. Years later, in a letter to Thomas Ward, James would characterize Agassiz as a "natural born *collector*," someone who hates "to lose *anything* in creation," like Goethe, but minus the "intelligent glance" that knows how to separate the wheat from the chaff (24 May 1868; *CWJ* 4: 306). Flipping through the pages of *Journey in Brazil* in his German *Pension* and thinking back to his Brazilian sojourn, James was surprised by how much *he*, the collector *malgré lui*, had already lost—in memory: "So many of those fish names wh. were familiar to me call up no image of what they represent now" (*CWJ* 4: 310).

If the nitty-gritty of fieldwork exasperated and, less frequently, amused James—"the everlasting old story, *fish*"; he complained to his mother on 9 December 1865; *CWJ* 4: 131)—the persona of the naturalist as such, exemplified to almost comical perfection by Louis Agassiz himself, retained its fascination for him. In James's verbal portraits, Agassiz appears touching in his childishness and appalling in his ruthlessness, stubbornly averse to the slipstream of change and yet continually on the move himself, amiable yet dangerous, tolerant yet greedy, a friendly but callous carnivore with barrels of putrefying fish by his side. But James's ambivalence about Agassiz stopped short of one issue, the crucial nature of which he had sensed the minute he set foot on the Brazilian shore: "Almost every one is a negro or a negress," he wrote to his mother, "which words I perceive we dont know the meaning of with us" (21 April 1865; *CWJ* 4: 102).

Agassiz and Alexandrina

Travelers from the United States recognized with surprise and satisfaction that Brazilians approached the problem of slavery with much more equanimity than their own fellow citizens. Given that runaway slaves in Brazil were hunted down by the police and army and that those who stayed were often tortured and mutilated by their masters,[80] it seems remarkable indeed that visitors such as William Edwards felt compelled to make a point of the strong bond of "affection" (*V* 71) between the slaves and their owners that he thought he had witnessed on Brazilian plantations. "Scattered here and there were neat-looking houses of the blacks," Edwards remembered about his visit to the plantation of Senhor Antonio, on the river Guamá. Many of the slaves were "about, and all as fat and happy as their master" (*V* 61). Wherever he looked, smiles lit up "coal-black" countenances (*V* 64) and "little ebonies" performed their happy dances, their eyes expressive of "fun and frolic" (*V* 61). No wonder that the masters here did not have to punish their slaves (*V* 72). In short, Brazilian slavery was "little more than slavery in name" (*V* 204). One of the reasons for this was perhaps that the hot Brazilian sun burned everything and everyone into "pretty nearly the same tint" (*V* 23) or, put more seriously, that "prejudice against colour is scarcely known" (*V* 204). In fact, educated Brazilian blacks were, wrote Edwards, "just as talented and just as gentlemanly as the whites" (*V* 204).

Elizabeth Agassiz, too, saw with relief that Brazil did not seem to need a civil war to solve its racial problems. Although she was well aware of the morally "enfeebling" aspects of slavery, she was no friend of immediate equality for blacks either and was impressed by the gradual nature of the transition in Brazil: "It seems to me that we may have something to learn here in our own perplexities respecting the position of the black race among us, for the Brazilians are trying gradually and by installments some of the experiments which are forced upon us without previous preparation" (*JB* 128). Each year larger and larger numbers of blacks either received or bought their freedom. These free blacks, generally speaking, compared favorably, in terms of "intelligence and activity," with Brazilian whites.[81] But then, according to Elizabeth, they also had a very different category of whites to compete with: the Portuguese, members of "a less energetic and powerful race than the Anglo-Saxon" (*JB* 129). If Alfred Russel Wallace had emphatically denied that it could be proper and "right to keep a number of our fellow-creatures in a state of adult infancy" (*TA* 83), Louis and Elizabeth Agassiz subscribed to a philosophy that easily blended racial prejudice and Anglo-Saxon exceptionalism.

For Louis Agassiz, as for a whole slew of North American naturalists before him, the proper study of natural history included the study of "man," and Brazil was the ideal place where one of Agassiz's favorite topics, racial amalgamation or "miscegenation," could be studied in vitro, as it were. Agassiz conceded that a thorough study of the "different nations and cross-breeds inhabiting the Amazonian Valley" would have required more empirical data than he had been able to assemble during his travels. But he also emphasized that a country like Brazil, "where the uncultivated part of the population go half naked, and are frequently seen entirely undressed," provided an ideal testing ground for the efficiency of what he called "the natural history method," that is, "the comparison of individuals of different kinds with one another, just as naturalists compare specimens of different species" (*JB* 529). Agassiz's results, as sketched out in appendix 5 to *Journey in Brazil*, are impressive indeed. For example, he notes that, like "long-armed monkeys," the "negroes" of Brazil have a "comparatively short body" but long limbs, while the Brazilian Indians sport a long, heavy trunk and short arms and legs. Indian women have conical breasts, whereas black women's breasts seem more cylindrical. And so forth.

Before his move to the United States, Agassiz had shown little interest in racial typologies, but, after a visit to Dr. Samuel Morton's "American Golgotha" with its six hundred skulls, he quickly "Americanized" himself by arguing that different races were, in fact, different species, "zoologically distinct," and that there really was no gap wider than the one separating black from white.[82] At a meeting of the American Association for the Advancement of Science in 1850, Agassiz declared that people of different skin color "did not originate from a common centre, nor from a single pair."[83] Charleston naturalists had taken Agassiz on a tour of the nearby plantations, where he examined, from a distance, the different tribes of Africans sweating in the fields and concluded that "human affairs with reference to the colored races would be far more judiciously conducted if, in our intercourse with them, we were guided by a full consciousness of the real differences existing between us and them . . . rather than by treating them on terms of equality."[84]

Apart from considerations of geographical range, the absence of interbreeding was regarded as the best test of specific distinctness,[85] and Agassiz was a bit hard pressed to account for the existence of offspring of parents that belonged to different races. Human species, like animal species, could not mix; confronted with plenty of evidence that in fact they did, Agassiz suggested that, well, perhaps they shouldn't. In 1863, the year of the Emancipation Proclamation, which was also the year during which the term "miscegenation" first entered the American language,[86] Agassiz touched upon this controversial topic in an exchange of letters with Samuel Gridley Howe, a member of the American Freedmen's Inquiry Commission. Howe had approached the "Prof" with a specific set of questions about the future of the "African race" in the United States.[87] Agassiz used the opportunity to lash out, from a "physiological" as well as "from a high moral point of view," against racial amalgamation. Relations between partners of different races were a violation of the law of the preservation of species as well as a "sin against nature," the product predominantly of the unnatural situation in the American South, where the "better instincts" of lusty southern gentlemen were stunted and blunted by the constant availability of sexually responsive "mulatto" housemaids. From their servants, lamented Agassiz on 9 August 1863, these hotspurs would soon graduate to "more spicy partners, as I have often heard the full blacks called by fast young men" (9 August 1963). Enter the "halfbreed": degenerate, sickly, sterile, a combination of the vices and defects of the races of his or her parents, destined to perish (10 August 1863).

There was, said Agassiz, no doubt in his mind that racial mixing, as personally repugnant to him as it seemed nationally unacceptable, had to be considered the real reason for the Civil War. What had caused the conflict was not the issue of slavery; it was the shocked "recognition of our own type in the offspring of Southern gentlemen, moving among us as negroes, which they are not" (9 August 1863). Agassiz conjured up a nightmarish scenario in which the progress of racial amalgamation in the United States, if not checked, would transform the country from a "manly population descended from cognate nations" into "the effeminate progeny of mixed races, half indian, half negro, sprinkled with white blood," He added, "I shudder from the consequences" (10 August 1863). Let us therefore, he demanded, "put every possible obstacle to the crossing of the races" (9 August 1863), and let us implement legislation intended to accelerate the disappearance of "halfbreeds" from our midst. Louis Agassiz opposed slavery, if only because he hoped that legal freedom for blacks in the South would encourage and foster the geographical separation of the races, with northern blacks migrating in droves to the South, where they would join their brothers and toil in a humid climate for which their natural indolence had made them eminently suitable. "Halfbreeds," impaired by their deficient physique and their lack of "fecundity" (11 August 1863), would then soon become extinct. And if Agassiz "rejoiced" at the prospect of universal emancipation, he did so chiefly "because hereafter a physiologist or ethnographer will be able to discuss the question of the races and to advocate a discriminating policy respecting them, without seeming to support legal inequality" (10 August 1863). Agassiz's argument is interestingly circuitous: free the blacks in the South so that we can control their freedom.[88]

The data Agassiz thought he had assembled among the scantily clad "colored" population of Brazil were meant to support this convoluted line of reasoning. In many ways, the slaveholding society of Brazil was the living embodiment of Agassiz's worst fears. Here, where masters freely cohabited with their slaves and the slaves mingled with the natives, where white children grew up "in the constant companionship of their blacks" (*JB* 481), there was "absolutely no distinction of color," as Elizabeth noted primly, adding that it was "rare to see a person in society who can be called a genuine negro; but there are many mulattoes and mamelucos, that is, persons having black or Indian blood" (*JB* 280). In the northern provinces especially, Mrs. Agassiz was shocked to behold the "enfeebled character of the population," which gave her occasion to rehearse her husband's theories of the detriments of racial mixing: "The variety of color in every society where slavery prevails tells the same story of amalgamation of race; but here this mixture of races seems to have had a much more unfavorable influence on the physical development than in the United States. It is as if all clearness of type had been blurred, and the result is a vague compound lacking character and expression" (*JB* 292–293).

Here, Elizabeth's husband, in one of his inevitable footnotes, for once did not elaborate scientifically but merely chimed in: "Let any one who doubts the evil of this mixture of races, and is inclined, from a mistaken philanthropy, to break down all barriers between them, come to Brazil" (*JB* 293). In Agassiz's skewed reading, miscegenation in Brazil, a country in which "mulattoes," "mamelucos," and "cafuzos" roamed freely, proved that mixing did not enhance but rather diluted the product, making it unpalatable. Whoever had been to Brazil could not deny, exclaimed Agassiz, "the deterioration consequent upon an amalgamation of races, more widespread here than in any other country in the world." In Brazil, in fact, racial mixing was "rapidly effacing the best qualities of the white man, the negro, and the Indian, leaving a mongrel nondescript type, deficient in physical and mental energy." It was not difficult for Agassiz to turn this harrowing experience into a cautionary tale to benefit his fellow Americans:

> At a time when the new social status of the negro is a subject of vital importance in our statesmanship, we should profit by the experience of a country where, though slavery exists, there is far more liberality toward the free negro than he has ever enjoyed in the United States. Let us learn the double lesson: open all the advantages of education to the negro, . . . but respect the laws of nature, and let all our dealings with the black man tend to preserve, as far as possible, the distinctness of his national characteristics, and the integrity of our own. (*JB* 293)

What was behind Agassiz's duplicitous reasoning was nothing else but the once radical belief that humans were part and parcel of the animal kingdom, which here had turned into a justification for separating humans from humans.[89] Interestingly enough, Agassiz inadvertently undermined his own dark vision of a mixed-race America by asserting that at least in Brazil "halfbreeds" always evidenced a tendency to revert to their "original stocks," which, of course, raised hopes that what was mixed would one day be completely pure again.[90] "Children between mameluco and mameluco, or between cafuzo and cafuzo, or between mulatto and mulatto, are seldom met with where the pure races occur; while offspring of mulattoes with whites, Indians and negroes, or of mamelucos with whites, Indians, and negroes, or of cafuzos with whites, Indians, and negroes, form the bulk of these mixed populations" (*JB* 298).

In *Journey in Brazil*, only one of these Brazilian "halfbreeds," who are normally merely object lessons of the dangers of miscegenation, gains a flicker of individuality. The mixed-blood girl Alexandrina appeared in the Harvard naturalist's Brazilian household on or around 29 September 1865, at the same time that Elizabeth Agassiz—in spite of her general reservations against hiring Indians, whose "irregular habits" and preference for liquor made them ill suited to be domestic servants—also acquired

59. "Head of Alexandrina." Woodcut from a drawing by William James. From L. and E. Agassiz, *A Journey in Brazil*, 1868.

an "Indian lad," called Bruno, whom she hoped to "break . . . in gradually" (*JB* 223–224). No such caution was needed with Alexandrina, since, with "a mixture of Indian and black blood in her veins," she was a so-called cafuzo and therefore, in Agassiz's homespun racial typology, combined "the intelligence of the Indian with the greater pliability of the negro" (*JB* 224). Here, for once, in the view of the not-so-doctrinaire Elizabeth, miscegenation had improved the product, for the consumer's comfort and convenience.

By all appearances, Alexandrina indeed did not let the Agassizs down. She proved to be a worthwhile investment, "not only from a domestic, but also from a scientific point of view," as Elizabeth noted on 9 October (*JB* 235). In addition to assisting Louis in the lab, preparing and cleaning his "skeletons of fish very nicely," she also accompanied Elizabeth on her neighborhood nature walks: "With the keen perceptions of a person whose only training has been through the senses, she is far quicker than I am in discerning the smallest plant in fruit or flower, and now that she knows what I am seeking, she is a very efficient aid" (*JB* 236). Her ultimate merit lay in her ability to climb trees, which proved helpful in the rainforest but also revealed who her real relatives were: "Nimble as a monkey, she thinks nothing of climbing to the top of a tree to bring down a blossoming branch" (*JB* 236).

Nimble Alexandrina appears for the last time in Elizabeth's narrative in the form of a portrait, a woodcut based on a sketch made by none other than William James (ill. 59). Here, Alexandrina once again proves her scientific value—this time not on account of her native expertise, as the subject of knowledge, but as an especially interesting and curious object in Louis Agassiz's collection of racial hybrids. It was only, added Elizabeth, "after a good deal of coy demur" (*JB* 246) that Alexandrina consented to sitting for her portrait—an attitude we are invited to think is derived from the native suspicion of likenesses, though this fear more usually took the form of a distrust of photography, in which the identification between the image and the person was even more complete: "There is a prevalent superstition among the Indians and Negroes that a portrait absorbs into itself something of the vitality of the sitter, and that any one is liable to die shortly after his picture is taken. This notion is so deeply rooted that it has been no easy matter to overcome it" (*JB* 276).

If in his sketch William James manages to give Alexandrina a modicum of individuality, this says more about his talents as an artist than about his suitability as Agassiz's "scientific" illustrator. "Cafuzos," Agassiz believed, did not have the "delicate" features of the mulatto; they were dark-complexioned and had "long, wiry, and curly" hair (*JB* 532). James's drawing duly emphasizes Alexandria's enormously expansive head of hair, a thicket of drawn-out curls radiating in all directions. It is, of course, impossible to enter into Alexandria's experience of the story, but, upon closer inspection, there indeed seems to be little "coyness" in the picture. In James's rendering, Alexandrina's full lips under the refined nose are slightly pursed, and her almond eyes look skeptically and, if not sadly, then at least without joy at the viewer. And why shouldn't she have resisted William James's portrait-making? Given her knowledge of what the whites were "seeking," Alexandrina also would have known that the finished product would not have been hers to keep; that instead it would be removed for external analysis[91] and made part of an inventory of interesting specimens, like the flowers she had helped Mrs. Agassiz collect in the forest. Indeed, in *Journey in Brazil,* having briefly gained individuality and a form of modest personal distinction as the scientific helpmate of Elizabeth and Louis Agassiz, Alexandrina quickly sinks back into the anonymity of a type, the generic value of an "illustration."

Normally, in his indefatigable pursuit of scientific data, Agassiz relied less on painting than on photography, a medium that seemed more suited to capture the contours as well as the essence of racial difference and to document the weakness of racial "hybrids." Walter Hunnewell, the son of a wealthy father, had brought with him a new photographic contraption that took pictures with less trouble than the daguerreotype, and Manáos presented Agassiz with the perfect opportunity to document his prejudices: "Our picturesque barrack of a room, which we have left for more comfortable quarters in Mr. Honorio's house, serves as a photographic saloon, and here Mr. Agassiz is at work half the day with his young friend Mr. Hunnewell, who spent almost the whole time of our stay in Rio in learning photography, and has become quite expert in taking likenesses" (*JB* 276). Agassiz had, in a sense, pioneered the use of the camera for the purposes of anthropological fieldwork when, in 1850, he commissioned the Charleston photographer Joseph T. Zealy to take daguerreotypes of fifteen slaves in the nude—a gallery of racial "types" intended to teach viewers a lesson in racial inferiority.[92] The photographic sessions in Brazil had a similar purpose. They elicited a "very complete series" of racial hybrids (*JB* 296).

Agassiz's photographic collection survives in three albums, crammed full with images of Brazilians in various stages of undress. A sad cornucopia of "racial types," this strange picture gallery featured men and women of African descent (their tribal affiliations were noted on the frame), children, "mulattoes," even Chinese living in Brazil. The albums themselves still bear the accession slips of the Library of Agassiz's Museum of Comparative Zoology (pasted over by more recent slips that indicate that they were received, as a gift from Louis Agassiz's son Alexander, by the Library of the Peabody Museum of American Archeology and Ethnology on 28 June 1910).

Agassiz lets most of his images speak for themselves. The majority of them are without captions; those that are identified bear the most generic titles, such as "Congo," "Monjola," or "Mina Ebu."[93] The greater number of the pictures present their subjects in the order Agassiz himself prescribed in *Journey in Brazil,* namely "in full face, in perfect profile, and from behind" (*JB* 529). It is fair to assume that all of the cabinet photographs that show a sequence of poses were produced by Agassiz's assistant Walter Hunnewell, experimenting with his father's new camera. They are often badly focused, uncertain in their use of background and foreground, stiff and amateurish in the way the subjects are posed, and generally far less accomplished than the full-face, expertly posed portraits that make up part of the second and the entire third album. These were the work of photographic studios, such as Stahl & Wahnschaffe in Rio, whose stamp appears on some of them (*JB* 529).

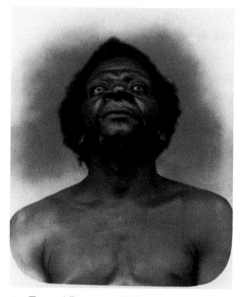

60. Frontal Portrait of Man, by Augusto Stahl, from L. Agassiz's Brazilian Albums, Albumen print, 1865. Courtesy of the Peabody Museum of Archaeology and Ethnology, Harvard University. PM # 2004.1.436.1.97.

One of the more professionally executed shots in Agassiz's albums, a bust portrait of a man by the accomplished German-Brazilian photographer Augusto Stahl (ill. 60), depicts the sitter with his head dramatically tilted back, the eyes directed upward. But here the dynamic pose, enhanced by the effective studio background, does not serve to heroicize or even individualize the subject. Rather, it seems intended to exaggerate and distort the subject's physiognomy, widening the nostrils, enlarging the nose, emphasizing the lower facial area and reducing the forehead, thus producing what white American viewers—especially those familiar with the profile views of apes, Hottentots, and African-Americans offered in Gliddon and Nott's *Types of*

Mankind (1854)—were supposed to interpret as an image of racial inferiority.

However, the amateurish photographs made by Hunnewell are, in their own way, no less composed than the pictures produced by Augusto Stahl. In all of Hunnewell's shots, taken in the same shabby courtyard, the same rickety chair appears. It seems to be there not only for the photographer's convenience—as a prop, as something he could direct his subject to sit on or lean against—but also as a basis for comparing the individuals depicted in different photographs and thus as a way of facilitating the "successful" application of Agassiz's "natural history method" (ill. 61)

Another criterion for comparison is supplied by the several stereoscopic postcards of marble statues Agassiz inserted at strategically important points in his albums (ill. 62). They were meant to serve not only as "substitutes, alleviating the need for photographs of nude Whites,"[94] but also established a fairly direct, if deplorable, iconographical context

61. Profile Portrait of a Woman. Photograph by Walter Hunnewell, 1865. From L. Agassiz's Brazilian Albums. Courtesy of the Peabody Museum of Archaeology and Ethnology, Harvard University. PM# 2004.1.436.1.40.

in which to read the photographs of black Brazilians. Several of the statues contrast the idealized, beautiful forms of the nude human body with the carefully draped folds of various items of clothing. Thorvaldsen's Venus, included in album 3, coyly clutches her dress with her left hand, letting it slowly sink to the floor; Apollo, also inserted into the third album, extends his left arm, thus unfolding the loose garment fastened around

62. "Gems of Statuary by Eminent Sculptors." Stereoscopic card of Egeria, sculpture by J. H. Foley (1818–1874). From L. Agassiz's Brazilian Albums. Courtesy of the Museum of Archaeology and Ethnology, Harvard University. PM# 2004.1.436.1.39.

his neck in one smooth, dramatic gesture of unveiling. Gracing the first album, Egeria, created by Irish sculptor John Henry Foley, emerges leisurely from the dress that is carelessly draped around her hips.

In Agassiz's images of Brazilian slaves, however, disorderly white heaps of discarded clothing piled on the floor or dropped on the chair appear like ironical leitmotifs—quotations of an iconographical tradition supposedly alien to the world of their subjects, indices of the alleged differences between the white classical standard and the black deviation from the rule. In the photograph immediately following the postcard with Egeria, for example, the contrast between the nymph's undulating body and Agassiz's subject is enhanced by the rigid profile view in which the woman is presented to the viewer (ill. 61). Standing in front of a dilapidated building, paint peeling off the walls, she is pressing her arms stiffly against the sides of her body; her dress lies crumpled behind her on the seat of the inevitable chair. The cloth the photographer has arranged over the fence in the background, perhaps to achieve better lighting, is a far cry from the elegant drapes behind Egeria. Hunnewell's picture is clearly not meant to seduce the viewer; if it served to stimulate the imagination at all, it was supposed to *repel* as it titillated. A similar impulse to denigrate the sitter is apparent in a tripartite image, also by Stahl, from Agassiz's third album, captioned "Mulâtre," which was evidently intended to support Agassiz's warnings about the dangers of miscegenation and the inevitable enfeeblement it brings to the human "races" (the last shot in the series drew specific attention to the subject's genitals, which appeared unusually formed). Ultimately, however, the photograph, a rigid side view of a man with curly hair, captured only the image of someone who did not have his picture taken for fun (ill. 63).

Elizabeth Agassiz noted that, after some initial problems with native misgivings about her husband's portrait-taking, it soon became easier for them to find models: "The desire to see themselves

63. Portrait of a Man (frontal, profile, and back), detail ("Mulâtre"), by Augusto Stahl. From L. Agassiz's Brazilian Albums. Courtesy of the Museum of Archaeology and Ethnology, Harvard University. PM# 2004.1.436.1.158.

in a picture is gradually gaining the ascendant, the example of a few courageous ones having emboldened the more timid, and models arc much more easily obtained now" (*JB* 277). When William James accidentally witnessed one of the professor's photographic sessions, he was unpleasantly surprised by the liberties Agassiz took with his native models whom he persuaded to pose in the nude. Before reluctantly embarking on one of his collecting trips,[95] James had sought out the "Prof" himself in his "photographic establishment." Hunnewell "with his black hands" catiously let him in. On entering the room, James "found Prof, engaged in cajoling 3 mocas whom he called pure indians but who, I thought and as afterwards appeared, had white blood. They were very nicely dressed in white muslin & jewelry with flowers in their hair & an excellent smell of pripioca. Apparently refined, at all events not sluttish, they consented to the utmost liberties being taken with them and two without much trouble were induced to strip and pose naked." Even members of the white Brazilian establishment found Agassiz's activities in the darkroom unusual, as James reported: "While we were there Sr. Tavares Bastos came in and asked me mockingly, if I was attached to the Bureau d'Anthropologie."[96] Exploiting the mystique that surrounded the photographic process itself (note Hunnewell's "black hands"), Agassiz was obviously taking advantage of young women ("môças") who under normal circumstances would have been too "refined" to expose themselves. The episode seemed all the more embarrassing since Tavares Bastos (1839–1875), a national deputy from the province of Alagoas, was one of the most passionate Brazilian admirers of the United States.[97]

What happened in Hunnewell's laboratory, where the camera looked on while the naturalist fingered the bodies of native women, could be regarded as a poignant conclusion, a damning footnote, to the tradition of American natural history that has been the subject of this book. In a way, James's portrait of Alexandrina and his story of Agassiz and Hunnewell in their "laboratory" encapsulate the contradictions that had marked American natural history from its beginnings—characterized as it was by the daring attempt to treat humans as part of nature and by its ultimate pathetic failure to do so in any other way than through a sharp division of those fully human from those just a little less human than human.

I have not been able to locate the original of James's sketch of Alexandrina, but his Brazilian portfolio also includes one drawing that offers an even more interesting comment on Louis Agassiz's obsession with racial mixing. I am referring to a sketch, my final image in this book, of two young Brazilians, one light-skinned, straight-haired boy, his companion dark and curly-haired, which shows that the former art student has not lost his grasp of visual detail. His painter's eye has captured the bold outlines of the boys' heads and bodies. Their symmetrically arranged backs turned to the viewer, they seem to look in different directions. However, as their position indicates, what separates them—like what unites them—is none of the viewer's business, but exclusively theirs (ill. 64). James's drawing records not physical facts but an inner mood—something that we can only speculate on, never ascertain. No image could be further from Agassiz's intrusive portraits of half- or completely undressed Brazilian girls. Small wonder that James felt the mockery when Tavares Bastos asked him if he was a member of the "Bureau of Anthropologic."

After his return from Brazil, it would take William James another ten years of agonizing headaches and back pain, despair, and depression before he was able to "select a vocation that stemmed from his sense of who he was,"[98] but his Brazilian notes already suggest the productive direction in which he would be looking. In 1868, James reviewed Darwin's *The Variation of Animals and Plants under Domestication*. While he disagreed with some of Darwin's "hypotheses," he still appreciated that the book "harrows and refreshes, as it were, the whole field of which it treats." And, thinking of "that scoundrel" Agassiz, he informed his novelist brother, "The more I think of Darwin's ideas the more weighty do they appear to me."[99] His former teacher's efforts notwithstanding, William James soon embraced Darwinism, but he used it to his own ends and began to develop a provisional and always revisable theory he said was "no science . . . only the hope of a science."[100] Among other things, James proposed the paradox that what made human beings fit for natural survival was in fact the one quality in them that apparently was *not* natural, a quality that made the mind independent of the machinery of the brain—personal consciousness. Mind and world "have been evolved together," insisted James, and mental facts could not be "properly studied apart from the physical environment of which they take cognizance."[101] In James's psychology, consciousness is a *"fighter for ends."*[102] Novel ideas are spontaneous mental variations, like the chance variations in nature pointed out by Darwin; they have to be adapted to the intellectual and social habitat into which they are born in order to survive.[103]

Significantly, when Agassiz appeared again in James's own work, it was not as an influence but as an object of study. In "Great Men and Their Environment," a lecture given before the Harvard Natural History Society and published in the *Atlantic Monthly* in October 1880, James applied Darwinian theory and argued

64. *Two Seated Brazilian Youths.* Sketch by William James, ca. 1865. Houghton Library, MS Am 1092.2 (7b). By permission of Houghton Library, Harvard University, and Bay James, literary executor for the James family.

that "great men" could be most usefully understood as "spontaneous variations" in the social organism. Their environment adopts or rejects, preserves or destroys, in short, *selects* them. Thus James arrived at the following analogy that superficially asserts Agassiz's greatness yet reduces it to size by turning him into an illustration of the validity of the very theory against which Agassiz had fought. The "great man," declared James, is selected by his environment, and the environment, in turn, becomes modified by his influence and presence. So Agassiz in America seems like the European rabbit in New Zealand. In James's view, the "great man" acts as a ferment, and changes its constitution, just as the advent of a new zoölogical species changes the faunal and floral equilibrium of the region in which it appears. We all recollect Mr. Darwin's famous statement of the influence of cats on the growth of clover in their neighborhood. We all have read of the effects of the European rabbit in New Zealand, and we have many of us taken part in the controversy about the English sparrow here—whether he kills most cankerworms, or drives away most native birds, just so the great man, whether he be an importation from without like Clive in India or Agassiz here, or whether he spring from the soil like Mahomet or Franklin, brings about a rearrangement, on a large or a small scale, of the pre-existing social relations.[104] Louis Agassiz remained unrepentant and incorrigible right up to the end in his iron opposition to Darwinian theory. A praiser of his own past, he died in 1873, "sad and intellectually isolated."[105] His onetime student William James, in his own work, went on to "overhaul the very idea of truth," developed a philosophy meant to "unstiffen all our theories," and became critical of what had been the implicit basis not just of Agassiz's work but of the project of natural history as such: "the simple duplication by the mind of a ready-made and given reality."[106] Twenty-three years after Agassiz's death, as the century was drawing to a close, William James, now himself in his fifties, spoke at a reception of the American Society of Naturalists given by the president and fellows of Harvard College at Cambridge and offered a tribute to his former teacher that memorably captures the poetry and the pathos of natural history in America: "I had the privilege of admission to his society during the Thayer expedition to Brazil. I well remember at night, as we all swung in our hammocks in the fairy-like moonlight, on the deck of the steamer that throbbed its way up the Amazon between the forests guarding the stream on either side, how he turned and whispered, 'James, are you awake?' and continued, 'I cannot sleep; I am too happy; I keep thinking of these glorious plans.'"

Encapsulated in this little anecdote is all the excitement and excitability that kept naturalists going: the dreamlike nature of the project that made natural history an affair of the senses as much as of the mind, a holiday of the imagination as well as an adventure in the world of empirical fact, an enterprise in which the cheerful solipsism of the naturalist (Agassiz apparently did not wait for an answer from James before he continued to talk about himself) was as much a given as was the pliability of the outside world, softly amenable, at least in the subject's joyful anticipation, to his "glorious plans." In passing, with the studied casualness of the good stylist, William James offered an assessment of Louis Agassiz that might very well count as the unacknowledged premise behind the "poetics of natural history" as I have attempted to describe it in this book. Agassiz, said William James, had "looked on the world as if it and he were made for each other, and on the vast diversity of living things as if he were there with authority to take mental possession of them all."[107]

NOTES

1 Charles Darwin, *On the Origin of Species by Means of Natural Selection, or the Preservation of Favoured Races in the Struggle for Life.* London: Murray, 1859, 310–311.
2 John James Audubon, "Labrador Journal," *The Audubon Reader*, New York: Everyman's Library, 2006, 401.

INTRODUCTION

1 Walt Whitman, *Complete Poetry and Collected Prose*, ed. Justin Kaplan (New York: Library of America, 1982) 30.
2 Linnaeus, *A General System of Nature, through the Three Grand Kingdoms of Animals, Vegetables, and Minerals, Systematically Divided into their Several Classes, Orders, Genera, Species, and Varieties, with their Habitations, Manners, Economy, Structure, and Peculiarities*, trans. William Turton, M.D., 7 vols. (London: Lachington, Allen, 1806) 1: 2–3.
3 "Extract of a letter, written by John Winthrop" (1670), *The Puritans: A Sourcebook of Their Writings*, ed. Perry Miller and Thomas H. Johnson, rev. ed., 2 vols. (New York: Harper & Row, 1963) 2: 740. The "otherness" of North American nature, which is unlike everything else in the world and everything-we-are-not, is a standard topic even in much contemporary writing on American natural history; see Peter Fritzell, *Nature Writing and America: Essays upon a Cultural Type* (Ames: Iowa State University Press, 1990).
4 See Susan M. Pearce, "The Urge to Collect," *Interpreting Objects and Collections*, ed. S. M. Pearce (London: Routledge, 1994) 157–159; James Clifford, "Collecting Ourselves," in Clifford, *The Predicament of Culture: Twentieth-Century Ethnography, Literature, and Art* (Cambridge, MA: Harvard University Press, 1988) 216–230.
5 Bruno Latour, *Science in Action* (Cambridge, MA: Harvard University Press, 1987) 225.
6 See Brenda Danet and Tamar Katriel, "No Two Alike: Play and Aesthetics in Collecting," Pearce, *Interpreting Objects and Collections*: "To relate to an object or an experience as a collectable is to experience it aesthetically" (225).
7 I have borrowed this phrase from Richard Jenkyns, "Child's Play," *New York Review of Books*, 17 July 1997:42–44.
8 Susan Stewart, On Longing: Narratives of the Miniature, the Gigantic, the Souvenir, the Collection (1984; Durham, NC: Duke University Press, 1993) 151.
9 Gaston Bachelard, *The Poetics of Space*, trans. Maria Jolas (1964; Boston: Beacon Press, 1994), see esp. 16. The French edition, *La Poétique de l'éspace*, appeared in 1958.
10 While in Philadelphia, Manasseh Cutler also visited Bartram's garden and complained that the plants appeared to have been arranged neither "ornamentally nor botanically" but seemed "to be jumbled together in heaps" (*LJC* 1: 273).
11 William Hedges, "Toward a National History," *The Columbia Literary History of the*

United States, gen. ed. Emory Elliott (New York: Columbia University Press, 1988) 190–191.

12 Edmund Wilson, *Patriotic Gore: Studies in the Literature of the Civil War* (1964; New York: Norton, 1994) 486.

13 Van Wyck Brooks, *The Flowering of New England* (New York: E. P. Dutton, 1952) 457; Robert J. Richardson Jr., *Emerson: The Mind on Fire* (Berkeley: University of California Press, 1995) 467, 545.

14 Ian F. A. Bell, "Divine Patterns: Louis Agassiz and American Men of Letters. Some Preliminary Explorations," *Journal of American Studies* 10 (1976): 349–381.

15 Christopher Looby, "The Constitution of Nature: Taxonomy as Politics in Jefferson, Peale, and Bartram," *Early American Literature* 22 (1987): 257, A similar argument is presented more succinctly by Ellen Sacco, "Racial Theory, Museum Practice: The Colored World of Charles Willson Peale," *Museum Anthropology* 20.2 (1997): 26–32.

16 Mary Louise Pratt, *Imperial Eyes: Travel Writing and Transculturation* (London: Routledge, 1992) 31, 27.

17 See Looby, "The Constitution of Nature" 256.

18 Paul Semonin, "'Nature's Nation': Natural History as Nationalism in the New Republic," *Northwest Review* 30.2 (1992): 6.

19 Philip Marshall Hicks, "The Development of the Natural History Essay in American Literature," diss., University of Pennsylvania, 1924,125.

20 Pamela Regis, Describing Early America: Bartram, Jefferson, Crevecoeur, and the Rhetoric of Natural History (De Kalb: Southern Illinois University Press, 1992) xi.

21 David Scofield Wilson, *In the Presence of Nature* (Amherst: University of Massachusetts Press, 1978) 7–9.

22 Linnaeus's *Lachesis Lapponica* was first published in English in 1811; the Swedish version did not appear before 1888.

23 See, for example, Wolfgang Iser, *The Fictive and the Imaginary: Charting Literary Anthropology* (Baltimore: Johns Hopkins University Press, 1993).

24 See Robert J. Scholnick, "Permeable Boundaries: Literature and Science in America," *American Literature and Science*, ed. R. J. Scholnick (Lexington: University of Kentucky Press, 1992) 1–17; Ludmilla Jordanova, introduction, *Languages of Nature: Critical Essays on Science and Literature*, ed. L. Jordanova (London: Free Association, 1986), esp. 23–25.

25 William Wordsworth, "A Poet's Epitaph," *Selected Poems*, ed. John O. Hayden (Harmondsworth: Penguin, 1994) 91.

26 Daniel Boorstin, The Lost World of Thomas Jefferson, with a new preface (1948; Chicago: University of Chicago Press, 1993) 16; Benjamin Smith Barton, *A Discourse on Some of the Principal Desiderata in Natural History, and on the Best Means of Promoting the Study of This Science, in the United-States* (Philadelphia Denham & Town, 1807) 12.

27 Alan Gross, "Science and Culture," *American Literary History* 7.1 (Spring 1995); 169–186.

28 Lawrence Buell, *The Environmental Imagination: Thoreau, Nature Writing, and the Formation of American Culture* (Cambridge, MA: Belknap, 1995) 422.

29 Charles Darwin, *The Descent of Man, and Selection in Relation to Sex*, rev. 2nd ed. (1888; New York: D. Appleton, 1927) 634.

30 Gillian Beer, "'The Face of Nature': Anthropomorphic Elements in the Language of *The Origin of Species*," Jordanova, *Languages of Nature* 212–243.

31 Vladimir Nabokov, "Audubon's *Butterflies, Moths, and Other Studies*, Comp. and Edited by Alice Ford" (1952), in Nabokov, *Strong Opinions* (New York: McGraw-Hill, 1973) 330.

32 See the surveys offered by William M. Smallwood and Mabel C. Smallwood, *Natural History and the American Mind* (New York: Columbia University Press, 1941); Charlotte M. Porter, *The Eagle's Nest: Natural History and American Ideas, 1812–1842* (Tuscaloosa: University of Alabama Press, 1986); Robert Bruce, *The Launching of Modern American Science, 1846–1876* (New York: Knopf, 1987). An essential tool for research is Max Meisel, *A Bibliography of American Natural History: The Pioneer Century, 1769–1865*, 3 vols. (1924–1929; New York: Hafner, 1967). For an extensive annotated bibliography of primary material and secondary studies, see *This Incomparable Lande: A Book of American Nature Writing*, ed. Thomas J Lyon (1989; New York: Penguin, 1991)399–476.

33 Danet and Katriel, "No Two Alike" 234.

34 Mark Kipperman, "The Rhetorical Case against a Theory of Literature and Science," *Philosophy and Literature* 10 (1986): 82

I. "AMERICA TRANSPLANTED": JOHN AND WILLIAM BARTRAM

1 On Breintnall, one of the members of Franklin's Junto Club for "mutual Improvement" and an unjustly neglected figure in colonial intellectual history, see Stephen Bloore, "Joseph Breintnall, First Secretary of the Library Company," *Pennsylvania Magazine of History and Biography* 59 (1935): 42–56.

2 To John Bartram, 25 February 1740/41, *CJB* 145. The English style of counting years was reformed in 1752, when the Julian calendar was abandoned in favor of the Gregorian calendar. The beginning of the year was set at 1 January, rather than at 25 March. In the century prior to this reform, writers, including John Bartram, would often refer to both the old and the new year for dates between 1 January and 25 March, and I have respected this practice here. Selections from the correspondence between Collinson and Bartram were first printed in William Darlington's *Memorials of John Bartram and Humphry Marshall, with Notices of Their Botanical Contemporaries* (Philadelphia: Lindsay & Blakiston, 1849). Thomas P. Slaughter, in a review published in *William and Mary Quarterly* 3rd ser. 50 (1993): 440–443, has leveled serious criticism against the only modern edition of Bartram's correspondence, prepared by Edmund Berkeley and Dorothy Smith Berkeley (*CJB*). Scholars curious about John Bartram should by all means consult the Bartram Papers at the Historical Society of Pennsylvania; however, since it is my aim to interest a wider audience in this remarkable exchange of letters, all my references are to the Berkeleys' edition, which is easily accessible. The editors of *CJB*, in order to remedy the frequent lack of punctuation in Bartram's letters and for the sake of better readability, have chosen to leave additional space between sentences, and it is thus that the letters appear here, too.

3 Quoted in Norman G. Brett-James, *The Life of Peter Collinson* ([London:] Edgar G.

Dunstan, 1926) 26. The standard biography of John Bartram is Edmund Berkeley and Dorothy Smith Berkeley, *The Life and Travels of John Bartram: From Lake Ontario to the River St. John* (Tallahassee: University Presses of Florida, 1982).

4 Quoted in Brett-James, *Life of Collinson* 52.

5 William Stork, *A Description of East-Florida with a journal, Kept by John Bartram of Philadelphia, Botanist to His Majesty for the Floridas; Upon a journey from St. Augustine up the River St. John's, as far as the Lakes. With Explanatory Botanical Notes. Illustrated with an accurate Map of East-Florida, and two plans; one of St. Augustine, and the other of the Bay of Espiritu Santo,* 3rd ed. (London: Sold by W. Nicoll . . . and T. Jefferies, 1769) pt. 2, ii.

6 Lewis Weston Dillwyn, *Hortus Collisonianus: An Account of the Plants Cultivated by the Late Peter Collinson* (Swansea: Printed by W. C. Murray and D. Rees, 1843).

7 Raymond Phineas Stearns, *Science in the British Colonies of America* (Urbana: University of Illinois Press, 1970) 516.

8 See Daniel McKinley, "Peter Collinson's Curious Amphibious Quadruped (1753)," *Archives of Natural History* 15.1 (1988): 73.

9 See Ann Leighton, *American Gardens in the Eighteenth Century: "For Use or for Delight"* (Boston: Houghton Mifflin, 1976) 129. Leighton briefly describes Collinson and Bartram as "staunch and understanding friends" (122), "two good and great men who never met but loved each other for years across the wide Atlantic, which they literally ploughed with their plants and seeds" (129). Thomas P. Slaughter acknowledges that the two collectors "sustained a closeness that was more intense than face-to-face encounters generally are in our time" but remains a little biased in his treatment of Collinson, whom he describes as an "English merchant who helped John market his plants"; *The Natures of John and William Bartram* (New York: Knopf, 1996) 85,13.

10 See Anne Larsen, "Equipment for the Field," *Cultures of Natural History,* ed. N. Jardine, J. A. Secord, and E. C. Spray (Cambridge: Cambridge University Press, 1996) 367.

11 Brenda Danet and Tamar Katriel, "No Two Alike: Play and Aesthetics in Collecting," *Interpreting Objects and Collections,* ed. Susan M. Pearce (London: Routledge, 1994) 228.

12 See *Historisches Wörterbuch der Rhetorik,* ed. Gert Ueding, 3 vols, to date (Tübingen: Niemeyer, 1992–) 2:67.

13 Ian Watt, *The Rise of the Novel: Studies in Defoe, Richardson, and Fielding* (London: Chatto & Windus, 1957) 189.

14 For a chronological list of John Bartram's letters to Collinson from 1740 to 1763 that were printed in the *Philosophical Transactions of the Royal Society,* see *CJB* 793–794.

15 Linnaeus's *Systema naturae* first appeared in 1735. Collinson's letter to Bartram, 14 December 1737, shows that his admiration for the Swedish naturalist remained mixed with skepticism: "The Systema Naturae is a Curious pformance for a young man but His Coining a set of new names for plants tends but to Embarrass & perplex the study of Botany. . . . be that as it will He is certainly a very Ingenious Man & a great naturalist" (*CJB* 72). In his *Letters from an American Farmer* (1782), J. Hector St. John de Crevecoeur devotes an admiring chapter to "Mr. John Bertram [sic], the Celebrated Pennsylvanian Botanist" (letter 11), in which he commends Bartram for having acquired, within a pe-

riod of three months, "Latin enough to understand Linnaeus"; *Letters from an American Farmer and Sketches of Eighteenth-Century America,* ed. Albert Stone (Harmondsworth: Penguin, 1981) 195. Bartram's Latin was in fact patchy, and for his knowledge of classificatory detail he would often depend on secondhand information from friends such as the Leiden physician Frederick Gronovius, whom he asked to write down his observations "in English which I can understand much better than Latin" (*CJB* 267). But his own fieldwork in America soon led Bartram to be even more critical of Linnaeus than Collinson had been: "I find abundance of many plants that do not answer to any of his Genera & is really A new Genus," he wrote to Philip Miller in 1755 (*CJB* 380). Bartram knew that there were "no general rules without some exceptions so it is with Lineus sistem" (*CJB* 457).

16 Brett-James, *Collinson* 42.

17 Brett-James, *Collinson* 37.

18 See also Bartram to Collinson, 21 June 1743: "I have many small children & none yet grown up to take care of business" (*CJB* 217).

19 James Gordon (1708–1780) was a gardener at Mile End, London, credited with the introduction of *Ginkgo biloba* to England.

20 See Collinson's memorandum from 1763, published in Dillwyn's *Hortus Collinsonianus*: "I often stand with wonder and amazement when I view the inconceivable variety of flowers, shrubs, and trees, now in our gardens and what there were forty years ago; in that time what quantities from all North America have annually been collected. . . . Very few gardens, if any excel mine at Mill Hill, for the rare exotics which are my delight" (vii).

21 Watt, *Rise of the Novel* 195.

22 Francis Bacon, "Of Gardens" (1625), *The Essays,* ed. John Pitcher (Harmondsworth: Penguin, 1985) 197.

23 See *Gentleman's Magazine* 71 (1801): 199–200; Ann Shteir, *Cultivating Women, Cultivating Science: Floras Daughters and Botany in England, 1760 to 1860* (Baltimore: Johns Hopkins University Press, 1996) 35. Erasmus Darwin encouraged the readers of *The Loves of Plants* (1789) to think of his work as "diverse little pictures suspended over the chimney of a Lady's dressing-room, *connected only by a slight festoon of ribbons*" (Oxford: Woodstock, 1991, vi).

24 Also named "moosewood," after the moose's penchant for grazing on its branches and leaves.

25 In his *Elements of Botany; or, Outlines of the Natural History of Vegetables,* Benjamin Smith Barton praises the *Sarraccnia purpurea* or "Purple Side-Saddle-flower" as a "very singular plant," "a native of various parts of NorthAmerica" (Philadelphia: Printed for the author, 1803, pt. 3,17). Specimens of the sarracenia were brought to England by John Tradescant in 1640, but Dillwyn believes the plant was first cultivated with success by Collinson (*Hortus Collinsonianus* 49).

26 Alexander Garden to Cadwallader Colden, 4 November 1754, quoted in Slaughter, *Natures of John and William Bartram* 81.

27 See, however, Bonalda Stringher, who is obviously unaware of John Bartram's letters to Collinson, in "The South as a Garden: The *Travels* of William Bartram," *The United*

States South: Regionalism and Identity, ed. Valeria Gennaro Lerda and Tjebbe Westendorp (Firenze: Bulzoni, 1990) 114.

28 See Collinson's letter to John Bartram, 6/9 April 1759: "We are all much Entertained with thy draught of thy House and Garden the situation most delightful and that for our plants is well chosen" (*CJB* 463). For reasons why the sketch should be attributed to William rather than John Bartram (handwriting, perspective, artistic skill), see Slaughter, *Natures of John and William Bartram* 35.

29 Andrew Cunningham, "The Culture of Gardens," *Cultures of Natural History* 38.

30 Bacon, *Essays* 199.

31 Collinson liked to tease Bartram about his correspondence with a widow in Charleston, Martha Logan, who provided Bartram with seeds from her garden and apparently "spare[d] no pains or cost to oblige" him (22 May 1761; *CJB* 517).

32 The species of *Cypripedium* which Collinson introduced into English gardens are listed in *Hortus Collinsonianus* 17.

33 In plate 72 of the second volume of Mark Catesby's *Natural History of Carolina, Florida, and the Bahama Islands,* a picture of the "LADY'S SLIPPER of Pensilvania" serves as the unlikely background for a representation of a large bullfrog ("Rana maxima Americana aquatica," now known as *Rana catesbiana*). "This curious *Helleborine,*" explains Catesby, "was sent from *Pensylvania,* by *Mr. John Bertram,* who, by his Industry and Inclination to the Searches into Nature, has discovered and sent over a great many new Productions both Animal and Vegetable. This Plant flowered in *Mr. Collinson's* garden, in *April,* 1738" (*NHC* 2: 72). For John Bartram's opinion on Catesby, see *CJB* 376: "Indeed his birds snakes & fishes may be excelent as far as I know for I am not so acquainted with them as with plants of which he seems to know little of thair natural growth but hath done all at random except outlines of ye leaves" (16 December 1754).

34 John Bartram consoled Collinson on 3 December 1762: "Ye red & white calceolus & ye grassy plant white orchis & autumnal gentian & A pretty ornithogalon perticular to ye desert of Jersey is ye most difficult plants I know to prosper in a garden" (*CJB* 579).

35 Billy Bartram promptly obliged, and Collinson acknowledged receipt of the picture in September 1760: "Pray thank Billy for so Elegantly painting It—my Flowers are not quite so Large & Whiter than yours" (*CJB* 492).

36 Danct and Katriel, "No Two Alike" 226.

37 The third edition of *Medicina Britannica; or, A Treatise on Such Physical Plants, as Are Generally to be found in the Fields or Gardens of Great-Britain; Containing a Particular Account of their Nature, Virtues, and Uses,* by Thomas Short, of Sheffield, M.D. (Philadelphia: B. Franklin, 1751), sported "a PREFACE by Mr. John Bartram, Botanist of Pennsylvania, and his NOTES throughout the Work, shewing the Places where many of the described PLANTS are to be found in these Parts of America, their Differences in Name, Appearance and Virtue, from those of the same Kind in Europe; and an Appendix, containing a Description of a Number of PLANTS peculiar to *America,* their Uses, Virtues, & c." The "Dog Piss Weed" is discussed on page 287. Even though Bartram promises to show the "Places" where the plants that are described grow, most of his notes in fact record their *near-absence* from the wild in the American colonies, thus

contributing, at least indirectly, to the American or English reader's realization of the remarkable difference between the two countries: "It is a fine Aromatick Plant, but is difficult to raise in Pennsylvania" (21); "I never observed any of the true kind to grow in our colonies" (66); "Grows wild no where in our Colonies" (157); "Grows only in curious Gardens with us" (206); "We have not this Species growing in our Provinces" (187), "We have had it from Europe, but it soon went off" (291); "I have often had Seed from Europe, but it never came up" (291).

38 On Collinson's falling-out with the Library Company, see Austin K. Gray, *The First American Library* (Philadelphia: Library Company of Philadelphia, 1936) 17–18.

39 See *The Spectator,* ed. G. Gregory Smith, 8 vols. (London, 1897–1898) 7: 17 (no. 477, 6 September 1712).

40 See Stearns, *Science in the British Colonies* 579: "If Bartram ever resented Collinson's attitude, he never mentioned it. He accepted the gifts with humble thanks, performed his tasks as a collector with industry and skill, grew in knowledge under his patron's generous and considerate care—and profited from the subscriptions which Collinson collected to support his efforts." For a more accurate assessment, see Leighton, *American Gardens* 128: "Bartram was always able to give as well as he got, sometimes better."

41 John Bartram to Sir Hans Sloane, 14 November 1742 (*CJB* 207).

42 Charlotte M. Porter, in "Philadelphia Story: Florida Gives William Bartram a Second Chance," *Florida Historical Quarterly* 71 (1993): 310–323, takes a less generous view of their relationship and, especially, of Collinson's part in it. She makes much of the "petty wrath" Collinson vented when Bartram neglected to send him a copy of his Florida journal, which had been appended (without John's knowledge) to the second edition of William Stork's *Account of East-Florida* (1766). The title page of Stork's edition of the journal identified Bartram as "Botanist to His Majesty for the Floridas." I am more inclined to understand Collinson's hurt pride, especially in view of his frequent reminders to John how difficult it had been for him to secure this appointment from the king: "John thou knows nothing what it is to Solicit at Court any favour" (May 1765; *CJB* 646).

43 In his edition of Bartram's correspondence, Darlington confuses the Tipitiwitchet with the *Schrankia,* the "sensitive briar." The origin of the name is unclear, because the letter is lost in which Bartram himself is believed to have commented on it. Naturalist John Ellis, in a brief note addressed to the "curious Collectors of rare Plants about London" and published shortly after Collinson's death in the *St. James's Chronicle; or the British Evening-Post,* 1–3 Sept. 1768, no. 1172 [p. 4], credits Bartram with the discovery and naming of the plant. It was subsequently called *Dionaea* (after Aphrodite) by the Swedish botanist Daniel Carl Solander, allegedly for the beauty of its leaves and the elegance of its flowers. Ellis added the trivial name *muscipula* (which did not mean "flytrap," as he probably intended, but "mousetrap"): "About three years ago that diligent and indefatigable Botanist, Mr. John Bartram, sent a dried Specimen of this extraordinary Plant in Flower to the worthy Peter Collinson, Esq. of Mill Hill, F.R.S. the lately deceased Friend of all Botanists, by the Indian Name, either Cherokee or Catabaw, but which I cannot now recollect, of Tippitywitchit, which he said he had collected in the Swamps beyond the Blue-Mountains." Ellis's claims are strongly disputed by Daniel McKinley,

in "'Wagish Plant as Wagishly Described.' John Bartram's Tipitiwitchet: A Flytrap, Some Clams and Venus Obscured," included in E. Charles Nelson, *Aphrodite's Mouse-trap: A Biography of Venus's Flytrap* (Aberystwyth: Boethius, 1990) 125–145. Ellis also described the *Dionaea* in an appendix to his *Directions for Bringing Over Seeds and Plants, from the East-Indies and Other Distant Countries, in a State of Vegetation: Together with a Catalogue of such Foreign Plants as are worthy of being encouraged in our American Colonies, for the Purposes of Medicine, Agriculture, and Commerce. To which is added, The Figure and Botanical Description of a New Sensitive Plant, called Dionaea Muscipula: Or, Venus's Fly-Trap* (London: Davis, 1770).

44 Charles Darwin, *Insectivorous Plants*, vol. 12 of *The Works of Charles Darwin* (New York: Appleton, 1896) 286.

45 On jokes and their function for the reassertion of masculinity in men's friendships, see Karen Walker, "'I'm Not Friends the Way She's Friends': Ideological and Behavioral Constructions of Masculinity in Men's Friendships," *masculinities* 2.2 (Summer 1994): 38–55.

46 See Hans Werner Ingensiep, "Der Mensch im Spiegel der Tier und Pflanzenseele: Zur Anthropomorphologie der Naturwahrnehmung im 18. Jahrhundert," *Der ganze Mensch: Anthropologie und Literatur im 18. Jahrhundert*, ed. Hans-Jürgen Schings (Stuttgart: J. B. Metzler, 1994) 54–79.

47 Crèvecoeur, *Letters from an American Farmer* 69.

48 See also *T* 379–380. William Bartram's *Colocasia*, an uncanny representation of gigantic flowering American Lotus plants, features, in the lower left corner, a drawing of the Venus flytrap; plate 21 in William Bartram, *Botanical and Zoological Drawings, 1756–1788*, ed. Joseph Ewan (Philadelphia: American Philosophical Society, 1968).

49 William Bartram, *Drawings* 60.

50 A hundred years later, on 28 June 1873, Charles Darwin reported a similar disappointment to the botanist Joseph Dalton Hooker: "I cannot make the little creature grow well"; *More Letters of Charles Darwin: A Record of His Work in a Series of Hitherto Unpublished Letters*, ed. Francis Darwin, 2 vols. (London: John Murray, 1903) 1: 349.

51 Benjamin Smith Barton, *Elements of Botany*, 2nd ed. (1803; Philadelphia: Printed for the author, 1814), pt. 2, 22.

52 Charlotte M. Porter, "The Drawings of William Bartram (1739–1823), American Naturalist," *Archives of Natural History* 16.3 (1989): 289. In 1766, Collinson showed Billy Bartram's paintings to one of the most renowned botanical artists of his time, Georg Ehret, who admired their "elegance" (*CJB* 667).

53 John Fothergill to John Bartram, 1772 (*CJB* 754); Fothergill to Bartram, September or October 1772 (*CJB* 751).

54 *Chain of Friendship: Selected Letters of Dr. John Fothergill of London, 1735–1780*, ed. Betsy C. Corner and Christopher C. Booth (Cambridge, MA: Belknap, 1971) 409.

55 Robert McCracken Peck, "William Bartram and His Travels," *Contributions to the History of North American Natural History*, ed. Alwyne Wheeler (London: Society for the Bibliography of Natural History, 1983) 38.

56 See John Fothergill to John Bartram, September or October 1772: "He [William Bar-

tram] proposes to go to Florida It is a country abounding with great variety of plants and many of them unknown. To search for these will be of use to Science in general but I am a little selfish I wish to introduce into this country the more hardy American plants, such as will bear our winters without much shelter" (*CJB* 750).

57 William Bartram's "Travels in Georgia and Florida, 1773–74: A Report to Dr. John Fothergill" was first published by Francis Harper in the *Transactions of the American Philosophical Society* ns 23.2 (1943): 121–242.

58 William Hedges, "Toward a National Literature," *The Columbia Literary History of the United States,* gen. ed. Emory Elliott (New York: Columbia University Press, 1988) 190.

59 See Peter Collinson to John Bartram, 10 March 1759: "I received Billey's Letter I am pleased to see him Improved in his Writing I wish I could say as much in his spelling—which will be Easily attain'd with application I send Him 2 books to assist him" (*CJB* 460).

60 See *William Bartram on the Southeastern Indians,* ed. Gregory A. Waselkov and Kathryn E. Holland Braund (Lincoln: University of Nebraska Press, 1995) 25. In 1786, the Philadelphia publisher Enoch Story Jr. printed a broadside inviting submissions to a planned edition of Bartram's *Travels,* but the national advertising and sales campaign was launched by another publisher, James and Johnson, in 1790. James and Johnson secured subscriptions from, among other luminaries, President Washington, Vice President Adams, and Thomas Jefferson. See Peck, "William Bartram and His Travels" 41.

61 *William Bartram on the Southeastern Indians* 26.

62 For early reviews of *Travels,* see *Monthly Review* 10 (1793): 13–22; *Massachusetts Magazine, or Monthly Museum* 4 (November 1792): 686–687; *Universal Asylum and Columbian Magazine* 5 (April 1792): 195–197, 255–267.

63 *William Bartram on the Southeastern Indians* 30; see also Lester J. Cappon, "Retracing and Mapping Bartram's Southern Travels," *Proceedings of the American Philosophical Society* 118 (December 1974): 507–513; Charlotte M. Porter, "William Bartram's Travels in the Indian Nations," *Florida Historical Quarterly* 70 (1992): 434–450.

64 John Seelye, "Beauty Bare: William Bartram and His Triangulated Wilderness," *Prospers* 6 (1981): 46; Carlyle to Emerson, 8 July 1851, *The Correspondence of Thomas Carlyle and Ralph Waldo Emerson, 1834–1872,* ed. Charles Eliot Norton, 3rd ed., 2 vols. (Boston: James R. Osgood, 1883) 2:198.

65 L. Hugh Moore, "The Aesthetic Theory of William Bartram," *Essays in Arts and Sciences* 12.1 (March 1983): 17–35; Charles H. Adams, "William Bartram's *Travels*: A Natural History of the South," *Rewriting the South: History and Fiction,* ed. Lothar Hönnighausen and Valeria Gennaro Lerda (Tübingen: Francke, 1993) 112–120.

66 Pamela Regis, *Describing Early America: Bartram, Jefferson, Crèvecoeur, and the Rhetoric of Natural History* (De Kalb: Southern Illinois University Press, 1992) xi.

67 Douglas Anderson argues that *Travels* challenges the reader's "capacities of adjustment"; "Bartram's *Travels* and the Politics of Nature," *Early American Literature* 25 (1990): 3.

68 For a discussion of Bartram's style, see Frieder Busch, "William Bartrams bewegter Stil," *Literatur und Sprache der Vereinigten Staaten: Aufsätze zu Ehren von Hans Galinsky,* ed. Klaus Lubbers (Heidelberg: Winter 1969) 47–61.

69 Interestingly, S. T. Coleridge described *Travels* as "a series of poems, chiefly descriptive, *occasioned* by the Objects, which the Traveller observed"; inscription on the flyleaf of the copy Coleridge presented to Sara Hutchinson on 19 December 1801, S. T. Coleridge, *The Notebooks,* ed. Kathleen Coburn, 4 vols. (New York, 1957–), 1: pt. 2, "Notes," entry 218. For Bartram's considerable influence on Coleridge, see N. Bryllion Fagin, *William Bartram: Interpreter of the American Landscape* (Baltimore: Johns Hopkins Press, 1933) 128–149.

70 William Bartram, *Drawings* 62.

71 Christopher Looby, "The Constitution of Nature: Taxonomy as Politics in Jefferson, Peale, and Bartram," *Early American Literature* 22 (1987): 258.

72 John Bartram, "Diary of a Journey through the Carolinas, Georgia, and Florida, from July 1, 1765, to April 10, 1766," ed. Francis Harper, *Transactions of the American Philosophical Society* 33, pt. 1 (1942): 31. Punctuation mine.

73 See Francis Harper's comments in William Bartram, "Travels in Georgia and Florida, 1773–74" 174. In 1774, John Fothergill, through the nurseryman William Malcolm, presented seedlings of the Franklin tree to the Royal Gardens at Kew, but whether he really had received these specimens from Bartram or from another collector in his employ is unclear; see Charles F. Jenkins, "The Historical Background of Franklin's Tree," *Pennsylvania Magazine of History and Biography* 57 (1933): 202–203.

74 "CATALOGUE OF AMERICAN TREES, SHRUBS AND HERBACIOUS PLANTS, most of which are now growing, and produce ripe Seed in JOHN BARTRAM'S Garden, near Philadelphia. The Seed and growing Plants of which are disposed of on the most reasonable Terms," broadside, 1783, reproduced in Leighton, *American Gardens* 306.

75 Humphry Marshall, *Arbustum Americanum: The American Grove; or, An Alphabetical Catalogue of Forest Trees and Shrubs, Natives of the American United States, Arranged According to the Linnean System* (Philadelphia: Joseph Cruikshank, 1785). The title page of the first edition of Marshall's book misspells "Arbustum" as "Arbustrum."

76 See Joseph Ewan and Nesta Ewan, "John Lyon, Nurseryman and Plant Hunter, and His Journal, 1799–1814," *Transactions of the American Philosophical Society* 53.2 (1963): 5.

77 *Arbustum Americanum* v, vi, xi.

78 Although other botanists subsequently corrected the spelling of the plant's trivial name to *altamaha,* the rules of botanical nomenclature require that the name stand as it was first published; see William Bartram, *Drawings* 151.

79 See Barton, *Elements of Botany,* pt. 3, 90–91: "This vast and interesting class embraces those hermaphrodite vegetables, which have all their stamens, or male organs, united below, that is by their filaments, into one body, or cylinder, through which the pistil, or female organ, passes."

80 *Arbustum Americanum* 48–50.

81 Sir Joseph Banks to Humphry Marshall, 6 May 1789, in Darlington, *Memorials of John Bartram and Humphry Marshall* 562. See Marshall's *Arbustum* 50: "It [the *Franklinia*] seems really allied to the Gordonia, to which it has, in some late Catalogues, been joined."

82 Charles Sprague Sargent, *Manual of the Trees of North America,* 2nd ed., 2 vols. (1922; New York: Dover, 1965) 2: 751.

83 See Rom Harrè, "Some Narrative Conventions of Scientific Discourse," *Narrative in Culture: The Uses of Storytelling in the Sciences, Philosophy, and Literature,* ed. Cristopher Nash (London: Routledge, 1994) 81–101.

84 Regis, *Describing Early America* 25.

85 Regis, *Describing Early America* 56. Regis reprints Bartram's description with crucial (and unmarked!) omissions. It may be that Bartram's *Travels* presents "few textual problems" (Regis 164, n. 17), but it *does* require scrupulously attentive reading.

86 See Richard Mabey, *The Flowers of Kew* (New York: Atheneum, 1989) 28.

87 Gavin D. R. Bridson, "From Xylography to Holography: Five Centuries of Natural History Illustration," *Archives of Natural History* 16 (1989): 127.

88 See Stephen A. Spongberg, *A Reunion of Trees: The Discovery of Exotic Plants and Their Introduction into North American and European Landscapes* (Cambridge, MA: Harvard University Press, 1990) 39.

89 William Cronon, "Telling Tales on Canvas," *Discovered Lands, Invented Pasts: Transforming Visions of the American West,* exhibition held at the Buffalo Bill Historical Center, Cody, WY, 15 June-16 August 1992 (New Haven: Yale University Press, 1992) 46–47.

90 See Sheila Connor, *New England Natives: A Celebration of People and Trees* (Cambridge, MA: Harvard University Press, 1994) 113–114.

91 François André Michaux, *The North American Sylva; or, A Description of the Forest Trees of the United States, Canada, and Nova Scotia, considered particularly with respect to their use in the Arts, and their introduction into Commerce: to which is added a description of the most useful of the European Forest Trees.* Trans. Augustus L. Hillhouse. 3 vols. (Paris: C. d'Hautel, 1817–1819).

92 Michaux, *North American Sylva* 1:298.

93 See Wilfrid Blunt and William T. Stearn, *The Art of Botanical Illustration,* new ed. (1950; Woodbridge, Suffolk: Antique Collectors' Club, 1994) 204–205.

94 See Bartram's note in "Mr. Barclay's Book," 1789; William Bartram, *Drawings* 164.

95 *Elements of Botany,* 1st ed., pt. 3, 98. See also the description of the *Franklinia* in the German-born botanist Frederick Pursh's *Flora Americae Septentrionalis; or, A Systematic Arrangement and Description of the Plants of North America,* 2 vols. (London: Printed for White Cochrane, 1814): "This elegant tree, whose large white flowers with yellow anthers have a most agreeable appearance, though a native of a very southern latitude, is able to stand a considerable northern climate" (2: 451).

96 Francis Harper and Arthur N. Leeds, "A Supplementary Chapter on Franklinia alatamaha," *Bartonia: Proceedings of the Philadelphia Botanical Club* 19 (1937): 1–13.

97 William Bartram, *Drawings* 164.

98 Michaux, *North American Sylva* 1:298.

99 See Edgar T. Wherry, "The History of the Franklin Tree, *Franklinia alatamaha,*" *Journal of the Washington Academy of Sciences* 18.6 (19 March 1928): 175.

100 J. Ewan and N. Ewan, "John Lyon" 22–23.

101 Wallace Stevens, "Desire & the Object," *Collected Poetry and Prose,* ed. Frank Kermode and Joan Richardson (New York: Library of America, 1997) 594.

102 Gayther L. Plummer, "*Franklinia alatamaha* Bartram ex Marshall: The Lost *Gordonia*

(Theaceae)," *Proceedings of the American Philosophical Society* 121.6 (December 1977): 479–480.

103 See J. Ewan and N. Ewan, "John Lyon" 23.

104 Plummer, "*Franklinia*" 480.

105 Plummer, "*Franklinia*" 481. See also Harper and Leeds, "A Supplementary Chapter" 13: "The disappearance of *Franklinia* in a wild state is a mystery still awaiting solution."

106 Jenkins, "Historical Background of Franklin's Tree" 207.

107 This is how Rafinesque defines himself in his "scientific" autobiography, *A Life of Travels and Researches in North America and South Europe; or, Outlines of the Life, Travels, and Researches of C. S. Rafinesque* (Philadelphia: Printed for the author, 1836) 148. On Rafinesque, see Wayne Hanley, *Natural History in America: From Mark Catesby to Rachel Carson* (New York: Quadrangle, 1977) 126–142.

108 *Atlantic Journal and Friend of Knowledge: A Cyclopedic Journal and Review of Universal Science and Knowledge* 1.2 (1832): 79.

109 Blunt and Stearn, *Art of Botanical Illustration* 205.

110 The image reappears in slightly different form as plate 108 of the Royal Octavo edition of Audubon's *Birds of America*. Maria Martin, a gifted illustrator, was the sister-inlaw and later wife of the Charleston naturalist and minister John Bachman, who first collected the warblers portrayed in the plate. In the 1830s, Martin provided drawings of southern plants for Audubon. She also drew the blacksnake and other "Carolina Reptiles" for John Edwards Holbrook's *North American Herpetology* (see chapter 4).

111 *Travels: Poems by W. S. Merwin* (1993; New York; Knopf, 1994) 38–40.

112 See Slaughter, *Natures of John and William Bartram* 90–93.

113 Spongberg, *A Reunion of Trees* 40.

2. COLLECTION AND RECOLLECTION: CHARLES WILLSON PEALE

1 My definition of Peale's aims depends on his own, proud self-assessment in *SDC*: "The Museum is still in its infancy. . . . It is the fruit of the persevering industry, patience and zeal of an individual; without any aid or countenance of a public nature" (viii).

2 For a general introduction to Peale's museum, see Robert E. Schofield, "The Science Education of an Enlightened Entrepreneur: Charles Willson Peale and His Philadelphia Museum, 1784–1827" *American Studies* 30.2 (Fall 1989): 21–40. The standard biography of Peale is Charles Coleman Sellers, *Charles Willson Peale* (New York: Charles Scribner's Sons, 1969).

3 See Edgar P. Richardson, Brooke Hindle, and Lillian B. Miller, *Charles Willson Peale and His World* (New York: Harry N. Abrams, 1983) 80, 86.

4 James Clifford, *The Predicament of Culture: Twentieth-Century Ethnography, Literature, and Art* (Cambridge, MA: Harvard University Press, 1988) 218.

5 Christopher Looby, "The Constitution of Nature: Taxonomy as Politics in Jefferson, Peale, and Bartram," *Early American Literature* 22 (1987): 269. The approach to Peale's museum sketched out here differs not only from Looby's revisionist interpretation but also from David R. Brigham's recent analysis of the Philadelphia Museum as a "cultural production, rather than the expression of one man's ideas"; *Public Culture in the Early Republic: Peale's Museum and Its Audience* (Washington, DC: Smithsonian Institution

Press, 1995) 138. On the basis of Peale's receipts and accession lists, notices in newspapers, and written comments by visitors in diary or letter form, Brigham meticulously identifies the names of those who patronized Peale's museum and donated objects to its collections. He provides a fifteen-page list of subscribers to Peale's museum in 1794, grouped by occupation and ranked by wealth (short-term visitors were not recorded for posterity). Not surprisingly, these visitors included few women and no African-Americans: "The museum's openness to people of different social ranks was . . . limited" (7). For Brigham, Peale's museum therefore embodies the tension in early American society between the lip service paid to democratic ideas and the persistence of social and racial inequality, which helped maintain a hierarchical order benefiting the prosperous. Ironically, in spite of his methodological premises Brigham himself almost inevitably reverts to the language of individual agency even in those chapters that attempt to shift, as he says, the "viewpoint from Peale to his audience." While offering his constituencies the "structures" within which they could express their own economic, social, and intellectual interests, Peale obviously retained his "authoritarian" stance over his collection, a fact Brigham himself nicely illustrates when in his epilogue he points out Peale's use of real and painted curtains as a means of exerting "audience control within a structured display" (147), While the present chapter is indebted to Brigham's scholarship, I am less interested in the institutional reality of the museum than in the artistic "structures" Peale invented, which I think are more complex, more brilliantly intricate and idiosyncratic, than Brigham's theoretical model or Looby's ideological reading allows us to realize.

6 See Sidney Hart and David C. Ward, "The Waning of an Enlightenment Ideal: Charles Willson Peale's Philadelphia Museum, 1790–1820," *New Perspectives on Charles Willson Peale: A 250th Anniversary Celebration,* ed. Lillian B. Miller and David C, Ward (Pittsburgh: University of Pittsburgh Press, 1991)219–235.

7 Looby, "The Constitution of Nature" 268. Looby argues that Peale's "Exhibition of Perspective Views" ritually removed visitors from, and then again returned them to, the "ordered environment" of the museum. However, Peale abandoned these exhibitions the year he started his museum and resuscitated his show only for special visitors, as a letter addressed to General Weedon, 25 September 1786, indicates: "I have now intirely laid aside my Exhibition of Transparent Paintings" (*PP* 1: 456). Looby's critical view of Peale has been adopted by other scholars. See Charles H. Adams, "William Bartram's *Travels: A Natural History of the South," Rewriting the South: History and Fiction,* ed. Lothar Hönnighausen and Valeria Gennaro Lerda (Tubingen: Francke, 1993) 114: "Peale's museum reflects the relatively static conception of both nature and history typical of . . . literary, scientific, and political intellectuals who made their home in Philadelphia's American Philosophical Society." David Brigham, too, argues that the "impulse toward order" in Peale's museum was "suggestive of a broader need to establish economic, social, and political structures within a country recently independent of monarchical rule and colonial status" (*Public Culture in the Early Republic* 149).

8 Peale complained in a letter to his son Rembrandt (28 October 1809) that visitors would sometimes stand on benches and "sully the Glass [of the display cases]" with their fingers (*PP* 2:1237). See Joel J. Orosz, *Curators and Culture: The Museum Movement in*

America, 1740–1870 (Tuscaloosa: University of Alabama Press, 1990) 81–83; Brigham, *Public Culture* 28–29; Brigham, "Social Class and Participation at Peale's Philadelphia Museum," *Mermaids, Mummies, and Mastodons: The Emergence of the American Museum,* ed. William T. Alderson (Washington, DC: American Association of Museums, 1992) 87. However, there is no evidence that such behavior was restricted to the lower orders of society, and Peale's rationale for disciplining ill-mannered visitors seems less ideological than pragmatic: "It must be obvious to every thinking being that these rules are necessary to preserve the articles of a Museum formed for the instruction and the amusement of the present as well as future generations" (PP 2:1237).

9 William H. Goetzmann, New Lands. New Men: America and the Second Great Age of Discovery (1986; New York: Penguin, 1987) 95–96.

10 Charles Coleman Sellers, *Mr. Peale's Museum: Charles Willson Peale and the First Popular Museum of Natural Science and Art* (New York: Knopf, 1980) 42. Peale received "articles of *lusus naturae*" with "caution" (*PP* 2: 18); some of those he owned he "never wished to exhibit" (*DI* 6).

11 Charles Catton Jr. (1756–1819) was an English landscape painter who had left his native country for the United States in 1804. The original painting of *Noah and His Ark* has not survived. On Peale and Catton, see Charles Coleman Sellers, *Charles Willson Peale with Patron and Populace: A Supplement to* Portraits and Miniatures by Charles Willson Peale *with a Survey of His Work in Other Genres, Transactions of the American Philosophical Society* ns 59.3 (1969): 44–45.

12 Roger B. Stein erroneously assumes that Peale had used a print as a model; "Charles Willson Peale's Expressive Design: The Artist in His Museum," *Prospects* 6 (1981): 165. See, however, Peale's letter to Rembrandt Peale, 16 December 1819: "Realy the *original* is an uncommon well executed piece of work" (*PP* 3: 781; my emphasis).

13 See Susan Stewart, "Death and Life, in That Order, in the Works of Charles Willson Peale," *The Cultures of Collecting,* ed. John Eisner and Roger Cardinal (Cambridge, MA: Harvard University Press, 1994) 219.

14 Roland Barthes, "The Plates of the *Encyclopedia*" (1964), trans. Richard Howard, *The Barthes Reader,* ed. Susan Sontag (New York: Hill & Wang, 1982) 222.

15 *The Historie of the World. Commonly called, The Naturall Historie of C. Plinius Secundits,* trans. Philemon Holland, 2 vols. (London: Adam Islip, 1601) 1:311 (bk. 11, ch. 2).

16 John Eisner and Roger Cardinal, introduction, *The Cultures of Collecting* 1.

17 See Joy Kenseth, "A World of Wonders in One Closet Shut," *The Age of the Marvelous,* ed. Joy Kenseth (Hanover, NH: Hood Museum of Art, Dartmouth College, 1991) 87–88.

18 See Sellers, *Mr. Peale's Museum* 244–245.

19 The wild turkey was brought home to Philadelphia by Titian Peale from Major Long's expedition to the Louisiana Territory in 1819–1820. See Charles Coleman Sellers, *Portraits and Miniatures by Charles Willson Peale, Transactions of the American Philosophical Society,* ns 42.1 (1952): 161, and Sellers, *Mr. Peale's Museum* 242.

20 See Rembrandt Peale, *An Historical Disquisition on the Mammoth, or Great American Incognitum, an Extinct, Immense, Carnivorous Animal, Whose Fossil Remains Have Been Found in North America* (London: E. Lawrence, 1803); in PP 2:550.

21 Peale invoked the Renaissance collector Aldrovandi whenever he sought public support; evidently he subscribed to the legend that Aldrovandi had died poor and neglected (*PP* 2: 177; *DI* 17).

22 Neil Harris, *Humbug: The Art of P. T. Barnum* (Boston: Little, Brown, 1973): "The Devil lurked only in playhouses; museums were out of his territory, and so safe for ordinary folk" (37).

23 Stein, "Peale's Expressive Design" 172.

24 Alan Trachtenberg, *Reading American Photographs: Images as History, Mathew Brady to Walker Evans* (New York: Noonday, 1989) 11.

25 See Paula Findlen, *Possessing Nature: Museums, Collecting, and Scientific Culture in Modern Italy* (Berkeley: University of California Press, 1994) 38; Amy Boesky, "'Outlandish-Fruits': Commissioning Nature for the Museum of Man," *ELH* 58 (1991): 317–319. For Roger Stein, *The Artist in His Museum* belongs firmly within the tradition of the "gallery picture" ("Peale's Expressive Design" 167).

26 A reference to the ancient belief, given new currency under the influence of Christianity, that the "pelican in her piety" tears its breast open so as to resuscitate its young with its own blood. The fable, outlandish as it sounds, has some basis in the observation that the pelican, while feeding its young, presses its beak, the tip of which is red as if tainted with blood, against its breast feathers; see Colin Clair, *Unnatural History* (London: Abelard-Schuman, 1967)133–134.

27 Thomas Morton, *The New English Canaan of Thomas Morton* (1632), ed. Charles Francis Adams, *Publications of the Prince Society* 14 (New York: Burt Franklin, 1967) 193.

28 See Doughty's *Cabinet of Natural History and American Rural Sports* 3.1 (1832): 12–13.

29 See Russell W. Belk and Melanie Wallendorf, "Of Mice and Men: Gender Identity in Collecting," *Interpreting Objects and Collections,* ed. Susan M. Pearce (London: Routledge, 1994) 245–246.

30 Sellers, *Mr. Peale's Museum* 89; see also *W* 46.

31 Peale's eagle "had in Gold letters on his cage *Feed me daily 100 years.*" Unfortunately, it lived for only fifteen years, "long enough," as its frugal owner couldn't help adding, "to cost Peale at least 300$ for its food" (*A* 221). Alexander Wilson's drawing of Peale's eagle would in turn inspire John James Audubon's watercolor for plate 31 of *The Birds of America.* See Carole Ann Slatkin's notes in John James Audubon, *The Watercolors for* The Birds of America, ed. Annette Blaugrund and Theodore E. Stebbins Jr. (New York; Villard, 1993) 161.

32 See Orosz, *Curators and Culture* 36- 42; Paul Ginsberg Sifton, "A Disordered Life: The American Career of Pierre Eugène Du Simitière," *Manuscripts* 25 (1973): 235–253.

33 Sellers, *Mr. Peale's Museum* 107. See also *A* 314: "Some men, to show their profound knowledge, not infrequently lard their discourse with Latin terms, this to be sure may have a powerful effect on the minds of the illiterate, but to those who have actual experience on the matter in question can have little effect."

34 Richard Pulteney, *A General View of the Writings of Linnaeus,* ed. George Maton (1781; London: J. Mawman, 1805): "The rightly ascertaining of species is the great end of all method" (107).

35 Pulteney, 244. This is a view shared by Peale: "Linnaeus did not attempt to form a natural system, but an artificial one, only in order to facilitate the study of this science" (*SDC* 2 n).

36 Hans Blumenberg, *Die Lesbarkeit der Welt* (Frankfurt am Main: Suhrkamp, 1981) 35.

37 Catherine Fritsch, "Notes of a Visit to Philadelphia, Made by a Moravian Sister in 1810," trans. A. R. Beck, *Pennsylvania Magazine of History and Biography* 36 (1912): 359–360.

38 Sellers, *Mr. Peale's Museum* 32. On 6 July 1788, Peale wrote in his diary that he had "removed my Books . . . to make room to place my small Birds" (*PP* 1: 512).

39 Herman Melville, *Redburn, White-Jacket, Moby-Dick,* ed. G. Thomas Tanselle (New York: Library of America, 1983) 935–936, 940.

40 Louis Agassiz, *An Essay on Classification* (London: Longman, Brown, Green, Longmans & Roberts, 1859) 9.

41 Findlen, *Possessing Nature* 37.

42 Werner Hüllen, "Reality, the Museum, and the Catalogue: A Semiotic Interpretation of Early German Texts of Museology," *Semiotica* 80.3/4 (1990): 271–272.

43 Peale later suspected that the "constant" inhalation of arsenic could have been responsible for "every indisposition that befel him" *(A* 110).

44 See Donna Haraway, *Primate Visions: Gender, Race, and Nature in the World of Modern Science* (New York: Routledge, 1989) 37. See, however, Karen Wonders's comprehensive history of the concept, *Habitat Dioramas: Illusions of Wilderness in Museums of Natural History* (Uppsala: Almqvist & Wiksell, 1993).

45 See Wonders, *Habitat Dioramas* 14. In 1932, the *New York Times Magazine* defined the "diorama" as a "miniature stage in which animals or human beings were grouped in their proper surroundings" (Wonders 15).

46 Wonders, *Habitat Dioramas* 16. For a semiotic perspective on habitat groups, see Mieke Bal, "Telling, Showing, Showing Off," *Critical Inquiry* 18.3 (1992): 556–594.

47 It is not clear if Carl Akeley's "Muskrat Group" (1889), usually regarded as the first example of a successful habitat diorama, originally had a painted background. In the museum's annual report, William Morton Wheeler only praised Akeley's use of "real" landscape elements in the foreground: "The great difficulties in accurately imitating the boggy earth, the half dead vegetation and the stagnant water have been very successfully overcome by Mr. Akeley" (Wonders 134). Ironically, Wheeler's description reveals that Akeley's "lifelike" arrangement presented a natural scenery that itself was already "half dead." Like Peale, Akeley thought of his animals as "sculptures," but he took the analogy one step further. He made plaster moulds from his specimens, and from these he produced hollow papier-mâché replicas of the originals, over which he would then stretch the preserved skins. Akeley's method, which combined supreme craftsmanship with artistic creativity, inaugurated modern taxidermy.

48 Susan Stewart, *On Longing: Narratives of the Miniature, the Gigantic, the Souvenir, the Collection* (1984; Durham: Duke University Press, 1993) 153.

49 John D. Godman, *American Natural History,* 2 vols. (Philadelphia: Carey & Lea, 1826–1828) 1:131.

50 *The Journals of Zebulon Montgomery Pike, with Letters and Related Documents,* ed. Donald Jackson, 2 vols. (Norman: University of Oklahoma Press, 1966) 2: 293.

51 Godman, *American Natural History* 1: 133.

52 Ivan Karp, "Culture and Representation," *The Poetics and Politics of Museum Display*, ed. Ivan Karp and Steven D. Lavine (Washington, DC: Smithsonian Institution Press, 1991) 23.

53 Richard Brilliant, *Portraiture* (Cambridge, MA: Harvard University Press, 1991) 15.

54 Timothy Lenoir and Cheryl Lynn Ross claim that natural history museums provide semiotic markers of nature instead of "authentic" nature itself. Therefore, what we really see in these museums is, in Emerson's words, "the figure of a disguised man." I agree but would also point out that, as Peale's museum proves, this effect cannot be considered an accidental byproduct. What Lenoir and Ross present as their own shrewd deconstruction of the natural history display was, in fact, the central premise of Peale's collecting. See Lenoir and Ross, "The Naturalized History Museum," *The Disunity of Science: Boundaries, Contexts, and Power*, ed. Peter Galison and David J. Stump (Stanford: Stanford University Press, 1996) 397.

55 This claim is endorsed by Clifford in *The Predicament of Culture*, where it seems equally off the mark: "The objective world is given, not produced, and thus historical relations of power in the work of acquisition are occulted. The *making* of meaning in museum classification and display is mystified as adequate *representation*. The time and order of the collection erase the concrete social labor of its making" (220). The third chapter of Brigham's *Public Culture in the Early Republic* includes a selection of written responses from visitors to Peale's museum. The more positive accounts stressed Peale's active share, his "skill and energy," in the making of the museum. Even the critical comments show that "visitors did not simply absorb Peale's version of the museum" and that Peale's museum was widely perceived not as finished or "given" but as a collection in flux, as a form of production rather than mindless reproduction (66–67).

56 Craig Black, "Evolution and All That," *Museums Journal* 92.2 (February 1992): 16–17.

57 See Black, "Evolution and All That," 16: "What do museum dioramas tell us of the real world? All too little. They are valid but romanticized reminders of the past but they are seldom so labelled."

58 Georges Bataille, "Musée," in "Dictionnaire," *Documents: Archéologie, Beaux-Arts, Ethnographie, Variétés* 2 (1930): 300.

59 Brandon Brame Fortune, "Charles Willson Peale's Portrait Gallery: Persuasion and the Plain Style," *Word & Image* 6.4 (1990): 317, mentions Peale's interest in the polygraph as one further example of his pervasive desire for "perfect preservation or exact reproduction."

60 Charles Willson Peale, *George Washington at the Battle of Princeton* (Pennsylvania Academy of the Fine Arts).

61 Apparently, in 1786 and 1787 there was a regular traffic of birds and other animals between Mount Vernon and Paris; see Lafayette to Washington, 36 October 1786: "I hope your jackass, with two females, and a few pheasants and red partridges, have arrived safe"; Lafayette to Washington, 3 August 1787: "I thank you, my dear general, for the fine birds . . . you have sent to me. The poor ducks died at the Havre on their arrival; I beg you will send me some again, and beg leave to add a petition for an invoice of mocking

birds." *Memoirs, Correspondence, and Manuscripts of General Lafayette,* published by his family, 3 vols. (London: Saunders and Otley, 1837) 2:148,191.

62 For a summary of this amusing episode, see James Thomas Flexner, *Washington: The Indispensable Man* (1969; New York: New American Library, 1984) 191–192.

63 Elizabeth Hall and Max Hall, "George Washington's Pheasants," *Harvard Magazine* 77.6 (February 1975): 19–21.

64 See the entries for 18 September 1788: "Mr Sadler brought me a summer Drake" (a wood duck), "which from its beauty and being well killed I begin the preservation—work on the background of Mrs Culbreath's picture" (*PP* 1: 540–541); 2 November 1788: "took the flesh out of 2 Ducks & brant & put the powder into the Skins to preserve them, worked on Mr. Lamings Miniature" (*PP* 1: 542); 14–23 November 1788: "finishing drapery & back ground of Mrs Hutchingson miniature . . . chiefly employed in making banks & fitting up the wings of the bird case for putting the birds in it" (*PP* 1: 547).

65 Linnaeus, *Systema naturae per regna tria naturae, secundum classes, ordines, genera, species, cum characteribus, differentiis, synonymis, locis,* 10th ed, 2 vols. (Holmiae; Impensis Direct. L. Salvii, 1758–1759) 1: 20–24.

66 Pulteney, *General View of the Writings of Linnaeus* 176. See John C. Greene, *The Death of Adam: Evolution and Its Impact on Western Thought* (1959, Ames: Iowa State University Press, 1996) 184–187, and Stephen Jay Gould, *The Mismeasure of Man* (New York: Norton, 1981) 31–39.

67 See Linnaeus's defense of his classification in *Fauna Suecia* (1746), as summarized in Phillip Sloan, "The Gaze of Natural History," *Inventing Human Science: Eighteenth-Century Domains,* ed. Christopher Fox, Roy Porter, and Robert Wokler (Berkeley: University of California Press, 1995) 123.

68 See Edward Tyson's preface and the "Epistle Dedicatory" in *Orang-Outang, sive Homo Sylvestris; or, The Anatomy of a Pygmie Compared with that of a Monkey, an Ape, and a Man. To which is added, A Philological Essay Concerning the Pygmies, the Cynocephali, the Satyrs, and Sphinges of the Ancients* (London: Printed for Thomas Bennet . . . and Daniel Brown, 1699), For observations on the larynx of Tyson's "pygmie," see 51–52.

69 Georges Louis Leclerc, comte de Buffon, "Les Orangs-Outangs," *Ocuvres complètes de Buffon, avec les descriptions anatomique de Daubenton, son collaborateur,* nouvellc édition, ed. J. V. F. Lamouroux and A. G. Desmarest, 44 vols. (Paris: Ladrange et Verdière, 1824–1832) 18: 64, 73.

70 Denis Diderot, *Le Réve de d'Alembert,* ed. Paul Vernière (Paris: Marcel Didier, 1951) 165.

71 See Robert Wokler, "Tyson and Buffon on the Orang-Utan," *Studies on Voltaire and the Eighteenth Century* 165 (1976): 2315.

72 Linnaeus, *Systema naturae* 1: 21–22.

73 As the passages about his first wife's death in Peale's autobiography show, he was fascinated with the notion of life-in-death and the "custom of burying the dead too soon" (*A* 135).

74 *The Papers of Benjamin Franklin,* ed. Leonard W. Labaree et al., 31 vols, to date (New Haven: Yale University Press, 1959–) 30: 189–190.

75 See Rosamond Wolff Purcell and Stephen Jay Gould, *Finders, Keepers: Eight Collectors* (New York: W. W. Norton, 1992) 30–32.

76 See *PP* 2: 24, n. 4.

77 In 1793, French revolutionaries dragged the embalmed body of the vicomte de Turenne (1611–1675) from his tomb in Saint-Denis and displayed it in the cabinets of the Jardin des Plantes. On 2 August 1796, Demolard, a member of the Conseil des Cinq-Cent, took exception to the disgraceful location of the body, "placés [sic] entre ceux d'un éléphant et d'un rhinocéros!" Leo Armagnac, *Histoire de Turenne, maréchal de France*, 5th ed. (Tours: A. Mame et fils, 1886) 355. Bonaparte saw to it that Turenne's remains found their final resting place in the Dôme des Invalides. Another—particularly poignant—example is Julia Pastrana, a sideshow attraction of the 1850s, billed as the "Ugliest Woman Who Ever Lived." After her death, her body was embalmed and exhibited by her husband, alongside that of her son, who had died within days of his birth. See Robert Lifson, *Enter the Sideshow* (Bala Cynwyd, PA: Mason, 1983) n.p.

78 In 1823, John C. Warren, professor of anatomy and surgery at Harvard, published an essay in the *Boston Journal of Philosophy and the Arts* that would have delighted Peale: "Description of an Egyptian Mummy, Presented to the Massachusetts General Hospital; With an Account of the Operation of Embalming"; later published separately (Boston: Cummings, Hilliard, 1824). Warren points out that it had not been unusual for the bodies of kings and other dignitaries in Europe to be embalmed, a fact that he takes as an incentive to propose "different operations . . . for the preservation of the bodies of our departed friends." He suggests that preserved bodies, when properly dried and washed in turpentine, be clothed in "becoming robes," and placed in airtight glass cases, in this way, he concludes, "we might, for a moderate expence, preserve the appearance of those forms which have been associated in our minds with agreeable occurrences, and we might gratify the curiosity of a remote futurity, by transmitting to them the bodies and even the feature of those who have been distinguished among us" (20). Warren's Egyptian mummy went on display in Rubens Peale's Baltimore Museum and became a fad; a year later, the market was "glutted" with mummies. John W. Durel, "In Pursuit of a Profit," Alderson, *Mermaids, Mummies, and Mastodons* 44, 46.

79 Traveling up the Missouri in 1843, John James Audubon, usually not known for his interest in Native Americana, pried open the coffin of a "three-years-dead Indian chief" and made off with a terrible trophy: "The head had still the hair on, but was twisted off in a moment" (*SJ* 300). At the turn of the century, Franz Boas's collection comprised one hundred skeletons and two hundred crania; see Robert E. Bieder, "The Collecting of Bones for Anthropological Narratives," *American Indian Culture and Research Journal* 16.2 (1992): 29.

80 Brigham, *Public Culture in the Early Republic* 132–135.

81 On Peale's interest in artificial limbs, see *PP* 2: 513–514.

82 Flexner, Washington 342–343. See also Margaret Coffin, *Death in Early America: The History and Folklore of Customs and Superstitions of Early Medicine, Funerals, Burials, and Mourning* (Nashville: Nelson, 1976) 41–44.

83 See John Josselyn, *An Account of Two Voyages to New-England* (1674), in *John Josselyn: Co-*

lonial Traveler, ed. Paul J. Lindholdt (Hanover: University Press of New England, 1988) 128.

84 See Esther Forbes, *Paul Revere and the World He Lived In* (Boston: Houghton Mifflin, 1942) 125.

85 Pulteney, *General View* 175.

86 Sidney Hart, "To Encrease the Comforts of Life': Charles Willson Peale and the Mechanical Arts," Miller and Ward, *New Perspectives on Charles Willson Peale* 237.

87 In 1804, Peale was afflicted with hydrocele, an accumulation of serous fluid in the testis—a condition that he tried to heal himself by applying ice to the affected area. See his letter to Dr. Wistar, professor of anatomy, surgery, and midwifery at the University of Pennsylvania, 13 November 1804: "My cure of the Hydrocele is not only enteresting to myself but must be so to you Gentlemen of the Faculty, as perhaps I have been the first who has made the application" (*PP* 2: 780–781).

88 See Charles Willson Peale to Angelica Peale, 14 February 1815: "If the teeth does not intirely fit your Mouth, it must be laid to my account" (*PP* 3: 303). Peale helped his children battle the infirmities of age in more than one way: letters written in 1820 indicate that, for example, he was also working on a pair of usable spectacles for Angelica.

89 Hart, "'To Increase the Comforts of Life'" 244.

90 "Alio studio . . . dell'uomo in genere è principalmente diretto lo scopo di questa opera. E di qual uomo si può egli meglio e piú dottamente parlare, che di sé stesso?" Vittorio Alfieri, *Vita,* ed. Giulio Cattaneo (Milano: Garzanti, 1977) 5.

91 Henry David Thoreau, "Natural History of Massachusetts" (1842), *The Natural History Essays,* ed. Robert Sattelmeyer (Salt Lake City: Peregrine Smith, 1980) 5.

92 Henry Fielding, *The History of Tom Jones, a Foundling,* ed. Martin C. Battestin and Fredson Bowers (Middletown, CT: Wesleyan, 1975) 76.

93 See Lichtenberg's *Sudelbücher,* Heft J (659), in Lichtenberg, *Aphorismen, Schriften, Briefe,* ed. Wolfgang Promies and Barbara Promies (München: Hanser, 1974) 177.

94 Whereas I am describing a deliberate autobiographical technique here, a sarcastic little note written by Benjamin Henry Latrobe to Thomas Jefferson on the occasion of Peale's third marriage suggests that lack of development might have been a factor not only in Peale's *written* life: "Peale, is again married I find, This is his third ticket in the lottery of marriage. . . . He has now a fry of five or six little uneducated children about him . . . he is a *boy* in many respects & unfortunately also in this; that he has saved nothing" (*PP* 2: 889).

95 Paul Ricoeur, "Narrative Time," *On Narrative,* ed. W. J. T. Mitchell (Chicago: University of Chicago Press, 1981) 165–186.

96 Benjamin Franklin, *The Autobiography,* in *Writings,* ed. J. A. Leo Lemay (New York: Library of America, 1987) 1328.

97 See *A* 2: "Men as well as boys often desire things which they do not really want. . . . Doct'r Franklin has wisely said: 'It is easier to build two chimneys than keep one fire.'" See Franklin's *Poor Richard Improved* (1758): "Expence is constant and certain; and *'tis easier to build two Chimneys than keep one in Fuel*" (Franklin, *Writings* 1302; see also *PP* 3: 319). William Temple Franklin's edition of Franklin's autobiography came out in 1818.

98 Peale exaggerates. In his *Confessions* (bk. 2), Rousseau, who professes to have "little dread" of "carrying" his guilt with him to his grave, slyly leaves himself a convenient loophole by claiming that if "this is a crime to be expiated, (as I hope it is), all the misfortunes which overwhelm me in the decline of life must have done it, added to *forty* years of uprightness and honor on different occasions"; *The Confessions of J. J. Rousseau: With the Reveries of the Solitary Walker, Translated from the French*, 2 vols. (London: J. Bew, 1783) 1: 117 (my emphasis).

99 Rousseau, *Confessions* 1:19, 18, 16.

100 Rousseau, *Confessions* 1:152.

101 Rousseau, *Confessions* 1: 49, 45.

102 A later example of a similar autobiographical technique is Noah Webster's "Memoir of Noah Webster, LL.D.," written around 1832, in which the seemingly objective use of the third person singular serves the purpose of self-elevation rather than self-distancing. See *Autobiographies of Noah Webster: From the Letters and Essays, Memoir, and Diary*, ed. Richard M. Rollins (Columbia: University of South Carolina Press, 1989) 129–186. In the "Preamble" to his collector's autobiography, *A Life of Travels and Researches in North America and South Europe; or, Outlines of the Life, Travels, and Researches of C. S. Rafinesque* (Philadelphia: Printed for the author, 1836), naturalist C. S. Rafinesque deplored the egotism of the first-person point of view, adding that he "should have wished" to have his text changed "to a recital in the third person" (3).

103 "Linnaeus's Diary," Pulteney, *General View of the Writings of Linnaeus* 513, 565.

104 Peale's farm was first called Persevere; later he renamed it Belfield.

105 Franklin, *Writings* 1376.

106 See also *A* 169: "Peale's active mind kept his hands constantly employed."

107 It is interesting to compare this with Noah Webster's considerably more pessimistic definition of "happiness" in his roughly contemporaneous *American Dictionary of the English Language*, 2 vols. (New York: Converse, 1828): "Perfect *happiness*, or pleasure unalloyed with pain, is not attainable in this life."

108 Not surprisingly, Peale has a "scientific" explanation, too hilarious to be suppressed here, for what he thinks is a disturbingly high incidence of widows in Maryland: "It is a remarkable circumstance that in this part of the country . . . there are a great number of Widows. how is this to be accounted for? The men have to attend to the out-door business and [are] therefore exposed to get wet feet. The women staying in the house attending to their domestic concerns seldom go out except in visiting their neighbours, then in some sort of carriage" (*A* 142–143).

3. COLLECTING HUMAN NATURE: P. T. BARNUM

1 See *Mermaids, Mummies, and Mastodons: The Emergence of the American Museum*, ed. William T. Alderson (Washington, DC: American Association of Museums, 1992) 63, 65.

2 Charles Coleman Sellers, *Mr. Peale's Museum: Charles Willson Peale and the First Popular Museum of Natural Science and Art* (New York: Knopf, 1980) 250.

3 Daniel Aaron, *Cincinnati: Queen City of the West, 1819–1838* (Columbus: Ohio State University Press, 1992) 276–280.

4 Aaron, *Cincinnati: Queen City of the West* 277.

5 Sellers, *Mr. Peale's Museum* 275–276.

6 William H. Goetzmann, *New Lands, New Men: America and the Second Great Age of Discovery* (1986; New York: Penguin, 1987) 178.

7 Paul L. Farber, "The Transformation of Natural History in the Nineteenth Century," *Journal of the History of Biology* 15.1 (Spring 1982): 147.

8 See Sellers, *Mr. Peale's Museum* 314–318.

9 Barnum did not distinguish between "dwarfs" and "midgets," a term that, according to the *Oxford English Dictionary,* became current only after 1865.

10 See A. H. Saxon, *P. T. Barnum: The Legend and the Man* (New York: Columbia University Press, 1989) 278.

11 Emerson called Barnum's financial troubles "the gods visible again"; see his letter to Lidian Emerson, 30 December 1855, *The Letters of Ralph Waldo Emerson,* ed. Ralph L. Rusk, 6 vols. (New York: Columbia University Press, 1939) 4: 541.

12 *The Diary of George Templeton Strong,* ed. Allen Nevins and Milton Halsey Thomas, 4 vols. (New York: Macmillan, 1952) 4:18.

13 See the entries in Strong's diary for 13 July 1865 and 3 March 1868; *Diary of George Templeton Strong* 4:17, 195.

14 Michael Zuckerman, "The Selling of the Self: From Franklin to Barnum," *Benjamin Franklin, Jonathan Edwards, and the Representation of American Culture,* ed. Barbara B. Oberg and Harry S. Stout (New York: Oxford University Press, 1993) 152.

15 James M. Mayo, *The American Grocery Store: The Business Evolution of an Architectural Space* (Westport, CT: Greenwood, 1993) 59.

16 Edgar Allan Poe, *Doings of Gotham,* ed. Jacob E. Spannuth and Thomas Ollive Mabbott (Pottsville, PA: Spannuth, n.d.) 48.

17 Quoted in Eric Lott, *Love and Theft: Blackface Minstrelsy and the American Working Class* (New York: Oxford University Press, 1993) 65.

18 Posterbill, Barnum's American Museum, 21 March 1864: "PHOTOGRAPH CARTES DE VISITE of the principal Living Curiosities are furnished to visitors as *keepsakes,* at the low price of 15 cents each." All the posterbills from Barnum's Museum quoted in this chapter are from the holdings of the Harvard Theatre Collection.

19 Mayo, *American Grocery Store* 59.

20 In the publicity writing for the museum, the "implied visitor" is usually male.

21 Neil Harris, *Humbug: The Art of P. T. Barnum* (Boston: Little, Brown, 1973) 286.

22 See the *Book of Jumbo, Largest Elephant In or Out of Captivity* (Buffalo: Courier, 1882) 4.

23 "Life Sketch of P. T. Barnum," *Illustrated and Descriptive History of the Animals Contained in Barnum & Van Amburgh's Museum and Menagerie Combination* (New York: S. Booth, 1866) 63.

24 Lewis Carroll, *Alice in Wonderland,* ed. Donald J. Gray, 2nd ed. (1971; New York: Norton, 1992) 11.

25 Paula Findlen, *Possessing Nature: Museums, Collecting, and Scientific Culture in Early Modern Italy* (Berkeley: University of California Press, 1994) 303.

26 *Brief Biographical Sketch of the Living Curiosities and the Leading Wonders of Barnum's*

Traveling Museum and Menagerie (New York: Press of Wynkoop & Hallenbeck, 1871) 3, 17 (Houghton Library, Harvard University).

27 See Aline Mackenzie Taylor, "Sights and Monsters and Gulliver's *Voyage to Brobding-nag*," *Tulane Studies in English* 7 (1957): 29–82.

28 On 29 November 1863, Barnum announced the appearance of "the smallest of all small men; throwing Gen. Tom Thumb, Com. Nutt and all other descendants of King Lilliput entirely into the shade." Miss Reed was hailed as "a most perfect MODEL OF A WOMAN," and Barnum encouraged visitors to form a "very correct idea" of her size "by noticing her relative proportion to that of a common chair" (22 April 1861).

29 Jonathan Swift, *Gulliver's Travels*, ed. Christopher Fox (Boston: Bedford, 1995) 104.

30 *Gulliver's Travels* 102, 108, 123, 99.

31 *Gulliver's Travels* 94, 145, 108.

32 See Lennard J. Davis, *Enforcing Normalcy: Disability, Deafness, and the Body* (London: Verso, 1995) 24.

33 Barnum's "miniature" acquisitions after Tom Thumb, billed as even smaller than he, were *Commodore* Nutt and *Admiral* Dot.

34 In the weeks before the wedding, Lavinia Warren's diamond engagement ring was on display in Barnum's museum. On 2 February 1863, Barnum assured visitors that Lavinia would "never" be exhibited as Tom Thumb's wife, but a new poster published on 1 June 1863 announced that, for a limited time only, "GENERAL TOM THUMB and his beautiful LITTLE WIFE" would again be appearing at the museum.

35 A similar principle was operative in the "marriages of opposites" advertised by Barnum: in 1864, "Mrs. Battersby, weighing nearly 700 lbs." appeared alongside her "LIVING SKELETON" husband, who was said to weigh only 67 pounds: "THE MOST ASTONISHING MATRIMONIAL CONTRACT EVER SEEN" (28 January 1864).

36 *The Autobiography of Mrs. Tom Thumb [Some of My Life Experiences] by Countess M. Lavinia Magri*, ed. A. H. Saxon (Hamden, CT: Archon, 1979) 62.

37 Walt Whitman, "Song of Myself," *Complete Poetry and Collected Prose*, ed. Justin Kaplan (New York: Library of America, 1982) 30. In the interview, Barnum ("of Tom Thumb notoriety") celebrated "Yankeedom" and criticized England, from where he had just returned, as full of kings and "absolutely frozen," whereas in America there were "life," "freedom," and "men." Added Whitman, "A whole book might be written on that little speech of Barnum's." See Theodore A. Zunder, "Whitman Interviews Barnum," *Modern Language Notes* 48.1 (January 1933): 40.

38 Mark Twain, "How the Mighty Are Fallen," letter 11, *Mark Twain's Travels with Mr. Brown, Being Heretofore Uncollected Sketches Written by Mark Twain for the San Francisco Alta California in 1866 and 1867*, ed. Franklin Walker and G. Ezra Dane (New York: Knopf, 1940) 116–117.

39 *Illustrated and Descriptive Catalogue* vi.

40 This assessment is apparent even in a recent catalogue published by the American Association of Museums, *Mermaids, Mummies, and Mastodons* (ed. William T. Alderson, 1992) in which the contributors refer to Barnum's work variously as "crass commercialism" (12), "humorous hokum" (21), and "hoopla" (33).

41 Joel J. Orosz, *Curators and Culture: The Museum Movement in America, 1740–1870* (Tuscaloosa: University of Alabama Press, 1990) 229. Orosz maintains that Barnum, having embarked on a quest for respectability after 1850, assembled "a magnificent collection of natural history," which, had fire not destroyed it, could have provided a successful example of a "synthesis of popular education and professional science" (231)—what Orosz terms "the American compromise."

42 See Strong's journal entry for 8 September 1863: "Visited Barnum's this afternoon. His aquaria have just received a large accession of fish from Bermuda. Many of them very curious and beautiful" (*Diary of George Templeton Strong* 3: 355).

43 Barnum, *Selected Letters of P. T. Barnum,* ed. A. H. Saxon (New York: Columbia University Press, 1983) 141.

44 Orosz, *Curators and Culture* 230.

45 In their introduction to *P. T. Barnum: America's Greatest Showman* (New York: Knopf, 1995), Philip B. Kunhardt Jr., Philip B. Kunhardt III, and Peter W. Kunhardt announce that Barnum was "the great liberating force, chasing out old puritanical inhibitions and letting in the light of joy" (ix). In a similar vein, Barnum's biographer A. H. Saxon rhapsodizes that Barnum, almost "divinely called to be a showman," "let the sunlight into so many lives" (*Barnum* 331).

46 On the distinction between "collecting" and "hoarding," see Brenda Danet and Tamar Katriel, "No Two Alike: Play and Aesthetics in Collecting," *Interpreting Objects and Collections,* ed. Susan M. Pearce (London: Routledge, 1994): "To collect is to set up an agenda for future action for oneself. Hoarding too is future-oriented, but while the hoarder is interested in quantity, the collector is interested in quality" (224–225).

47 Barnum's autobiography offers a splendid instance of this curatorial facetiousness. Finding his museum crowded on St. Patrick's Day with an unusually large contingent of Irish visitors who had all brought their dinners along with the intention of "making a day of it," Barnum had one of his scene-painters paint a message on a piece of canvas, which he then put up over the door leading to the back stairs. It read: TO THE EGRESS. The Irish visitors, "pouring down the main stairs from the third story," stopped to look at the new sign: "Some of them read audibly: To the Aigress.' 'The Aigress,' said others, 'sure that's an animal we haven't seen,' and the throng began to pour down the back stairs only to find that the 'Aigress' was the elephant, and that the elephant was all out o'doors." Meanwhile, Barnum had already begun to accommodate "those who had long been waiting with their money at the Broadway entrance" (*ST* 1: 220).

48 Robert Harbison, *Eccentric Spaces* (1977; Boston: Nonpareil, 1988) 153.

49 Twain, *Travels with Mr. Brown* 117–18.

50 In the museum's lecture room "original dramas of natural incidents, well drawn characters, and unexceptionable moral" were performed; the selection of them was guided, Barnum claimed, "by the knowledge of what the public really want" (*AMI* 26). The highlight of Barnum's theatrical entertainments was *The Moral, Domestic Drama of the Drunkard; or, The Fallen Saved,* which luridly depicted the downfall of an alcoholic from "moderate drinking" to despair and suicide. After the performance, teetotal pledges could be signed at the box office; see M. R. Werner, *Barnum* (Garden City, NY: Garden City Publications, 1923) 111–113.

51 I borrow this phrase from Lawrence Weschler's delightful book *Mr. Wilson's Cabinet of Wonder* (New York: Pantheon, 1995).

52 See *AMI* 1: "The genius of Barnum is truly American. He has not rested content, as many other men would have done, after the great efforts he has already made, and the handsome fortune he has already realised; on the contrary, he is the same active spirit as ever."

53 See Saxon, *Barnum* 9.

54 Whitman, *Complete Poetry and Collected Prose* 50. In March 1855, *Knickerbocker Magazine* reviewed the autobiographies of both Barnum and Thoreau, under the title "Town and Rural Humbugs," as the most remarkable works published that season. See Barry Kritzberg, "Parallel Lives: Town and Rural Humbugs," *Concord Saunterer* 19.2 (1987): 37–40. Barnum's *Life* was simultaneously published in London, by Samson Low, and pirated by other English publishers. Translations into French, German, Swedish, and Dutch soon became available; see A. H. Saxon, *Barnumiana: A Select, Annotated Bibliography of Works by or Relating to P. T. Barnum* (Fairfield, CT: Jumbo's Press, 1995) 15.

55 Kunhardt et al., *Barnum: America's Greatest Showman* 120.

56 *The Autobiography of Petite Bunkum, the Showman; Showing His Birth, Education, and Bringing Up; His Astonishing Adventures by Sea and Land; His Connection with Tom Thumb, Judy Heath, the Woolly Horse, the Fudge Mermaid, and the Swedish Nightingale; Together with Many Other Strange and Startling Matters in His Eventful Career; All of Which Are Illustrated with Numerous Engravings. Written by Himself* (New York: P. F. Harris, 1855) 26.

57 *Autobiography of Petite Bunkum* 65; Saxon, *Barnumiana* 73.

58 P. T. Barnum, *The Humbugs of the World: An Account of Humbugs, Delusions, Impositions, Quackeries, Deceits, and Deceivers Generally, in All Ages* (New York: Carleton, 1866) 66.

59 In 1978, disregarding Neil Harris's *Art of Humbug* (1973), Leslie Fiedler egregiously suggested that "scarcely any critic (before me, at any rate)" had taken Barnum seriously; *Freaks: Myths and Images of the Secret Self* (1978; New York: Anchor, 1993) 279. Other treatments of Barnum's autobiography are few and far between; see James C. Austin, "Seeing the Elephant Again: P. T. Barnum and the American Art of Hoax," *Thalia: Studies in Literary Humour* 4.1 (1981): 14–18, and especially Lawrence Buell, "Autobiography in the American Renaissance," *American Autobiography: Retrospect and Prospect*, ed. Paul John Eakin (Madison: University of Wisconsin Press, 1991) 47–69.

60 Whitman, *Complete Poetry and Collected Prose* 188.

61 Harris, *Art of Humbug* 207.

62 George W. Haines, *Plays, Players, and Playgoers! Being Reminiscences of P. T. Barnum and His Museums. Also, a Graphic Description of the Great Roman Hippodrome and Lives of Celebrated Players* (New York: Bruce, Haines, 1874) 62.

63 Whitman, "So Long!" *Complete Poetry and Collected Prose* 611.

64 Haines, *Plays, Players, and Playgoers!* 60–61.

65 *Struggles and Triumphs; or, Forty Years' Recollections of P. T. Barnum. Written by Himself.* Author's Edition, revised, enlarged, newly illustrated and written up to April, 1875. by the Author (Buffalo, NY: Courier, 1875) vi.

66 Saxon, *Barnumiana* 72.

67 Harris, *Art of Humbug* 4–5.

68 A. Owen Aldridge, *Franklin and His French Contemporaries* (New York: New York University Press, 1957) 204.

69 An animal "made up of the Elephant, Deer, Horse, Buffalo, Camel, and Sheep" (L 350). Unbeknownst to Barnum, his "Woolly Horse" belonged to an existing rare breed of "curly horses"; see Saxon, *Barnumiana* 68.

70 See Barnum's letter to "Messrs. Griffin & Co.," 27 January 1860: "The Mermaid, Woolly Horse, Ploughing Elephants &c. were merely used by me as skyrockets, or advertisements, to attract attention and give notoriety to the Museum & such other really valuable attractions as I provided for the public" (*Selected Letters* 103).

71 As early as 1573, Ambroise Paré suggested that monsters such as "pig- or dog- headed boys" were produced by "the mixture of human and animal seed" (Fiedler, *Freaks* 234).

72 Stephen Jay Gould, *The Flamingo's Smile: Reflections in Natural History* (New York: Norton, 1985) 264. Under the temporalizing impact of evolutionary theory, "connecting links" were increasingly thought of as "*missing* links," i.e., as creatures who represented earlier stages of development and were "lost in time."

73 Arthur O. Lovejoy, *The Great Chain of Being: A Study of the History of an Idea* (Cambridge, MA: Harvard University Press, 1936) 236.

74 See Karen Wonders, *Habitat Dioramas: Illusions of Wilderness in Museums of Natural History* (Uppsala: Almqvist & Wiksell, 1993) 107.

75 Darwin, Notebook C 79 and Notebook B 232; *Charles Darwin's Notebooks, 1836–1844: Geology, Transmutation of Species, Metaphysical Enquiries,* ed. Paul H. Barrett et at. (Cambridge: Cambridge University Press, 1987) 264, 229.

76 Charles Darwin, *Journal of Researches into the Natural History and Geology of the Countries Visited during the Voyage of H.M.S.* Beagle *round the World*, new ed. (1845; New York: Appleton, 1902) 426, 430.

77 Jefferson, *Writings*, ed. Merrill D. Peterson (New York: Library of America, 1984) 265.

78 John Bachman, *An Examination of the Characteristics of Genera as Applicable to the Doctrine of the Unity of the Human Race* (Charleston: James, Williams & Gitsinger, 1845) 10–11.

79 Samuel George Morton, *Hybridity in Animals and Plants Considered in Reference to the Question of the Unity of the Human Species* (New Haven: B. L. Hamlen, 1847) 23.

80 Quoted in Stephen Jay Gould, *The Mismeasure of Man* (New York: Norton, 1981) 70. For Bachman's interest in human skulls, see the chilling letter addressed to him by John James Audubon from London, 4 October 1834: "My Friend W^{am}—MacGillivray wishes to posses Some Sculls of Negroes, *Africans if possible*—but at all events send him half a dozen or whatever you Can—he wishes also to have the *heads* of a few alligators of different Sizes—Turkey Buzzards and Carrion Crow heads indeed Sculls or heads of any description"; Audubon Papers, Houghton Library, bMS Am 1482 (63), quoted by permission of the Houghton Library, Harvard University.

81 Josiah Clark Nott and George R, Gliddon, *Types of Mankind; or. Ethnological Researches Based upon the Ancient Monuments, Paintings, Sculptures, and Crania of Races, and upon*

Their Natural, Geographical, Philological, and Biblical History (Philadelphia: Lippincott, Grambo, 1854).

82 William Carlos Williams noticed the "recurrent image of the ape" in Poe's works and looked for a more personal meaning: "Is it his disgust with his immediate associates and his own fears which cause this frequent use of the figure to create the emotion of extreme terror?" *In the American Grain* (Norfolk, CT: New Directions, 1925), 229.

83 Poe, who had moved to New York in 1844, was aware of and wrote about Barnum's museum. According to Thomas Ollive Mabbott, the "City Museum" in "Some Words with a Mummy" (1845) is probably a reference to the American Museum; *Collected Works of Edgar Allan Poe*, ed. Thomas Ollive Mabbott et al., vol. 3, *Tales and Sketches, 1843–1849* (Cambridge, MA: Belknap, 1978) 1195. On 1 November 1845, Poe suggested in the *Broadway Journal* that the rats known to scour the floors of the Park Theatre and said to disappear on hearing the "rhyming couplets" announcing the end of a performance should consider an engagement with Barnum's "celebrated dog Billy." Billy appeared in 1844 and 1845 in the American Museum as part of the act of contortionist William Cole, known as the "India Rubber Man" (see, for example, the posterbills for May 1844). Coincidentally, the first (posthumous) edition of Poe's *Works* (1855) came out just a few months after Barnum's *Life* from the same publisher, J. S. Redfield.

84 Quotations are from Patrick F. Quinn's edition of Poe's *Poetry and Tales* (New York: Library of America, 1984).

85 While Hop-Frog "hops," Trippetta "trips" or "moves nimbly," "with a quick light tread" (*Oxford English Dictionary*).

86 See, for example, John W. Robertson, *Bibliography of the Writings of Edgar Allan Poe* (1934; New York: Kraus, 1969), who writes that "Hop-Frog" lacks "constructive ingenuity, interest or humor. . . . How or why he conceived it is unexplainable" (259).

87 Barnum's collection also boasted "living monkeys," whose "burlesque similarity with mankind" was, the 1850 guidebook asserted, "a cause of considerable diversion." (Later catalogues were even more explicitly disparaging, pointing out, for example, that the monkeys in Van Amburgh's menagerie were "favorites with the crowd, who make invidious comparisons between them and the colored population"; *Illustrated and Descriptive History* 59).

88 Edgar Allan Poe to Annie Richardson, 8 February 1849, *The Letters of Edgar Allan Poe*, ed. John Ward Ostrom, 2 vols. (Cambridge, MA: Harvard University Press, 1948) 2:425.

89 *In Memoriam A.H.H.*, st. 118; Tennyson, *In Memoriam*, ed. Robert H. Ross (New York: Norton, 1973) 79.

90 The fact that Trippetta, though small, is of otherwise "exquisite proportions" makes Poe's story also a disturbing variation on the myth of "the Beauty and the Beast."

91 Thomas Wyatt admitted, though, that many "exaggerated descriptions of this resemblance have arisen"; see *A Synopsis of Natural History: Embracing the Natural History of Animals, with Human and General Animal Physiology, Botany, Vegetable Physiology and Geology, Translated from the Latest French Edition of C. Lemmonnier, Professor of Natural History in the Royal College of Charlemagne, with Additions from the Works of Cuvier, Dumaril, Lacepede, etc.; and Arranged as a Text Book for Schools* (Philadelphia: Thomas

Wardle, 1839) 31. On the Poe-Wyatt collaboration, see Stephen Jay Gould's recent essay "Poe's Greatest Hit," *Dinosaur in a Haystack: Reflections in Natural History* (New York: Crown, 1996) 173–186.

92 P. T. Barnum to Moses Kimball, 18 August 1846, *Selected Letters* 35.

93 Barnum's fondness for "connecting links" might seem rather difficult to reconcile with his abolitionism and his advocacy of racial equality, which induced him to argue vehemently in favor of black suffrage when he was a member of the Connecticut legislature. But it is not true that Barnum "dramatically and totally" changed his views on race (Kunhardt et al., *P. T. Barnum: America's Greatest Showman* 74). The often-cited speech Barnum delivered on 26 May 1865 luridly invokes African "thick skulls and lips, their woolly heads, their flat noses, their dull, lazy eyes" and finally does little more than offer the Lamarckian-sounding promise that under the influence of civilization and over the course of several generations the "low foreheads" of blacks will expand as a result of their newly awakened "active . . . brain" (*ST* 2: 580).

94 Winthrop D. Jordan, *White over Black: American Attitudes toward the Negro, 1550–1812* (Chapel Hill: University of North Carolina Press, 1968) 238.

95 An ironical reference to the conventional stage figure of the urban dandy in the minstrel show and "hero" of the song of that name ("Old Zip Coon is a very lamed scholar / He plays on the Banjo Cooney in de hollar"). In the 1920s, "What-Is-It" Johnson was still "employed" as a living curiosity at Coney Island in New York. At the time of his death, the *Times* speculated that Zip might have been seen by more people "than anybody that ever lived"; see Bernth Lindfors, "P. T. Barnum and Africa," *Studies in Popular Culture* 7 (1984): 23.

96 *New York Times*, 14 March 1860: 7.

97 The "gorilla" had been identified as a separate species in 1847 by Thomas Savage and Jeffries Wyman, "Notice of the External Characters and Habits of *Troglodytes* Gorilla, a New Species of Orang from the Gaboon River and Osteology of the Same," *Boston Journal of Natural History* 5 (1845–1847): 417–441.

98 The *New York Times* was quick to conclude from the physical "disabilities" of Zip that he was mentally impaired, too, arguing that no matter *what* the "What-Is-It" really was, it certainly exhibited the "cerebral peculiarities of an idiot" ("Amusements," 5 March 1860: 5).

99 See Harriet Ritvo, *The Animal Estate: The English and Other Creatures in the Victorian Age* (Cambridge, MA: Harvard University Press, 1987) 33–34. The term had been proposed by the eighteenth-century naturalist Johann Friedrich Blumenbach, who felt that apes, whose natural home was in the treetops, should not been seen as either bipeds or quadrupeds but as "four-handed," since their hind feet were furnished not with a big toe but with an opposable thumb.

100 Wyatt, *Synopsis* 30.

101 Johnson is the object of some remarkably disrespectful comments by Barnum's biographers. Saxon describes him as a "cone-headed Negro," grinning "imbecilely" (*Barnum* 99), while the Kunhardts, who so adamantly insist on Barnum's "basic decency," refer to Zip as a "microcephalic black dwarf" (*Barnum: America's Greatest Showman* 149). It

might not be amiss to point out that Lavinia Warren, Charles Stratton's wife, when quoting Barnum's *Life* in her autobiography, replaced references to her as a "dwarf" by a phrase she obviously preferred: "little woman." I leave it to the reader's judgment whether this and similar substitutions are just a matter of "more tender sensibilities" (thus editor Saxon in *Autobiography of Mrs. Tom Thumb* 9). Unwittingly or not, Barnum's recent biographers share a predilection for condescending language and jokes with the more popular literature on the subject of "freaks." See, for example, Robert Lifson's *Enter the Sideshow*, an illustrated inventory of human "disfigurement," in which the author never forgets to mention how "kindly" disposed most of these "creatures" were. Lifton writes about Zip (whose real name he mistakenly gives as "Jackson"): "Obviously of a gentle and child-like nature, Zip's smiling and comical presence made him legendary" (Bala Cynwyd, PA: Mason, 1983, n.p.).

102 *Diary of George Templeton Strong* 3:12.

103 See Gillian Beer, "Forging the Missing Link: Interdisciplinary Stories," in Beer, *Open Fields: Science in Cultural Encounter* (Oxford: Clarendon, 1996) 138.

104 Conveniently enough, according to Barnum's posterbill, Zip was the last survivor of the specimens captured by Barnum's exploring party. These "connecting links" couldn't have been particularly hardy, then.

105 See Haines, *Plays, Players, and Playgoers!* 39.

106 Fiedler, *Freaks* 34. Fiedler first contends that "freaks," because of their "strangeness," have a cathartic, "therapeutic" influence on audiences and later in the book contradicts himself, claiming that at the "freak show" we experience the "freakishness of the normal, the precariousness and absurdity of being, however we define it, fully human" (347).

107 The "What-Is-It," however, was more enduringly successful and still appears in posters for January, February, and March 1864.

108 James Joyce, *Finnegans Wake* (New York: Viking, 1939) 546. Joyce's "misspelling" is not accidental, as he obviously also wanted to play on the German *Fee* (fairy).

109 On the history of mermaids, see Colin Blair, *Unnatural History* (London: Abelard-Schuman, 1967) 225.

110 Barnum no doubt would have appreciated his biographer Saxon's attempts to locate the "right" mermaid. Saxon examined the two mermaids kept in a drawer at Harvard University's Peabody Museum, one of which had been identified by Charles Coleman Sellers as the "Fejee" specimen, and came to the devastating conclusion: "I remain unconvinced that either of them is the Fejee Mermaid. . . . The tiny creatures at Harvard [do not] correspond with Barnum's written account" (*Barnum* 365, n. 19).

111 Saxon, *Barnum* 121. This assessment is echoed by Sally Gregory Kohlstedt, "Entrepreneurs and Intellectuals: Natural History in Early American Museums," Alderson, *Mermaids, Mummies, and Mastodons* 34. An admiring portrait of John Bachman's contribution to natural history is painted by Jay Shuler in *Had I the Wings: The Friendship of Bachman and Audubon* (Athens: University of Georgia Press, 1995).

112 See Kunhardt et al., *Barnum: America's Greatest Showman* 43: "Bachman became incensed not only by the fake mermaid but by what he considered its obviously fraudulent partner, the 'ornithorhinchus,' or duck-billed platypus." But in fact Bachman writes:

"The Ornithorynchus is a true and really interesting animal; but it may be seen at the Museum in the Queen-street College gratis" (*Charleston Mercury,* 20 January 1843).

113 Barnum, *Humbugs of the World* 54.

114 In December 1861, Harvard professor Louis Agassiz (see chapter 6) issued Barnum a certificate as to the genuineness of the beluga whale. Referring to a specimen exhibited in Boston, Agassiz pompously informed his visitors that "the species known to Naturalists as the BELUGA" had afforded him *"the means of much valuable information, and I trust it may afford as much pleasure to many others, to see it turning round and round in its large tank, and now coming to the surface to breathe, or blow, as is the phrase"* (2 December 1861).

115 Barnum, *Humbugs of the World* 55.

116 Maureen Howard, *Natural History: A Novel* (New York: HarperPerennial, 1992) 225–226.

117 See Amy Boesky, "'Outlandish-Fruits': Commissioning Nature for the Museum of Man," *ELH* 58 (1991): 305–330.

118 Taylor, "Sights and Monsters" 30–31.

119 For some of these examples, I am indebted to Dennis Todd's essay "The Hairy Maid at the Harpsichord: Some Speculations on the Meaning of *Gulliver's Travels,*" *Texas Studies in Language and Literature* 34.2 (1992): 239–283.

120 See Bluford Adams, "'A Stupendous Mirror of Departed Empires': The Barnum Hippodromes and Circuses, 1874–1991," *American Literary History* 8.1 (1996): 34–56.

121 See the reproduction in Kunhardt et al., *Barnum: America's Greatest Showman* 253.

122 Ralph Waldo Emerson, Notebook XO, 155, *The Topical Notebooks of Ralph Waldo Emerson,* gen. ed. Ralph H. Orth, 4 vols. (Columbia: University of Missouri Press, 1990–1994) 1: 247.

123 William Leach, *Land of Desire: Merchants, Power, and the Rise of a New American Culture* (New York: Pantheon, 1993) 21.

124 Henry James, *A Small Boy and Others,* in James, *Autobiography,* ed. Frederick W. Dupee (New York: Criterion, 1956) 95, 89.

125 Barnum to "Messrs. R. Griffin & Co.," 27 January 1860, *Selected Letters* 103.

126 *The Diary of Alice James,* ed. Leon Edel (Harmondsworth: Penguin, 1982) 63.

4. THE POWER OF FASCINATION

1 Royal Society, London, Journal Books of Scientific Meetings (1660–1800), 17: 45, quoted by permission of the Royal Society, London.

2 Jonathan Edwards, *Images or Shadows of Divine Things,* ed. Perry Miller (New Haven: Yale University Press, 1948) 45 (no. 11).

3 Edwards, *Images or Shadows* 66–67 (no. 63).

4 On the distinction between "serpent" and "snake," see Edward O. Wilson, *Biophilia: The Human Bond with Other Species* (Cambridge, MA: Harvard University Press, 1984) 100–101: "Culture transforms the snake into the serpent, a far more potent creation than the literal reptile."

5 Oliver Wendell Holmes, "Jonathan Edwards," *Pages from an Old Volume of Life* (1891), *The Works of Oliver Wendell Holmes,* Standard Library Edition, 13 vols. (Boston: Houghton Mifflin, 1892) 8: 364.

6 *Newton's Philosophy of Nature: Selections from His Writings,* ed. H. S. Thayer (New York: Hafner, 1953) 174 (query 31).

7 *The Works of William Hogarth, Including* The Analysis of Beauty, 3 vols. (London: Black & Armstrong, 1837) title page; 3:70, xvii.

8 Benjamin Smith Barton, "A Memoir Concerning the Fascinating Faculty Which Has Been Ascribed to the Rattle-Snake, and Other American Serpents," *Transactions of the American Philosophical Society* 4 (1799): 75. The paper, which was also separately published in 1796, was presented to the American Philosophical Society on 4 April 1794.

9 11 April 1834; *The Journals and Miscellaneous Notebooks of Ralph Waldo Emerson,* ed. William H. Gilman et al., 16 vols. (Cambridge, MA: Harvard University Press, 1960–1982) 4: 272.

10 Perry Miller, *Errand into the Wilderness* (1956; Cambridge, MA: Belknap, 1984) 153.

11 Edwards, *Images or Shadows* 124 (no. 181).

12 *Travels and Works of Captain John Smith,* ed. Edward Arber, new ed. by A. G. Bradley, 2 vols. (Edinburgh: John Grant, 1910) 2: 954–955.

13 *William Byrd's Histories of the Dividing Line betwixt Virginia and North Carolina,* ed. William K. Boyd (Raleigh: North Carolina Historical Commission, 1929) 152–153.

14 *Hamilton's Itinerarium,* ed. Albert Bushnell Hart (St. Louis, MO: Privately printed, 1907) 94.

15 "An Account of Some Experiments on the Effects of the Poison of the Rattle-Snake, by Captain Hall," *Philosophical Transactions* 34 (1727): 309–315.

16 Benjamin Franklin, *The Autobiography,* in *Writings,* ed. J. A. Leo Lemay (New York: Library of America, 1987) 1361; "A Letter from Mr. *J. Breintal* to Mr. *Peter Collinson, F.R.S.,* Containing an Account of What He Felt after Being Bitten by a *Rattle-Snake,*" *Philosophical Transactions* 44 (1746): 150. On the misspelling of Breintnall's name, see Stephen Bloore, "Joseph Breintnall, First Secretary of the Library Company," *Pennsylvania Magazine of History and Biography* 59 (1935): 53, n. 43.

17 See Frederick B. Tolles, "A Note on Joseph Breintnall, Franklin's Collaborator," *Philological Quarterly* 21. 2 (April 1942): 248–249.

18 Benjamin Franklin, "Rattle-Snakes for Felons. To the Printers of the Gazette," *Pennsylvania Gazette,* 9 May 1751, *Writings* 359–361.

19 On the use of the rattlesnake as a colonial symbol, see Laurence M. Klauber, *Rattlesnakes: Their Habits, Life Histories, and Influence on Mankind,* 2nd ed., 2 vols. (1956, Berkeley: University of California Press, 1972) 2: 1240–1243; David Scofield Wilson, "The Rattlesnake," *American Wildlife in Symbol and Story,* ed. Angus K. Gillespie and Jay Mechling (Knoxville: University of Tennessee Press, 1987) 44.

20 *Histoire naturelle des quadrupédes ovipares et des serpents, par M. be Comte de la Cépède, Garde du Cabinet du Roi; des Academies & Sociétés Royales de Dijon etc.,* 2 vols. (Paris: Hôtel du Thou, 1788–1789) 2: 419–420: "Ne regrettons pas les beautés naturelles de ces climats plus chauds que le nôtre, leurs arbres plus touffus, leurs feuillages plus agréables, leurs fleurs plus suaves, plus belles: ces fleurs, ces feuillages, ces arbres cachent la demeure du Serpent à sonnette."

21 "Vipera Caudisona Americana; or, The Anatomy of a Rattle-Snake dissected at the Re-

pository of the Royal Society in January 1683, by *Edward Tyson*, M.D., Coll. Med. Lond, etc.," *Philosophical Transactions* 13 (1683): 25–44.

22 "An Extract of Several Letters from *Cotton Mather*, D.D. to *John Woodward*, M.D., and *Richard Waller*, Esq., S.R. Seer.," *Philosophical Transactions* 29 (1714–1717): 62–71.

23 C. Plinii Nat Hist. 8.21 and 29.4; see *The Historie of the World. Commonly called, The Natural Historie of C. Plinius Secundus,* trans. Philemon Holland, 2 vols. (London: Adam Islip, 1601) 1: 206–207, 2: 356. The connection between American myths about the rattlesnake's powers of enchantment and Pliny's description of the basilisk was not lost on the German translator of Barton's "Memoir," E.A.W. von Zimmermann; see his note in *Benjamin Smith Barton's Abhandlungen über die vermeinte Zauberkraft der Klapperschlange und anderer amerikanischen Schlangen; und über die wirksamsten Mittel gegen den Biss der Klapperschlange* (Leipzig: Reinicke und Hinrichs, 1798) 7–11 (note).

24 See Mather's letter to Richard Waller, the secretary of the Royal Society, March 1715: "Admission into that SOCIETY has been granted unto an American. . . . It will be a greater honor to be taken into the list of your servants, than to be mixed with the great men of Achaia"; *Selected Letters of Cotton Mather,* ed. Kenneth Silverman (Baton Rouge: Louisiana State University Press, 1971) 175. On the problematic question of Mather's membership and the ambiguous date of his election, see George Lyman Kittredge, *Cotton Mather's Election into the Royal Society* (Cambridge: John Wilson and Son, 1912).

25 "Capt Walducks Acc' of yᵉ Rattle Snake Read before yᵉ Royall Society, Jan. 7th, 1713/4." The manuscript of Walduck's letter is no. 21 of Sloane MS 3339 in the British Museum and was published by James R. Masterson in "Colonial Rattlesnake Lore, 1714," *Zoologica: New York Zoological Society* 23.9 (1938): 213–216. For information on Walduck, see Raymond Phineas Stearns, *Science in the British Colonies of America* (Urbana: University of Illinois Press, 1970) 351–356.

26 Byrd, *Histories of the Dividing Line* 158.

27 The background of the fatal-boot story is discussed by Laurence M. Klauber, *Rattlesnakes* 2: 1281–1283. In his version of the story, Crèvecoeur's Farmer James, who disingenuously claims that he has heard about the "deplorable" incident from the widow herself, substitutes for Walduck's luckless husbands a son (who dies) and a neighbor (who, thanks to an "eminent physician's" timely intervention, survives). "James" also condenses the time of the story to a mere week.

28 In a similar vein, Crèvecoeur later pointed out that "in the thick settlements" rattlesnakes have already become scarce: "in a few years there will be none left but on our mountains"; *Letters from an American Farmer and Sketches of Eighteenth-Century America,* ed. Albert E. Stone (Harmondsworth: Penguin, 1981) 183. See also John Bartram's letter to Peter Collinson, 17 July 1733/34: "Near *German Town,* about six miles from this City, we found a *Rattle Snake,* which is now become a Rarity so near our Settlements" (*CJB* 3; reprinted in *Philosophical Transactions of the Royal Society* 41 [1740]: 358–359).

29 "An Account of the Rattlesnake. By the Honourable *Paul Dudley,* Esq., F.R.S.," *Philosophical Transactions* 32 (1723): 292. On Dudley, a naturalist who felt equally at home discussing the habits of whales and the methods for making maple sugar, see David Scofield Wilson, *In the Presence of Nature* (Amherst: University of Massachusetts Press, 1978) 36–43.

30 See chapter 19 in part 4 of Beverley's work, first published in 1705, where the "Rattle-Snake" is listed under "troublesom Vermin," as one of the "Annoyances and Inconvenienc-es" of life in Virginia; *The History and Present State of Virginia, in Four Parts, by a Native and Inhabitant of the Place,* 2nd ed. (London: Printed for B. and S. Tooke . . . , 1722) 262–264.

31 See the first edition of Beverley's *History and Present State of Virginia* (London: Printed for R. Parker . . . , 1705), part 4, 65.

32 See his entry for 2 November 1749, *Peter Kalm's Travels in North America: The English Version of 1770,* ed. Adolph B. Benson, 2 vols. (1937; New York: Dover, 1964) 2: 613.

33 "Remarkable and Authentic Instances of the Fascinating Power of the Rattlesnake over Men and Other Animals, with Other Curious Particulars, communicated by Mr. *Peter Collinson* from a Letter of a Correspondent at *Philadelphia,*" *Gentleman's Magazine* 35 (November 1765): 511–514.

34 See, for example, Joseph Ewan's bibliography in *William Bartram: Botanical and Zoologi-cal Drawings, 1756–1788,* ed. Joseph Ewan (Philadelphia: American Philosophical Society, 1968)168.

35 See Peter Collinson to John Bartram, 20 September 1736: "I have Rec'd from my Ingenious frds. J. Breintnall & Doer Witt very pticular accts of the power it has over Creatures by Charming them Into its very Jaws" (*CJB* 35). In Klauber's *Rattlesnakes,* Breintnall is proposed as the author of the letter to Collinson; David Scofield Wilson concurs but points out that, for lack of solid evidence, "the ascription remains unproven" (Scofield, *In the Presence of Nature* 207, n. 22).

36 My emphasis; Royal Society, Journal Book 17: 44–46; quoted by permission of the Royal Society of London.

37 "Of Serpents very few I believe have escaped me, for upon shewing my Design of them to several of the most intelligent Persons, many of them confess'd not to have seen them all, and none of them pretended to have seen any other kinds" (*NHC* 1: x).

38 "An Account of the Most Effectual Means of Preventing the Deleterious Consequences of the Bite of the *Crolalus Horridus,* or *Rattle-Snake,* by *Benjamin Smith Barton,* M.D.," *Transactions of the American Philosophical Society* 3 (1793): 104.

39 In the 1758 edition of his *Systema naturae,* Linnaeus had introduced the scales and plates of the snake as a more secure basis on which to found the specific character of a snake than, for example, coloration; *Systema naturae per regna tria naturae, secundum classes, ordines, genera, species . . . ,* 2 vols. (Holmiae: Impensis Direct. L. Salvii, 1758–1759) 1:195.

40 A.M.F.J. Palisot de Beauvois, "Memoir on Amphibia," *Transactions of the American Philosophical Society* 4 (1799): 364, 371, 378.

41 Barton, "Memoir" 82, 75, 105, 107. In a letter Barton sent to Lacépède in 1802, he brags that he had "a number of living rattlesnakes under my immediate care" and adds that, "in the presence of several gentlemen," he once "ventured to apply a portion of the undi-luted venom of the rattlesnake, recently thrown from its fang, to my tongue." Barton's conclusion is somewhat anticlimactic: "I do not think I shall venture to repeat the experiment." "Letter to *M. Lacepede,* of Paris, on the Natural History of North America. By *Benjamin Smith Barton,* M.D., Professor of Materia Medica, Natural History, and Botany, in the University of Pennsylvania," *Philosophical Magazine* 22 (1805): 206.

42 *Niles' Weekly Register* 23 (14 September 1822): 28; *Niles' Weekly Register* 22 (17 August 1822): 387. Eventually, the enterprising Mr. Neal might have interacted a bit too closely with his perilous pets. At any rate, one can only hope that it wasn't he who was featured in a depressing news item published a year later in *Niles' Weekly Register.* Apparently, a man with a box of rattlers turned up in Genesee, New York, where he appalled the inhabitants by allowing his reptiles, as if they were "very harmless things," to "crawl on his face" and even touch his lips. Sure enough, one of his snakes bit him; he died the next morning, "a spectacle of indescribable horror," and was quickly buried by the alarmed villagers. *Niles' Weekly Register* 27 (9 October 1824): 83.

43 *North American Herpetology of the Reptiles Inhabiting the United States,* 5 vols. (Philadelphia: J. Dobson, 1842–1843). See Wayne Hanley, *Natural History in America: From Mark Catesby to Rachel Carson* (New York: Quadrangle, 1977) 214–223, and Ann Shelby Blum, *Picturing Nature: American Nineteenth-Century Zoological Illustration* (Princeton: Princeton University Press, 1993) 147–150.

44 Louis Agassiz at a meeting of the Natural History Society of Boston, quoted in Thomas Louis Ogier, *A Memoir of Dr. John Edwards Holbrook,* read before the Medical Society of South Carolina, Charleston, 1 November 1871 (privately printed) 11 (Museum of Comparative Zoology, Harvard University). Given its importance in the history of American science, Holbrook's work has received remarkably little attention from intellectual historians. In Michael O'Brien and David Moltke Hansen's collection, *Intellectual Life in Antebellum Charleston* (Knoxville: University of Tennessee Press, 1986), "John Holbrook's" *Herpetology* is mentioned just once, as a "notable" example of the expanding intellectual ambitions of Charleston (6).

45 Tom Beaumont observes: "The only thing we can offer to pass the time is just a drink. And that's the reason, by Jove, that we're always nipping"; John William De Forest, *Kate Beaumont* (State College, PA: Bald Eagle Press, 1963) 57.

46 See Theodore Gill, *Biographical Memoir of John Edwards Holbrook, 1794–1871* (Washington, DC: Judd & Detweiler, 1903) 54. According to Gill, Holbrook, unhappy about the many contradictions he had observed between the reptiles' superficial features and their anatomical characters, abandoned his intention to publish an anatomical supplement to his *Herpetology.*

47 See also Richard Harlan's *American Herpetology; or, Genera of North American Reptilia, with a Synopsis of the Species:* "The great obscurity and confusion peculiarly prevalent in the descriptions of Authors who have written on this subject [herpetology], though gradually dissipating, are by no means sufficiently cleared" (Philadelphia: n.p., 1827, 1).

48 Although Holbrook's second home was in Wrentham, Massachusetts, where his father came from, his lifestyle in Charleston in no way differed from that of his neighbors. See the effusive description in Harriott Horry Ravenel's *Charleston: The Place and the People* (New York: Macmillan, 1906), where the hospitality of Dr. Holbrook and his wife at Belmont, their country home situated about four miles from Charleston, is praised: "Few societies were more agreeable and intellectual than that gathered in the old house" (476). Similar compliments were extended to the Holbrooks in the memoirs of Louis Agassiz's wife, Elizabeth Cary Agassiz. In 1851, Agassiz had been invited to lecture at Holbrook's

college and also stayed at Belmont, "the centre of a stimulating and cultivated social intercourse." It was here that Agassiz aired his view that "different centres of growth" had made the "human family" what it was today, hence "we ought to consider the propriety of applying to man the same rules as to animals" and rank the races (*LC* 2: 497, 499).

49 As Klauber points out, for many years during the nineteenth century the Linnean names for the timber rattler, *Crotalus horridus,* and for the Central American rattlesnake, *Crotalus durissus,* were confused (Klauber, *Rattlesnakes* 1:34, n. 4).

50 Sera died in 1837, "halfway through the publication" of *North American Herpetology* (Blum, *Picturing Nature* 150).

51 *Herrn De la Cepede's Naturgeschichte der Amphibien oder der eyerlegenden vierfüßsigen Thiere and der Schlangen. Eine Fortsetzung von Büffon's Naturgeschichte. Aus dem Französischen übersetzt und mil Anmerkungen und Zusätzen versehen von Johann Matthäus Bechstein,* 5 vols. (Weimar: Verlag des Industrie-Comptoir's, 1801–1802).

52 Sec John Bartram to Collinson, 27 February 1736/37: "Yet we may justly admire the goodness of Providence in giveing this noxious Animal a Rattle in his Tail to give notice where he is—when in motion: for it happens that most that are bit is by accident, by treading on them at unawares as they lie coil'd up, or asleep" (*CJB* 40).

53 Charles Darwin, *The Origin of Species,* ed. Gillian Beer (Oxford: Oxford University Press, 1996) 164.

54 *Niles' Weekly Register* (19 June 1824): 252.

55 Hans Blumenberg, *Die Lesbarkeit der Welt* (Frankfurt am Main: Suhrkamp, 1981) 87.

56 See David Locke, *Science as Writing* (New Haven: Yale University Press, 1992).

57 Oliver Sacks, "The Poet of Chemistry," *New York Review of Books,* 4 November 1993: 56.

58 [T. B. Thorpe], "The Rattlesnake and Its Congeners," *Harper's New Monthly Magazine* 10 (1854–1855): 470–483.

59 Daniel Aaron, *The Unwritten War: American Writers and the Civil War* (1973; Madison: University of Wisconsin Press, 1987) 30.

60 The New York psychiatrist Clarence Oberndorf was so bothered by the escapades of the narrator that, in 1943, he doctored Doctor Holmes's novel, presenting an abridged version to the public that he hoped had separated the clinically valuable content—a story of neurosis—from its untidy and "dull" fictional packaging: *The Psychiatric Novels of Oliver Wendell Holmes* (New York: Columbia University Press, 1943).

61 *Elsie Venner* has in recent years received more critical attention, but apparently none of its modern readers has noticed just how much it continues the tradition of serpentine fascination stories. See Stanton Garner, "*Elsie Venner:* Holmes's Deadly 'Book of Life,'" *Huntington Library Quarterly* 37 (May 1974): 283–298; Alice Hall Petry, "The Ophidian Image in Holmes and Dickinson," *American Literature* 54.4 (1982); 598–601; Margaret Hallissy, "Poisonous Creature: Holmes's *Elsie Vernier,*" *Studies in the Novel* 17: 4 (Winter 1985): 406–419; Anne Dalke, "Economics, or the Bosom Serpent: Oliver Wendell Holmes's *Elsie Venner: A Romance of Destiny,*" *American Transcendental Quarterly* 2.1 (1988): 57–68.

62 For other references to Elsie's nonhuman "diamond eyes," see *EV* 1: 104, 131, 136, 186; 2: 142, 202, 229, 234, 244, 245, 248, 256, 263.

63 See *EV* 2: 11: "It was impossible not to see in this old creature a hint of the gradations by which life climbs up through the lower natures to the highest human developments."

64 This term is used, with reference to Holmes's novel, in an unpublished letter from Newton Arvin to Edmund Wilson, 28 August 1962, which was brought to my attention by Daniel Aaron. For Wilson's response ("Elsie herself is incredible"), see Edmund Wilson, *Letters on Literature and Politics, 1912–1972,* ed. Elena Wilson (New York: Farrar, Straus & Giroux, 1977) 629.

65 Jonathan Miller, "Going Unconscious," *Hidden Histories of Science,* ed. Robert B. Silvers (New York: New York Review of Books, 1995) 13.

66 James Braid, *Electro-Biological Phenomena Considered Physiologically and Psychologically* (Edinburgh: Sutherland & Knox, 1851) 9.

67 Braid, *Electro-Biological Phenomena* 5.

68 *Braid on Hypnotism: Neurypnology; or, The Rationale of Nervous Sleep Considered in Relation to Animal Magnetism or Mesmerism and Illustrated by Numerous Cases of Its Successful Application in the Relief and Cure of Disease,* ed. Arthur Edward Waite (1843; London: Redway, 1899) 86.

69 See *Braid on Hypnotism* 58.

70 John Newman, *Fascination; or, Philosophy of Charming: Illustrating the Principles of Life in Connection with Spirit and Matter* (New York: Fowlers & Wells, 1847) 20–21.

71 Jacob Baker, *Human Magnetism: Its Origin, Progress, Philosophy, and Curative Qualities, with Instruction for Its Application* (Worcester, MA: Privately printed, 1843).

72 Edgar Allan Poe, *Poetry and Tales,* ed. Patrick F. Quinn (New York: Library of America, 1984) 726–727.

73 *Life of P. T. Barnum Written by Himself, Including His Golden Rules for Money-Making, Brought up to 1888* (Buffalo: Courier, 1888) 70.

74 See Holmes's notes on Dr. Majolin's lectures, 14 November 1833 to 22 March 1834, lecture 22, n.p., Houghton Library, Harvard University, bMS Am 1240 (392).

75 *Homeopathy and Its Kindred Delusions* (1842) and *The Contagiousness of Puerperal Fever* (1843; rev. ed. 1855).

76 It was not rare for "mesmerized" patients to assert that they saw "blue or variegated sparks passing from the fingers of the operator" or "a blue fluid" issuing from his eyes; see Braid, *The Power of the Mind over the Body: An Experimental Inquiry into the Nature and Cause of the Phenomena Attributed by Baron Reichenbach and Others to the "New Imponderable"* (London: Churchill, 1841) 29.

77 See Thorpe's "Rattlesnake and Its Congeners" 480: "In more northern latitudes, it has been found that the rattlesnake will not live where the white ash grows in abundance. It has even been the practice among hunters who traverse the forests in summer, to stuff their pockets with white ash leaves, for the purpose of securing themselves against the snakes; and it is said that no person was ever bitten who resorted to this specific."

78 R.W.B. Lewis, *The American Adam: Innocence, Tragedy, and Tradition in the Nineteenth Century* (Chicago: University of Chicago Press, 1955) 38.

79 Holmes, "Jonathan Edwards," *Works of Oliver Wendell Holmes* 8: 401.

80 Holmes, "Border Lines in Medical Science," *Works of Oliver Wendell Holmes* 13: 211.

81 In his preface to the 1861 edition of *Elsie Venner*, Holmes did and did not commit himself: first he asserted that for him the "scientific doctrine" had been a "convenient medium of truth" rather than "an accepted scientific conclusion"; then he claimed that while writing the novel he "received the most startling confirmation of the possibility of the existence of a character like that which he had drawn as a purely imaginary conception in Elsie Venner" (*EV* 1: x). The term "medicated novel" was coined by one of Holmes's friends, as Holmes recalls in "Crime and Automatism" (*Works of Holmes* 8: 358).

82 *From a Darkened Room: The Inman Diary*, ed. Daniel Aaron (Cambridge, MA: Harvard University Press, 1996) 41–42.

5. AUDUBON AT LARGE

1 Audubon usually capitalizes the names of his birds, and I have followed his practice when paraphrasing and discussing his texts. Modern bird names, especially where they differ from the ones Aubudon knew, are given in lower case. See Susanne M. Low, *An Index and Guide to Audubon's* Birds of America: *A Study of the Double-Elephant Folio of John James Audubon's* Birds of America, *as Engraved by William H. Lizars and Robert Havell* (New York: Abbeville Press, 1988).

2 In his chapter "Les Pics," Buffon vividly describes the arduous conditions under which woodpeckers have to obtain their food and find their mates and concludes that their "narrow" and "crude" instincts have limited them to such sad and stunted lives: "Tel est l'instinct étroit et grossier d'un oiseau borné à une vie triste et chétive"; *Ouevres complètes de Buffon, aves les descriptions anatomiques de Daubenton, son collaborateur,* nouvelle édition, ed. J.V.F. Lamouroux and A. G. Desmarest, 44 vols. (Paris: Ladrange et Verdière, 1828) 37: 276. Commenting on Buffon's indictment of the woodpecker, Charles Willson Peale witheringly remarked: "a jargon of blindness . . . and folly, unworthy of the author" (*W* 64).

3 Audubon's wife, Lucy Bakewell Audubon, copied the manuscript of the first two volumes of *Ornithological Biography* and mailed them to the United States for publication and copyright, where they appeared, in an edition of five hundred copies, in Philadelphia and Boston.

4 Francis Hobart Herrick, *Audubon the Naturalist: A History of His Life and Time,* 2nd ed. 2 vols. (1938; New York: Dover, 1968) 1:1.

5 Ron Tyler, *Audubon's Great National Work: The Royal Octavo Edition of* Birds of America (Austin: University of Texas Press, 1993) 2.

6 See David Scofield Wilson, *In the Presence of Nature* (Amherst: University of Massachusetts Press, 1978) 141–147. For Catesby's place in the history of ornithology, see Elsa Guerdrum Allen, *The History of American Ornithology before Audubon, Transactions of the American Philosophical Society,* ns 41.1 (1951): 474–477.

7 George Ord, "Life of Wilson," *AO* 9: xviii. After Wilson's death in 1813, George Ord, member of the Academy of Natural Sciences of Philadelphia, pretending to be merely the editor, in fact *wrote* the entire ninth volume of *American Ornithology.*

8 See William Jardine, "Life of Alexander Wilson," in Wilson, *American Ornithology; or, The Natural History of the Birds of the United States, with a Continuation by Charles Lucian*

Bonaparte, ed. William Jardine, 3 vols. (London: Whittaker, Treacher & Arnot, 1832) 1: xxxix.

9 Jardine, "Life of Alexander Wilson" xl; *AO* 1: 3.

10 For a reproduction of this drawing, titled *Le Pic Noir a Bec Blanc,* see fig. 8 in Theodore E. Stebbins Jr., "Audubon's Drawings of American Birds, 1805–38," *Audubon, The Watercolors for* The Birds of America, ed. Annette Blaugrund and Theodore E. Stebbins Jr. (New York: Villard, 1993) 9. Selections from the Harvard collection of Audubon's drawings were made available in *The Bird Biographies of John James Audubon,* ed. Alice Ford (New York: Macmillan, 1957).

11 See Herrick, *Audubon, the Naturalist* 1: 359–360.

12 Ann Shelby Blum, *Picturing Nature: American Nineteenth-Century Zoological Illustration* (Princeton: Princeton University Press, 1993)106.

13 Audubon later recalled that when he was living in Kentucky, he "seldom passed a day without drawing a bird, or noting something respecting its habits" (*AJ* 1: 29).

14 See Audubon, "Louisville in Kentucky," *AJ* 2: 202. This sketch was one of sixty "episodes," or personal essays (called "delineations of American scenery and character"), included in the first three volumes of *Ornithological Biography,* which Audubon later deleted from the text of the Royal Octavo edition of *Birds of America.* They were republished, separately, in Maria Audubon's edition of the journals (1897).

15 Bernard Taper, *Balanchine: A Biography* (Berkeley: University of California Press, 1987) 388.

16 John Burroughs, *John James Audubon* (Boston: Small, Baynard, 1902) 126.

17 The two standard biographies of Audubon are Herrick, *Audubon the Naturalist,* and Alice Ford, *John James Audubon: A Biography* (1964; New York: Abbeville Press, 1988), but see also Shirley Streshinsky's recent, spirited *Audubon: Life and Art in the American Wilderness* (New York: Villard, 1993). For a concise profile, see Mary Durant, "The Man Himself," *The Bicentennial of John James Audubon,* ed. Alton A. Lindsey (Bloomington: Indiana University Press, 1985) 96–114.

18 Wayne Hanley, *Natural History in America: From Mark Catesby to Rachel Carson* (New York: Quadrangle, 1977) 65.

19 References to his "old master" David crop up time and again in Audubon's writings; see especially Audubon's "Introductory Address" to *Ornithological Biography (OB* 1: viii: "David had guided my hand in tracing objects of large size") and *The 1826 Journal of John James Audubon,* ed. Alice Ford (1967; New York: Abbeville Press, 1987) 227. When we think of what Walter Friedlaender has identified as the elements of David's heroic vision, his "mastery of space composition, of group arrangement, and of balanced movement," the fusion of the monumental with the precisely observed realistic detail, it is easy to imagine parallels between Audubon's work and the style of "*le premier peintre*" of the French Empire; *David to Delacroix,* trans. Robert Goldwater (Cambridge, MA: Harvard University Press, 1952) 28–29. However, as Alice Ford points out, Audubon's name is not among the abundant references to pupils in David's papers (*1826 Journal* 104–105, n. 15). Of equal longevity was the legend, never actively discouraged by Audubon and still kept alive in Constance Rourke's biography, that he might be the Lost Dauphin of France; *Audubon* (New York: Harcourt, Brace, 1936).

20 John Keats to George and Georgians Keats, 17–27 September 1819, *The Letters of John Keats, 1814–1821*, ed. Hyder Edward Rollins, 2 vols. (Cambridge, MA: Harvard University Press, 1958) 2:185.

21 John Chancellor, *Audubon: A Biography* (New York: Viking, 1978) 10. For a different view, see Barbara J. Cicardo, "From Palette to Pen: Audubon as Writer," *Audubon: A Retrospective*, ed. James H. Dorman and Allison H. De Pena (Lafayette: Center for Louisiana Studies, University of Southwestern Louisiana, 1990) 74–91.

22 In his journal, Audubon described himself as a "man who never looked into an English grammar and very seldom, unfortunately, into a French or a Spanish one" (12 December 1826; *1826 Journal* 388). It is perhaps an indication of the extent of the changes made by Audubon's granddaughter in her edition of the journals that she felt compelled to rewrite even this innocent sentence, perhaps to ward off suspicions that Audubon couldn't speak French properly. In her version, Audubon saw himself as a man "who never looked into an English grammar and who has *forgotten* most of what he learned in French and Spanish ones" (*AJ* 1: 181; my emphasis).

23 Audubon Papers, Houghton Library, bMS Am 1483 (179), quoted by permission of the Houghton Library, Harvard University.

24 Audubon himself must have appreciated MacGillivray's solicitous editing, as he disliked people who, in "improv[ing] the style," "destroyed the matter" (*1826 Journal* 401). This objection, incidentally, could be applied to Alice Ford's editing of one of Audubon's manuscripts in *Audubon, by Himself: A Profile of John James Audubon*, ed. Alice Ford (Garden City, NY: Natural History Press, 1969). Ford rewrites Audubon's "Pitting of Wolves," from *Ornithological Biography*, arguing that MacGillivray, a "writer only of necessity," had failed to remove the "cascading redundancies, *non sequiturs*, misleading allusions, indefinite antecedents, and desperate participles" in Audubon's manuscript (vi). In Ford's version (62–64), gone are Audubon's appeals to the "kind reader" as well as such wonderfully cadenced sentences of his as "Winter once more had come dreary, sad, cold and forbidding" (Ford rephrases: "Winter had set in, cold, dark and forbidding"). The effect of Ford's sanitized sample is not to convince us that this is how "Audubon would have written if English had been his native tongue" (foreword, vi) but to make us grateful for all those occasions in which MacGillivray decided to preserve rather than to change Audubon's original.

25 28 October 1826; *1826 Journal* 318.

26 This is briefly suggested by Burroughs: "In writing the biography of his birds he wrote his autobiography as well" (*John James Audubon* 120).

27 Among several instances of "sagacity of the brute creation" mentioned by White, see especially his letter 16 to Daines Barrington (on the nest-building of the martins), in *The Natural History and Antiquities of Selborne, in the County of Southhampton*, ed. Richard Mabey (Harmondsworth: Penguin, 1977) 147–151.

28 See Paul L. Farber, *The Emergence of Ornithology as a Scientific Discipline, 1760–1850* (Dordrecht: D. Reidel, 1982) 4.

29 Jean de La Fontaine, "Le Héron," *Fables* 7.4; *Fables et oeuvres diverses de J. La Fontaine*, ed. C. A. Walckenaer (Paris: Firmin Didot Frères, 1865) 205. For Audubon's love of La

Fontaine, a topic that deserves further study, see his *Mississippi River Journal* ("I had the satisfaction of ransacking the *Fables* of La Fontaine, with Engravings"; 29 December 1820; *SJ* 80) and his biography of the Pewee Flycatcher, where La Fontaine is mentioned along with the popular novelist Maria Edgeworth (1767–1849) as one of the authors whose works young Audubon would regularly peruse: "My paper and pencils, with now and then a volume of Edgeworth's natural and fascinating Tales or Lafontaine's Fables, afforded me ample pleasures" (*BA 1:* 224).

30 *BA* 1: 101; 4: 179; 5: 11; 6: 380, 446. See also *BA* 6: 190 and 7: 10, "with the swiftness of thought," and *BA* 7: 120, "with the velocity of thought."

31 Edward A. Muschamp estimates that, in the course of his work on *Birds of America*, Audubon must have traveled "between 30,000 and 35,000 miles." Audubon drew more than "1,000 birds representing close to 500 species" and wrote "nearly 1,000,000 words of text." *Audacious Audubon: The Story of a Great Pioneer, Artist, Naturalist, and Man* (New York; Brentano's, 1929) 263.

32 See also Audubon's "Introductory Address" to *Ornithological Biography*, where he claims that he "received life and light in the New World" (*OB* 1: v).

33 See Harriet Ritvo, "Learning from Animals: Natural History for Children in the Eighteenth and Nineteenth Centuries," *Children's Literature* 13 (1985): 86.

34 Peter Sitt, "An Interview with Robert Penn Warren" (1977), *Talking with Robert Penn Warren*, ed. Floyd C. Watkins et al. (Athens: University of Georgia Press, 1990) 244.

35 See Robert Penn Warren's essay "The Love and the Separateness in Miss Welty" (1944), *Critical Essays on Eudora Welty*, ed. W. Craig Turner and Lee Ending Harding (Boston: G. K.Hall, 1989) 42–51.

36 *The Collected Stories of Eudora Welty* (New York: Harcourt, Brace, Jovanovich, 1980) 197.

37 See Chancellor, *Audubon* 7: "Audubon was not a latter-day St Francis of Assisi and he indulged in what appears to us today to be revolting and wantonly indiscriminate slaughter of birds." For a critical view of Audubon's "incursion[s] into natural wholeness," mainly with reference to the "episodes" included in *Ornithological Biography*, see Annette Kolodny, *The Lay of the Land: Metaphor as Experience and History in American Life and Letters* (Chapel Hill-University of North Carolina Press, 1975) 74–88.

38 Paul Binding, *The Still Moment: Eudora Welty* (London: Virago, 1994) 156–157.

39 John Burroughs, "The Art of Seeing Things," *Leaf and Tendril* (Boston: Houghton Mifflin, 1908) 10.

40 Ben Forkner, "Writing the American Woods: The Journals and Essays of John James Audubon," *SJ* xxi.

41 John Burroughs, "In the Hemlocks" (1865), in Burroughs, *Wake-Robin* (Boston: Houghton Mifflin, 1904) 49.

42 Mary Durant and Michael Harwood, *On the Road with John James Audubon* (New York: Dodd, Mead, 1980) 211. Even the ecologist Alton A. Lindsey confirms that "a naturalist in this period really needed a good series of skins" for any species he was working on ("Saving the Pieces," Lindsey, *Bicentennial of Audubon* 118).

43 Audubon's scientific curiosity in this case did not prevent him from eating his specimens: "The flesh is excellent, and they are generally very fat" (*BA* 5:12).

44 Audubon takes care, however, to distinguish his own culinary preferences from those of the "epicure": "Young birds from the nest afford tolerable eating; but the flesh of the old birds is by no means to my taste, nor so good as some epicures would have us to believe, and I would at any time prefer that of a Crow or young Eagle" ("The Great Blue Heron"; *BA* 6: 126); "Indeed I know none, excepting what is called *an Epicure,* who could relish a Scaup Duck" (*BA* 6: 317).

45 See *BA* 6: 294: "Repeated inspection of the stomach has shewn me that the Shoveller is not more *nice* as to the quality of its food than the Mallard or any other of the Duck tribe, for I have found in it leeches, small fishes, large ground-worms, and snails" (my emphasis).

46 Audubon shared his antipathy with Benjamin Franklin, who called the bald eagle "a Bird of bad moral Character," "lousy," and "a rank Coward." Franklin facetiously recommended that Americans adopt the wild turkey as an emblem, "a much more respectable Bird, and withal a true original Native of America," albeit "a little vain and silly"; to Sarah Bache, 26 January 1784, *Writings,* ed. J. A. Leo Lemay (New York: Library of America, 1987) 1088. The last part of Franklin's statement, especially, would not have amused Audubon, who had chosen the wild turkey and the motto "America My Country" for his personal crest (see *1826 Journal* 243, 340).

47 La Fontaine, "Les Vautours et les pigeons" (*Fables* 7.8), *Fables et oeuvres diverses de J. La Fontaine* 210.

48 Forkner, "Writing the American Woods," *SJ* xvii,

49 Forkner, "Writing the American Woods," *SJ* xxxvii.

50 Durant, "The Man Himself," *Bicentennial of Audubon* 98.

51 Peter Matthiessen, *Wildlife in America* (New York: Penguin, 1987) 157–158.

52 See White, letters 30 and 21 to Barrington, *Natural History of Selborne* 190, 168; Richard Mabey, *Gilbert White* (London: Century, 1987) 86–87.

53 Streshinsky, *Audubon* 162.

54 James Fenimore Cooper, *The Leatherstocking Tales,* ed. Blake Nevius, 2 vols. (New York: Library of America, 1985) 1: 247.

55 *Leatherstocking Tales* 1: 249–250.

56 John Muir, *Nature Writings,* ed. William Cronon (New York: Library of America, 1997) 29–30, 78, 44.

57 The analogy between the slaughter of the Indians and the killing of birds is also inherent in Cooper's *The Pioneers:* the cannon used to fire volleys into the flocks of the arriving birds was once employed "by a war-party of the whites, in one of their inroads into the Indian settlements" (*Leatherstocking Tales* 1: 247).

58 The last Carolina parakeet died in 1914, during the same year that "Martha," the last passenger pigeon on earth, "blinked a final time in the Cincinnati Zoo" (Matthiessen, *Wildlife* 181).

59 See *BA* 1: 164: "Since the progress of civilization in our country has furnished thousands of convenient places for this Swallow to breed in, safe from storms, snakes, or quadrupeds, it has abandoned, with a judgment worthy of remark, its former abodes in the hollows of trees, and taken possession of the chimneys." According to Audubon, the

cardinal, too, "relishes . . . the neighbourhood of cities" and is "constantly found in our fields, orchards and gardens; nay, it often enters the very streets of our southern towns and villages" (*BA* 3:199).

60 See *Mississippi River Journal*, 27 October 1821: "Dressed all new, Hair Cut, my appearance altered beyond My expectations, fully as much as a handsome Bird is when robbed of all its feathering, the Poor thing Looks, Bashfull dejected and is either entirely Neglected or Look^d upon With Contempt; such was my situation Last Week—but When the Bird is Well fed, taken care of, sufered to Enjoy Life and dress himself; he is cherished again" (*SJ* 137). In a letter to John Bachman of 25 August 1834, Audubon complained about his friend's laziness as a correspondent and threatened that he would "keep hammering" at Bachman's door "like a Woodpecker on the bark of some tough tree, the inside of which it longs to see"; Audubon Papers, Houghton Library, bMS 1482 (61), quoted by permission of the Houghton Library, Harvard University.

61 The numerous literary treatments of Audubon's life include a whimsical poem by Stephen Vincent Benét, "John James Audubon" (1933), which describes the naturalist "Scrambling through a wilderness /. . . . / All around America, in a feathered dream"; *Selected Works of Stephen Vincent Benét*, 2 vols. (New York: Farrar & Rinehart, 1942) 1: 400–401. Other notable examples are Robert Penn Warren's cycle *Audubon: A Vision* (1969) and Pamela Alexander's book-length *A Commonwealth of Wings* (Hanover, NH: Wesleyan University Press, 1991).

62 Sarah Orne Jewett, *Novels and Stories*, ed. Michael Davitt Bell (New York: Library of America, 1994) 674, 679.

63 *Collected Stories of Eudora Welty* 198.

64 *1826 Journal* 369.

65 See *Mississippi River Journal*, 21 December 1820: "Drawing nearly all day I finished the Carion Crow, it stunk so intolerably, and Looked so disgusting that I was very glad when I through it over Board" (*SJ* 73); 27 March 1821: "Drawing at My Heron yet, it Smelt so dreadfully bad" (*SJ* 104). Disagreeable as these experiences must have been, Audubon apparently still preferred the company of dead birds to that of recently deceased humans. While at Oakley, James Pirrie's plantation on Bayou Sara, Louisiana, he had to accompany his host's wife to a dying neighbor's house and then had "the displeasure of Keeping his body's Company the remainder of the Night. On such Occasions time flys very slow indeed, so much so that it looked as if it Stood Still like the Hawk that Poises in the air over its prey" (*SJ* 125). In 1826, a shocked Audubon sat in on the dissections performed by Dr. Robert Knox in Edinburgh and was glad to "leave this charnel and breathe again the salubrious atmosphere of the streets" (*1826 Journal* 371). Watching Dr. John Lizars operate "on a beautiful dead body of a female, quite fresh," Audubon was overcome with the suffocating "horrible stench" of the dissecting rooms: "I soon made my escape, I assure thee, and went home" (*1826 Journal* 379).

66 Hereafter abbreviated as DEF. On 2 June 1863, Audubon's widow sold the original drawings of the plates of *The Birds of America* for a mere four thousand dollars to the New York Historical Society, which did not mount a full exhibition of the works for 108 years. See the complete edition annotated by Edward H. Dwight, *The Original Water-Color*

Paintings of John James Audubon for The Birds of America, 2 vols. (New York: American Heritage, 1966).

67 The quotation is from the beginning of La Fontaine's fable of the hornets and the bees, "Les Frelons et les mouches à miel" (*Fables*, 1.21), *Fables et oeuvres diverses de J. La Fontaine* 74.

68 *Letters of John James Audubon, 1826–1840*, ed. Howard Corning, 2 vols. (Boston: Club of Odd Volumes, 1930) 1; 3.

69 Rudolf Arnheim, *The Power of the Center: A Study of Composition in the Visual Arts*, new ed. (Berkeley: University of California Press, 1988) 47.

70 *1826 Journal* 347. In his "Introductory Address" to *Ornithological Biography*, Audubon defended himself against those who would criticize him for the large-paper format he was using: "It could not be avoided without giving up the desire of presenting to the world those my favourite objects in nature, of the size which nature has given to them" (*OB* 1: xvii).

71 See Audubon, *Watercolors for* The Birds of America 86.

72 Adam Gopnik, "Audubon's Passion," *New Yorker*, 25 February 1991:100.

73 Audubon to William Rathbone, 24 November 1826; *1826 Journal* 345. For Audubon, the Double Elephant Folio was obviously not just a jumble of plates; characteristically, he refers to it as his "great *Book of Nature*" (27 December 1826; *1826 Journal* 420) and his "dear Book" (*AJ* 1:258).

74 See Robert Henry Welker, *Birds and Men: American Birds in Science, Art, Literature, and Conservation, 1800–1900* (Cambridge, MA: Belknap, 1955) 88.

75 Along with the elimination from the plate of material too dramatically subjective, the story of how Audubon procured and painted the golden eagle, originally published in the second volume of *Ornithological Biography*, was also deleted from the Royal Octavo edition.

76 See Waldemar H. Fries, *The Double Elephant Folio: The Story of Audubon's* Birds of America (Chicago: American Library Association, 1973) 156.

77 Edgar Allan Poe, *Poetry and Tales*, ed. Patrick F. Quinn (New York: Library of America, 1984) 555.

78 See Carole Anne Slatkin's comment in Audubon, *Watercolors for* The Birds of America 232: "By juxtaposing an image of the eagle's hard-won but triumphant seizure of the rabbit . . . with a depiction of his own struggle to capture the eagle physically (and, by extension, to capture it on the page)—the artist may be referring to the difficulties he underwent to create this picture."

79 John Berger, "Why Look at Animals?" in Berger, *About Looking* (New York: Vintage, 1980) 3–28.

80 *The Autobiography of Charles Darwin, 2809–2882*, ed. Nora Barlow (1958; New York: Norton, 1969) 51.

81 Darwin, *The Descent of Man, and Selection in Relation to Sex*, rev. 2nd ed. (1888; New York: Appleton, 1927) 373, 374.

82 Darwin, *Descent* 424–425; see *BA* 6: 179–180.

83 Darwin, *Descent* 507, 468.

84 Darwin, *Descent* 632, 613.
85 See Gillian Beer, " 'The Face of Nature'; Anthropomorphic Elements in the Language of *The Origin of Species*," *Languages of Nature: Critical Essays on Science and Literature*, ed. Ludmilla Jordanova (London: Free Association, 1986) 222.
86 Darwin, *Descent* 621; Adrian Desmond and James Moore, *Darwin* (New York: Norton, 1991) 579.
87 Walt Whitman, *Complete Poetry and Collected Prose*, ed. Justin Kaplan (New York: Library of America, 1982) 690.
88 See Frank Stewart, *A Natural History of Nature Writing* (Washington, DC: Island, 1995) 67.
89 Whitman, *Complete Poetry and Prose* 904–905.
90 Edward Lurie, *Louis Agassiz: A Life in Science* (1960; Baltimore: Johns Hopkins University Press, 1988) 197–199.
91 *Boston Daily Advertiser*, 29 and 30 May 1855; Agassiz, *Contributions to the Natural History of the United States of America*, 4 vols. (Boston: Little, Brown, 1857–1862) 1: x, ix; see Lee Rust Brown, *The Emerson Museum: Practical Romanticism and the Pursuit of the Whole* (Cambridge, MA: Harvard University Press, 1997) 134.

6. AGASSIZ AGONISTES

1 Louis Agassiz to Sir Charles Lyell, 5 May 1856, Agassiz Papers, Houghton Library, bMS Am 1419 (131), quoted by permission of the Houghton Library, Harvard University.
2 Edward Lurie, *Louis Agassiz: A Life in Science* (1960; Baltimore: Johns Hopkins University Press, 1988) 198.
3 Louis Agassiz to Charles Sumner, 8 February 1856, quoted in Lurie, *Agassiz* 198.
4 In "Textbooks and Texts from the Brooks: Inventing Scientific Authority in America," *American Quarterly* 49 (1997): 1–25, Laura Dassow Walls argues that Agassiz's authority benefited from "the wide circulation of a set narrative" about his teaching technique, "a story about the triumph of fact-based modern science" (4). The best-known instance of Agassiz's lingering pedagogical influence is, of course, Ezra Pound's declaration in *The ABC of Reading* that "no man is equipped for modern thinking until he has understood the anecdote of Agassiz and the fish." In Pound's version, a postgraduate student who had approached Agassiz to receive "the final and finishing touches" is asked to describe a common sunfish and comes up with the conventional definitions and even a four-page essay, all of which are rejected by Agassiz, who "told him to look at the fish." Finally, the fish is "in an advanced stage of decomposition," but the student has learned his lesson—he "knew something about it" (London: Faber, 1934, 17). See also James David Teller, *Louis Agassiz: Scientist and Teacher* (Columbus: Ohio State University Press, 1947), and Guy Davenport's introduction to *The Intelligence of Louis Agassiz: A Specimen Book of Scientific Writings*, ed. Guy Davenport (Boston: Beacon, 1963) 1–30.
5 *The Education of Henry Adams* (1906), Henry Adams, *Novels, Mont Saint Michel, The Education* (New York: Library of America, 1983) 774.
6 See the entry for 30 November 1858 in *The Journal of Henry David Thoreau*, ed. Bradford Torrey and Francis Allen, 14 vols. (Boston: Houghton Mifflin 1906) 11: 359–360.

7 William James, "Louis Agassiz" (1896), *Essays, Comments, and Reviews: The Works of William James*, gen. ed. Frederick H. Burkhardt (Cambridge, MA: Harvard University Press, 1987) 48.

8 For biographical information on William James, see Ralph Barton Perry, *The Thought and Character of William James, as Revealed in Unpublished Correspondence and Notes, together with His Published Writings*, 2 vols. (Boston: Little, Brown, 1935); Gay Wilson Allen, *William James: A Biography* (New York: Viking, 1967); Gerald E. Myers, *William James: His Life and Thought* (New York: Yale University Press, 1986).

9 Henry James, *Notes of a Son and Brother*, in James, *Autobiography*, ed. Frederick W. Dupee (New York: Criterion, 1956) 308.

10 Louis Agassiz, *Twelve Lectures on Comparative Embryology, Delivered before the Lowell Institute, in Boston, December and January, 1848–49* (Boston: Henry Flanders, 1849) 26. My summary of Agassiz's scientific status in the United States is indebted to Ernst Mayr's seminal essay, "Agassiz, Darwin, and Evolution" (1959), now revised in Mayr, *Evolution and the Diversity of Life: Selected Essays* (Cambridge, MA: Belknap, 1997) 250–276.

11 *The Autobiography of Charles Darwin, 1809–1882*, ed. Nora Barlow (1958; New York: Norton, 1969) 122.

12 Louis Agassiz, *Contributions to the Natural History of the United States of America*, 4 vols. (Boston: Little, Brown, 1857–1862) 3: 4.

13 Darwin to Agassiz, 11 November 1859, *The Correspondence of Charles Darwin*, ed. Frederick H. Burkhardt and Sydney Smith, 10 vols. (Cambridge: Cambridge University Press, 1986–) 7: 366. Agassiz and Darwin stayed in touch, even though the tone of their correspondence had cooled off since 1848, when they first exchanged notes about barnacles. On 12 April 1864, Darwin gratefully acknowledged receiving one of Agassiz's books ("I know well how strongly you are opposed to nearly everything I have written & it gratifies me deeply that you have not for this cause taken, like a few of my former English friends, a personal dislike to me"). After his return from Brazil, Agassiz informed even his scientific and ideological opponent about the results of his ichthyological investigations, which Darwin said interested him "*extremely*" and told him "exactly what I wanted to know" (19 August 1868). Darwin's letters are in the Agassiz Papers at the Houghton Library, bMS Am 1419 (277–278), and are quoted with the permission of the Houghton Library, Harvard University.

14 See Agassiz's marginal note on page 194 (chapter 6) in his presentation copy of *On the Origin of Species by Means of Natural Selection* (London: Murray, 1859), Special Collections, Museum of Comparative Zoology, Harvard University. In the same chapter of *Origin*, Darwin illustrated the concept of "natural selection" by referring to geese with webbed feet that live on dry land as an example of how animals take possession of the habitats of other species over which they have gained an advantage, "however different it may be from [their] own place." Retorted Agassiz, again in the margin, "Why Sir, there is room enough for all the Ducks of the world in the forests of North America" (186).

15 Charles Darwin, *The Origin of Species*, ed. Gillian Beer (Oxford: Oxford University Press, 1996) 397.

16 Louis Agassiz, *Methods of Study in Natural History*, 2nd ed. (Boston: Ticknor & Fields, 1864) 13–14.

17 Agassiz, *Contributions to the Natural History of the United States of America* 3: 4.

18 A phrase I have borrowed from Armand Marie Leroi, "A Duck Folded in Half," *London Review of Books*, 19 June 1997:19–20.

19 "We were much amused by their evident *verdancy* in regard to life in the woods," wrote the geologist Josiah Dwight Whitney, adding that he should "love to see them in camp and watch their proceedings" (Lurie, *Agassiz* 149).

20 *Selecta genera et species piscium: quos in itinera per Brasiliam annis MDCCCXVII–MDCCCXX jussu et auspiciis Maximiliani Josephi I . . . peracto collegit et pingendos curavit Dr. J. B. de Spix . . . digessit, descripsit et observationibus anatomicis illustravit Dr. L. Agassiz, praefatus est et edidit itineris socius Dr. F. C. Ph. de Martins* (München: C. Wolf, 1829–[1831]).

21 See Johann Baptist Spix and Carl Friedrich Philipp von Martius, *Reise in Brasilien, auf Befehl Sr. Majestät Maximilian Joseph I., Königs von Baiern, in den Jahren 1817 bis 1820 gemacht und beschrieben*, 3 vols. (München: M. Lindauer, 1823–1831).

22 Agassiz to Rose Mayor Agassiz, 22 March 1865; *LC* 2: 626.

23 Theodore Roosevelt, *Through the Brazilian Wilderness* (New York: Charles Scribner's Sons, 1914) 324.

24 Brazil had recognized the belligerency of the Confederacy in the Civil War and, "much to the chagrin of the government in Washington," allowed Confederate ships to use Brazilian harbors; see E. Bradford Burns, *A History of Brazil*, 3rd ed. (New York: Columbia University Press, 1993) 196.

25 Oliver Wendell Holmes to Louis and Elizabeth Agassiz, 5 January 1868, Agassiz Papers, Houghton Library, bMS Am 1419 (405), quoted by permission of the Houghton Library, Harvard University.

26 See George Pendle, *A History of Latin America*, rev. ed. (Harmondsworth: Penguin, 1976) 123.

27 Thomas Jefferson, *Writings*, ed. Merrill D. Peterson (New York: Library of America, 1984) 1409.

28 See Thomas E. Skidmore, *Black into White: Race and Nationality in Brazilian Thought* (New York: Oxford University Press, 1974) 14. As late as 1845, no fewer than twenty thousand slaves were "imported," the protests of Britain notwithstanding. The same year, Britain declared its intention of treating as pirates all Brazilian slaveships encountered at sea; see Pendle, *History of Latin America* 154. Total emancipation was not granted before 1888, and then only because Brazilian planters realized that the replacement of slavery by free labor had become economically inevitable (Skidmore 17).

29 Skidmore, however, emphasizes that Brazil had never been as rigidly bi-racial as the United States, since there was always the "middle category . . . of racial mixtures" (*Black into White* 40).

30 In October 1865, Elizabeth Agassiz was exhilarated to learn that the first steamer of the line recently opened between New York and Brazil had stopped in Pará on its way to Rio (*JB* 252).

31 See Roger D. Stone, *Dreams of Amazonia* (1985; New York: Penguin, 1986) 56.

32 Charles Darwin, *Journal of Researches into the Natural History and Geology of the Countries Visited during the Voyage of H.M.S. Beagle round the World*, new ed. (1845; New York: Appleton, 1902) 11.

33 28 February 1832, *Charles Darwin's* Beagle *Diary*, ed. Richard Darwin Keynes (New York: Cambridge University Press, 1988) 42.

34 Darwin, *Journal of Researches* 31.

35 28 February 1832, *Darwin's* Beagle *Diary* 42.

36 1 March and 6 April 1832; *Darwin's* Beagle *Diary* 43, 52.

37 See Paul Russell Cutright, *The Great Naturalists Study South America* (New York: Macmillan, 1940) 22–23.

38 "This little book was so clearly and brightly written, described so well the beauty and grandeur of tropical vegetation, and gave such a pleasing account of the people . . . that Bates and myself at once agreed that this was the very place for us to go to"; see Alfred Russel Wallace, *My Life: A Record of Events and Opinions*, 2 vols. (New York: Dodd, Mead, 1905) 1: 264. On Bates and Wallace, see *JB* 443.

39 See Darwin's letter to H. W. Bates, 18 April 1863, *The Life and Letters of Charles Darwin, Including an Autobiographical Chapter*, ed. Francis Darwin, 3 vols. (London: John Murray, 1888) 2: 381. Darwin was "a *little* disappointed by Wallace's book on the Amazon," which he thought had "hardly facts enough" (to Bates, 3 December 1861; *Life and Letters* 2: 380).

40 Charles Darwin to Charles Lyell, 18 June 1858, *The Correspondence of Charles Darwin* 7: 107. On Wallace and Darwin, see Adrian Desmond and James Moore, *Darwin* (New York: Norton, 1991) 467–470.

41 Darwin, *Autobiography* 122.

42 See Agassiz's marginalia on page 184 of his copy of *The Origin of Species*, Museum of Comparative Zoology, Harvard University.

43 See, for example, Charles Darwin, *The Descent of Man, and Selection in Relation to Sex*, rev. 2nd ed. (1888; New York: Appleton, 1927) 329–330.

44 Lurie, *Agassiz* 357.

45 *New-York Daily Tribune*, 23 January 1868: 6; Boston *Evening Transcript* 8 January 1868: n.p.; *Springfield Republican*, 15 January 1868: 6.

46 Oliver Wendell Holmes to Louis and Elizabeth Agassiz, 5 January 1868, Agassiz Papers, Houghton Library, bMS Am 1419 (405), quoted by permission of the Houghton Library, Harvard University.

47 Ralph Waldo Emerson to Louis and Elizabeth Agassiz, 23 February 1868, Agassiz Papers, Houghton Library, bMS Am 1419 (1419), quoted by permission of the Houghton Library, Harvard University.

48 Because as a child Elizabeth Cary suffered from a chronic cough, her parents kept her at home, except for a brief period during which she attended Elizabeth Peabody's Historical School; see Louise Hall Tharp, *Adventurous Alliance: The Story of the Agassiz Family of Boston* (Boston: Little, Brown, 1959) 28. Elizabeth's sister Caroline later remembered: "It was curious enough in consideration of the life before her that she had not as solid an education as the others. But this was balanced by her love of reading in general . . . wider

than it would have been if she had had to bring lessons home from school to study"; see Lucy Allen Paton, *Elizabeth Cary Agassiz: A Biography* (Boston: Houghton Mifflin, 1919) 17.

49 See Actaea [Elizabeth Cary Agassiz], *A First Lesson in Natural History* (Boston: Little, Brown, 1859); Elizabeth Agassiz and Alexander Agassiz, *Seaside Studies in Natural History* (Boston: Ticknor & Fields, 1865).

50 See Linda S. Bergmann, "A Troubled Marriage of Discourses: Science Writing and Travel Narrative in Louis and Elizabeth Agassiz's *Journey in Brazil*," *Journal of American Culture* 18.2 (Summer 1995): 83–89. I should add that the present chapter, even where I disagree with her conclusions, has profited immensely from Professor Bergmann's scholarship.

51 On Agassiz's "collecting mania," see Edward Lurie, *Nature and the American Mind* (New York: Science History Publications, 1974) 104.

52 Louis Agassiz to Pedro II, 23 July 1863, Houghton Library, bMS Am 1282 (7), quoted by permission of the Houghton Library, Harvard University. See also *JB* 238–239: "The story of the Acaras, the fish which carries its young in its mouth, grows daily more wonderful."

53 Bergmann, "Troubled Marriage of Discourses" 85.

54 See Luigi Monga, "Travel and Travel Writing: An Historical Overview of Hodoeporics," *Annali d'italianistica* 14 (1996): 12, n. 22.

55 See Alexander von Humboldt, *Personal Narrative of Travels to the Equinoctial Regions of America, during the Years 1799–1804*, trans, and ed. Thomasina Ross, 3 vols. (London: Henry G. Bohn, 1852–1853) 2: 420, 467, 371. Humboldt's work first appeared in French in 1814; Elizabeth Agassiz relies on Ross's translation.

56 See *JB* 324, 342, 386.

57 Fourteen of the twenty woodcuts included in *Journey in Brazil* are based on photographs, and only four were made from watercolors and drawings by the expedition's official illustrator, Jacques Burkhardt.

58 What Nancy L. Stepan describes as Wallace's failure "to engage narratively with the romantic aspects of the tropical nature genre" is in fact a complex exercise in authorial self-irony; "Tropical Nature as a Way of Writing," *Mundialización de la ciencia y cultura national*, ed. A. Lafuente et al. (Madrid: Doce Calles, 1993) 501.

59 Wallace believed that "travellers who crowd into one description all the wonders and novelties which it took them weeks and months to observe, must produce an erroneous impression on the reader" (*TA* 4).

60 See also Elizabeth Agassiz's later assessment in *LC* 2: 632: "From beginning to end this journey fulfilled Agassiz's brightest anticipations."

61 George Levine, "Darwin and the Problem of Authority," *Raritan* 3 (1984): 47.

62 Claude Lévi-Strauss, *Tristes Tropiques: An Anthropological Study of Primitive Societies in Brazil*, trans. John Russell (1955; New York: Atheneum, 1967) 18.

63 Wallace had in fact been forwarding specimens to his sales agent in England throughout his stay, even though the larger portion of his shipments—the collections made during his last two years in Brazil—got delayed at Pará and therefore disappeared with the ship

that was to take him home; see Wallace, *My Life* 1: 284. From eight years of travel in Indonesia and Malaysia, Wallace returned in 1862 with a staggering 125,660 specimens.

64 The only extended treatment of James's Brazilian experience is Carleton Sprague Smith's chronological account, "William James in Brazil," *Four Papers Presented in the Institute for Brazilian Studies, Vanderbilt University* (Nashville: Vanderbilt University Press, 1951) 97–138.

65 William James's attitude to the Civil War receives short shrift in Gay Wilson Allen's biography; see, however, George Cotkin, *William James: Public Philosopher* (1989; Urbana: University of Illinois Press, 1994) 29–35, and, more generally, Kim Townsend, *Manhood at Harvard: William James and Others* (New York: Norton, 1996) 38–39.

66 Alfred Habegger, *The Father: A Life of Henry James Sr.* (New York: Farrar, Straus & Giroux, 1994) 442.

67 Howard M, Feinstein, *Becoming William James* (Ithaca: Cornell University Press, 1984) 170.

68 See John Owen King, *The Iron of Melancholy: Structures of Spiritual Conversion in America from the Puritan Conscience to Victorian Neurosis* (Middletown, CT: Wesleyan University Press, 1983) 149–153.

69 See William James to Mary Robertson Walsh James, 23 August 1865: "I find it impossible to tear myself away & this morning I told the Prof, that I wd. see this Amazon trip through at any rate" (*CWJ* 4:110–111).

70 Linda Simon, *Genuine Reality: A Life of William James* (New York: Harcourt Brace, 1998) 162.

71 See Walter Hunnewell's comment on the picture in an unpublished letter to Samuel Henshaw, 19 September 1918: "Now in the matter of the Agassiz photograph Monsieur Bourget is the Frenchman who sold bird skins to the Professor. The Brazilian interpreter standing behind Burkhardt with his hands on Burkhardt . . . I can't remember his name, he was a sort of courier & we picked him in Rio Janeiro & he was only with us a short time" (quoted by permission of the Ernst Mayr Library, Museum of Comparative Zoology Archives, Harvard University, bMU 1498.10.1).

72 James, Brazilian Diary, William James Papers, Houghton Library, bMS Am 1092 (9), quoted by permission of Bay James, literary executor for the James family, and the Houghton Library, Harvard University.

73 Other collections made by William James were more "satisfactory to the Prof"; see his letter to his family of 21 October 1865 about his exploration of the Icá and the Jutai, two tributaries of the Solimões (*CWJ* 4:126).

74 Dexter is a frequent object of James's mockery in the Brazilian notebooks as well as in the portfolio of drawings from the Brazilian journey. One sketch shows a thick, hairy neck protruding from an open-necked shirt. James's comment: "Mr. N. Dexter as he appears entering the dining room rear elevation."

75 Lord Byron, *Childe Harold's Pilgrimage*, canto 2, st. 28: "Pass we the joys and sorrows sailors find / . . . . / The foul, the fair, the contrary, the kind, / As breezes rise and fall and billows swell, / Till on some jocund morn—lo, land! and all is well."

76 Brazilian Diary (Houghton) [pp. 10–11].

77 In her recent biography of James, Linda Simon, apparently unimpressed by the textual evidence, argues that in Brazil James "developed the kind of filial relationship with Agassiz that made both men comfortable" (*Genuine Reality* 94).

78 James, Diary and Sketches from the Brazilian Expedition [pp. 9,13], quoted by permission of Bay James, literary executor for the James family, and the Ernst Mayr Library, Museum of Comparative Zoology Archives, Harvard University, bMn 1556.41.1.

79 Louis Agassiz to William James, 8 December 1865, William James Papers, Houghton Library, bMS Am 1092 (11), quoted by permission of the Houghton Library, Harvard University.

80 Skidmore, *Black into White* 5, See Darwin's *Journal of Researches* 492: "Near Rio de Janeiro I lived opposite to an old lady, who kept screws to crush the fingers of her female slaves. I have stayed in a house where a young household mulatto, daily and hourly, was reviled, beaten, and persecuted enough to break the spirit of the lowest animal." Either Darwin was more clear-sighted than others or his North American colleagues traveling through Brazil must have blinked constantly. In 1851, the U.S. government ordered Naval Lieutenant William Lewis Herndon to reconnoiter the Amazon valley. In a cheerful book he published about his travels, the Virginian officer encouraged those of his fellow Southerners who were concerned about "the state of affairs as regards slavery at home" to pack up their belongings as well as their slaves and remove to the Amazon to partake of its wealth and resources. In Herndon's experience, uncontaminated by close acquaintance with the facts, the "negro slave," constantly singing and chattering merrily, "seems very happy in Brazil." See William Lewis Herndon, *Exploration of the Valley of the Amazon* (Washington, DC: Taylor and Maury, 1854), especially 341–342.

81 See Skidmore, *Black into White* 42: "By the time of final abolition Brazil had already had long experience with millions of free colored; and it had an even longer tradition stretching into earlier centuries of upward mobility by a small number of free colored."

82 See *JB* 297: "With reference to their off-spring, the races of men stand . . . to one another in the same relation as different species among animals; and the word *races*, in its present significance, needs only to be retained till the number of human species is definitely ascertained and their true characteristics fully understood." See also Agassiz's letter to Samuel Gridley Howe, 9 August 1863, Houghton Library, bMS Am 1419 (153).

83 See the summary of Louis Agassiz's remarks on Josiah H. Nott's paper, "An Examination of the Physical History of the Jews," *Proceedings of the American Association for the Advancement of Science. Third Meeting, Held at Charleston, S.C., 1850* (Charleston: Walker & James, 1850) 106–107.

84 Louis Agassiz, "The Diversity of Origin of the Human Races," *Christian Examiner* 49 (July 1850): 144.

85 See chapter 7 in Darwin's *Descent of Man, and Selection in Relation to Sex*, where Agassiz is one of the ghosts whispering into the ears of a composite character called by Darwin "our supposed naturalist." Darwin's naturalist first considers arguments in "favour of classing the races of man as distinct species" (171) and refers to Agassiz, who had argued that "the different races of man are distributed over the world in the same zoological provinces, as those inhabited by undoubtedly distinct species and genera of mam-

mals" (172). But, continues Darwin, "if our supposed naturalist were to enquire whether the forms of man keep distinct like ordinary species, when mingled together in large numbers in the same country, he would immediately discover that this was by no means the case. In *Brazil* [my emphasis], he would behold an immense mongrel population of Negroes and Portuguese; in Chiloe, and other parts of South America, he would behold the whole population consisting of Indians and Spaniards blended in various degrees. In many parts of the same continent he would meet with the most complex crosses between Negroes, Indians, and Europeans; and judging from the vegetable kingdom, such triple crosses afford the severest test of the mutual fertility of the parent forms." Disconcerted, Darwin's naturalist concludes that apparently "the races of man are not sufficiently distinct to inhabit the same country without fusion." As "the absence of fusion affords the usual and best test of specific distinctness," the races of man cannot be specifically distinct (177).

86 The word was created, during the Lincoln reelection campaign, out of the Latin words *miscere* and *genus* (for "race") by George Wakeman and David Goodman Croly as a Democratic Party trick. The best accounts of the historical contexts and literary consequences of miscegenation are Daniel Aaron, "The 'Inky Curse': Miscegenation in the White Literary Imagination," in Aaron, *American Notes: Selected Essays* (Boston: Northeastern University Press, 1994) 113–133, and Werner Sollors, *Neither Black nor White yet Both: Thematic Exploration of Interracial Literature* (Oxford: Oxford University Press, 1997).

87 Agassiz's letters appear, together with Howe's replies, in heavily expurgated form in *LC* 2: 591–517 and are here quoted from the transcripts of the originals in the Agassiz Papers at the Houghton Library, bMS Am 1419 (150–154), with the permission of the Houghton Library, Harvard University.

88 An unfinished letter to Howe among Agassiz's manuscripts in the Houghton Library shows him plotting a policy "which may retain the whole country under the control of our race, without wronging the blacks" (15 August 1863). This and similar statements explain why the use of Agassiz's ideas by antiabolitionists cannot be considered as "tragic"; Lothar Schott, "Der Polygenismusbegriff vor Darwin, dargestellt in Verbindung mit einer kritischen Würdigung der phylogenetischen Vorstellungen von Louis Agassiz," *Archäologie als Geschichtswissenschaft: Studien und Untersuchungen*, ed. Joachim Herrmann (Berlin-Ost: Akademie Verlag, 1977) 486.

89 See Darwin's remark in Notebook C (1838), no. 154: "Has not the white Man, who has debased his Nature . . . by making slave of his fellow black, often wished to consider him as other animal . . . I believe those who soar above *Such prejudices*"; *Charles Darwin's Notebooks, 1836–1844: Geology, Transmutation of Species, Metaphysical Enquiries*, ed. Paul H. Barrett et al. (Cambridge: Cambridge University Press, 1987) 286.

90 On the ideal of "whitening" in Brazilian racist thought, see Skidmore, *Black into White* 64–65 and passim.

91 See Marina Warner, "Lost Shadows, Lost Souls: Body and Soul in Photography," *Raritan* 15.2 (1995): 54.

92 See Brian Wallis, "Black Bodies, White Science: Louis Agassiz's Slave Daguerreotypes," *American Art* 9.2 (1995): 39–61.

93 Gwyniera Isaac, in her incisive essay "Louis Agassiz's Photographs in Brazil: Separate Creations," *History of Photography* 21.1 (Spring 1997): 3–12, claims that we can "only imagine what race or mix of races Agassiz believed he was documenting." However, quite a few of the photographs (which were taken not only in Manáos but also in Rio) do carry captions listing the alleged race of the subjects.

94 Isaac, "Louis Agassiz's Photographs" 8.

95 James left on 9 November and returned on 6 December 1865; see *JB* 282, 294.

96 James, Brazilian Diary (Houghton) [p. 8]. Since the Bureau of Ethnology at the Smithsonian Institution was established only in 1879, the reference to the "Bureau d'Anthropologie" in the quotation remains mysterious, if intriguing. The term "anthropology" in its comprehensive sense (as "the natural history of man") did not appear in the annual reports of the Smithsonian before 1870; see Patricia J. Lyon, "Anthropological Activity in the United States, 1865–79," *Kroeber Anthropological Society Papers* 40 (1969): 8–37. However, some of the data gathered by the agents of the Bureau of Indian Affairs, first established in 1824 and transferred from the War Department to the Interior Department in 1846, were ethnographic descriptions of sorts.

97 Bastos, however, was also vehemently in favor of immigration from the United States and argued that it was "necessary for the pure blood of the Northern races" to "come to develop and regenerate our degenerate race." He might have had some sympathy with Agassiz's theories, even though, for obvious reasons, he would not have shared the latter's fear of racial mixing (sec Burns, *History of Brazil* 181). On Bastos, sec also *JB* 255.

98 James William Anderson, "'The Worst Kind of Melancholy': William James in 1869," *Harvard Library Bulletin* 30 (October 1982): 373.

99 William James, *Essays, Comments, and Reviews* 235. James's review appeared in the *North American Review* in July 1868; a second, shorter review appeared during the same month in the *Atlantic Monthly.* For William James's letter of 4 March 1868 to Henry James, see *CWJ* 1: 39.

100 William James, *Psychology: Briefer Course* (1892), in James, *Writings, 1878–1899,* ed. Gerald E. Myers (New York: Library of America, 1992) 433.

101 James, *Psychology: Briefer Course* 13.

102 William James, *The Principles of Psychology* (1890; Cambridge, MA: Harvard University Press, 1983) 144.

103 My summary is indebted to Robert J. Richards, "The Personal Equation in Science: William James's Psychological and Moral Uses of Darwinian Theory," *Harvard Library Bulletin* 30.4 (1982): 387–425.

104 William James, "Great Men and Their Environment," *Writings, 1878–1899,* 625–626. See Darwin, *Origin of Species* 62.

105 See Stephen Jay Gould, "Agassiz in the Galápagos," *Hen's Teeth and Horse's Toes: Further Reflections in Natural History* (New York: Norton, 1983) 108.

106 William James, *Pragmatism: A New Name for Some Old Ways of Thinking* (1907), in James, *Writings, 1902–1910,* ed. Bruce Kuklick (New York: Library of America, 1987) 510, 570.

107 William James, *Essays, Comments, Reviews* 49, 46.

SELECTED BIBLIOGRAPHY

Aaron, Daniel. *American Notes: Selected Essays.* Boston: Northeastern University Press, 1994.

Adams, Charles H. "William Bartram's *Travels:* A Natural History of the South." *Rewriting the South: History and Fiction,* ed. Lothar Hönnighausen and Valeria Gennaro Lerda. Tübingen: Francke, 1993. 112–120.

Agassiz, Elizabeth Cary. *Louis Agassiz: His Life and Correspondence.* 2 vols. Boston: Houghton, Mifflin, 1885.

Agassiz, Louis. *Contributions to the Natural History of the United States of America.* 4 vols. Boston: Little, Brown, 1857–1862.

———. "The Diversity of Origin of the Human Races." *Christian Examiner* 49 (July 1850): 110–145.

———. *An Essay on Classification.* London: Longman, Brown, Green, Longmans & Roberts, 1859.

———. *Lake Superior: Its Physical Character, Vegetation, and Animals, Compared with Those of Other and Similar Regions. With a Narrative of the Tour by J. Elliot Cabot and Contributions by Other Scientific Gentlemen.* Boston: Gould, Kendall & Lincoln, 1850.

———. *Methods of Study in Natural History.* 2nd ed. Boston: Ticknor & Fields, 1864 (¹1863).

Agassiz, Louis, and Elizabeth Agassiz. *A Journey in Brazil.* 6th ed. Boston: Ticknor & Fields, 1869.

Alderson, William T., ed. *Mermaids, Mummies, and Mastodons: The Emergence of the American Museum.* Washington, D.C.: American Association of Museums, 1992.

Allen, Elsa Guerdrum. *The History of Ornithology before Audubon. Transactions of the American Philosophical Society,* ns 41.1 (1951): 387–590.

Allen, Gay Wilson. *William James: A Biography.* New York: Viking, 1967.

Anderson, Douglas. "Bartram's Travels and the Politics of Nature." *Early American Literature* 25 (1990): 3–17.

Anderson, James William. "'The Worst Kind of Melancholy': William James in 1869." *Harvard Library Bulletin* 30 (October 1982): 367–386.

Arnheim, Rudolf. *The Power of the Center: A Study of Composition in the Visual Arts.* New version. Berkeley: University of California Press, 1988.

Audubon, John James. *Audubon, by Himself: A Profile of John James Audubon.* Ed. Alice Ford. Garden City, NY: Natural History Press, 1969.

———. *The Birds of America, from Drawings Made in the United States and Their Territories.* 7 vols. New York: J. J. Audubon; Philadelphia: J. B. Chevalier, 1840–1844.

———. *The 1826 Journal of John James Audubon.* Ed. Alice Ford, New York: Abbeville Press, 1987 (¹1967).

———. *Letters of John James Audubon, 1826–1840.* Ed. Howard Corning. 2 vols. Boston: Club of Odd Volumes, 1930.

———. *Ornithological Biography; or, An Account of the Habits of the Birds of the United States of America, Accompanied by Descriptions of the Objects Represented in the Work Entitled The Birds of America, and Interspersed with Delineations of American Scenery and Manners*. 5 vols. Edinburgh: Adam and Charles Black, 1831–1839.

Audubon, John James. *Selected Journals and Other Writings*. Ed. Ben Forkner. New York: Penguin, 1996.

———. *The Watercolors for* The Birds of America. Ed. Annette Blaugrund and Theodore E. Stebbins Jr. New York: Villard, 1993.

Audubon, Maria R., ed. *Audubon and His Journals*. 2 vols. New York: Dover, 1986 (¹1896).

Bachelard, Gaston. *The Poetics of Space*. Trans. Maria Jolas. Boston: Beacon Press, 1994 (¹1964).

Bachman, John. *An Examination of the Characteristics of Genera as Applicable to the Doctrine of the Unity of the Human Race*. Charleston: James, Williams & Gitsinger, 1845.

Bal, Mieke. "Telling, Showing, Showing Off." *Critical Inquiry* 18.3 (1992): 556–594.

Barnum, Phineas Taylor. *The Humbugs of the World: An Account of Humbugs, Delusions, Impositions, Quackeries, Deceits, and Deceivers Generally, in All Ages*. New York: Carleton, 1866.

———. *The Life of P. T. Barnum Written by Himself*. New York: Redfield, 1855.

———. *Selected Letters of P. T. Barnum*. Ed. A. H. Saxon. New York: Columbia University Press, 1983.

———. *Struggles and Triumphs; or, The Life of P. T. Barnum, Written by Himself*. Ed. George S. Bryan. 2 vols. New York: Knopf, 1927.

Barnum's American Museum Illustrated. New York: William Van Norden & Frank Leslie, 1850.

Barton, Benjamin Smith. "An Account of the Most Effectual Means of Preventing the Deleterious Consequences of the Bite of the *Crotalus Horridus*, or *Rattle-Snake*, by *Benjamin Smith Barton, M.D.*" *Transactions of the American Philosophical Society* 3 (1793): 100–115.

———. *A Discourse on Some of the Principal Desiderata in Natural History, and on the Best Means of Promoting the Study of This Science, in the United-States*. Philadelphia: Denham & Town, 1807.

———. *Elements of Botany; or, Outlines of the Natural History of Vegetables*. Philadelphia: Printed for the author, 1803.

———. "Letter to M. *Lacepede*, of Paris, on the Natural History of North America. By *Benjamin Smith Barton*, M.D. Professor of Materia Medica, Natural History, and Botany, in the University of Pennsylvania." *Philosophical Magazine* 22 (1805): 97–103, 204–211.

———. "A Memoir Concerning the Fascinating Faculty Which Has Been Ascribed to the Rattle-Snake, and Other American Serpents." *Transactions of the American Philosophical Society* 4 (1799): 74–133.

Bartram, John. *The Correspondence of John Bartram, 1734–1777*. Ed. Edmund Berkeley and Dorothy Smith Berkeley. Gainesville: University Presses of Florida, 1992.

———. "Diary of a Journey through the Carolinas, Georgia, and Florida, from July 1, 1765, to April 10, 1766." Ed. Francis Harper. *Transactions of the American Philosophical Society* 33, pt. 1 (1942): 1–120.

Bartram, William. *Travels and Other Writings*. Ed. Thomas P. Slaughter. New York: Library of America, 1996.

____. *William Bartram: Botanical and Zoological Drawings, 1756–1788*. Ed. Joseph Ewan. Philadelphia: American Philosophical Society, 1968.

———. *William Bartram on the Southeastern Indians*. Ed. Gregory A. Waselkov and Kathryn E. Holland Braund. Lincoln: University of Nebraska Press, 1995.

Bates, Henry Walter. *The Naturalist on the River Amazons: A Record of Adventures, Habits of Animals, Sketches of Brazilian and Indian Life, and Aspects of Nature under the Equator, during Eleven Years of Travel*. 2 vols. London: Murray, 1863.

Beauvois, Ambrose Marie Francois Joseph Palisot de. "Memoir on Amphibia." *Transactions of the American Philosophical Society* 4 (1799): 362–381.

Beer, Gillian. "'The Face of Nature': Anthropomorphic Elements in the Language of *The Origin of Species*." *Languages of Nature: Critical Essays on Science and Literature*. Ed. Ludmilla Jordanova. London: Free Association, 1986. 212–243.

———. *Open Fields: Science in Cultural Encounter*. Oxford: Clarendon, 1996.

Belk, W. Russell, and Melanie Wallendorf. "Of Mice and Men: Gender Identity in Collecting." *Interpreting Objects and Collections*. Ed. Susan M. Pearce. London: Routledge, 1994. 240–253.

Berger, John. "Why Look at Animals?" Berger, *About Looking*. New York: Vintage, 1980. 3–28.

Bergmann, Linda S. "A Troubled Marriage of Discourses: Science Writing and Travel Narrative in Louis and Elizabeth Agassiz's *Journey in Brazil*." *Journal of American Culture* 18.2 (Summer 1995): 83–89.

Berkeley, Edmund, and Dorothy Smith Berkeley. *The Life and Travels of John Bartram: From Lake Ontario to the River St. John*. Tallahassee: University Presses of Florida, 1982.

Beverley, Robert. *The History and Present State of Virginia, in Four Parts, by a Native and Inhabitant of the Place*. 2nd ed. London: Printed for B. and S. Tooke . . . , 1722 ('1705).

Bloore, Stephen. "Joseph Breintnall, First Secretary of the Library Company." *Pennsylvania Magazine of History and Biography* 59 (1935): 42–56.

Blum, Ann Shelby. *Picturing Nature: American Nineteenth-Century Zoological Illustration*. Princeton: Princeton University Press, 1993.

Blumenberg, Hans. *Die Lesbarkeit der Welt*. Frankfurt am Main: Suhrkamp, 1981.

Blunt, Wilfrid, and William T. Stearn. *The Art of Botanical Illustration*. New ed. Woodbridge, Suffolk: Antique Collectors' Club, 1994 ('1950).

Boesky, Amy. "'Outlandish-Fruits': Commissioning Nature for the Museum of Man." *ELH* 58 (1991): 305–330.

Braid, James. *Braid on Hypnotism: Neurypnology; or. The Rationale of Nervous Sleep Considered in Relation to Animal Magnetism or Mesmerism and Illustrated by Numerous Cases of Its Successful Application in the Relief and Cure of Disease*. Ed. Arthur Edward Waite. London: Red way, 1899 ('1843).

Breintnall, Joseph. "A Letter from Mr. J. Breintal to Mr. Peter Collinson, F.R.S., Containing an Account of What He Felt after Being Bitten by a Rattle-Snake." *Philosophical Transactions of the Royal Society* 44 (March/ April 1746): 147–150.

———. "Remarkable and Authentic Instances of the Fascinating Power of the Rattlesnake over Men and Other Animals, with Other Curious Particulars, Communicated by Mr. Peter Collinson, from a Letter of a Correspondent at Philadelphia." *Gentleman's Magazine* 35 (November 1765): 511–514.

Brett-James, Norman G. *The Life of Peter Collinson*. [London:] Edgar G. Dunston, 1926.

Brigham, David R. *Public Culture in the Early Republic: Peale's Museum and Its Audience*. Washington, D.C.: Smithsonian Institution Press, 1995.

Buell, Lawrence. "Autobiography in the American Renaissance." *American Autobiography: Retrospect and Prospect.* Ed. Paul John Eakin. Madison: University of Wisconsin Press, 1991. 47–69.

———. *The Environmental Imagination: Thoreau, Nature Writing, and the Formation of American Culture.* Cambridge, MA: Belknap, 1995.

Buffon, comte de [Georges Louis Leclerc], *Oeuvres completes de Buffon, avec les descriptions anatomique de Daubenton, son collaborateur,* nouvelle édition, ed. J. V. F. Lamouroux and A. G. Desmarest, 44 vols. Paris: Ladrange et Verdière, 1824–1832.

Burns, E. Bradford. *A History of Brazil.* 3rd ed. New York: Columbia University Press, 1993.

Burroughs, John. *John James Audubon.* Boston: Small, Baynard, 1902.

Busch, Frieder. "William Bartrams bewegter Stil." *Literatur und Sprache der Vereinigten Staaten: Aufsatze zu Ehren von Hans Galinsky.* Ed. Klaus Lubbers. Heidelberg: Winter 1969. 47–61.

Byrd, William. *William Byrd's Histories of the Dividing Line betwixt Virginia and North Carolina.* Ed. William K. Boyd. Raleigh: North Carolina Historical Commission, 1929.

Catesby, Mark. *The Natural History of Carolina, Florida and the Bahama Islands: Containing the Figures of Birds, Beasts, Fishes, Serpents, Insects, and Plants: Particularly the Forest-Trees, Shrubs, and other Plants, Not Hitherto Described, or Very Incorrectly Figured by Authors. Together with their Descriptions in English and French. To Which, are Added Observations on the Air, Soil, and Waters: With Remarks upon Agriculture, Grain, Pulse, Roots, &c. To the Whole, is Prefixed a New and Correct Map of the Countries Treated Of.* 2 vols. London: Printed at the expence of the author, 1731, 1743. Appendix, 1748.

Chancellor, John. *Audubon: A Biography.* New York: Viking, 1978.

Cicardo, Barbara J. "From Palette to Pen: Audubon as Writer." *Audubon: A Retrospective.* Ed. James H. Dorman and Allison H. De Pena. Lafayette: Center for Louisiana Studies, University of Southwestern Louisiana, 1990. 74–91.

Clair, Colin. *Unnatural History.* London: Abelard-Schuman, 1967.

Clifford, James. *The Predicament of Culture: Twentieth-Century Ethnography, Literature, and Art.* Cambridge, MA: Harvard University Press, 1988.

Cooper, James Fenimore. *The Leatherstocking Tales.* Ed. Blake Nevius. 2 vols. New York: Library of America, 1985.

Cotkin, George. *William James: Public Philosopher.* Urbana: University of Illinois Press, 1994 (¹1989).

Crèvecoeur, J. Hector St. John de. *Letters from an American Farmer and Sketches of Eighteenth-Century America.* Ed, Albert E. Stone. Harmondsworth: Penguin, 1981.

Cutler, Manasseh. *Life, Journals, and Correspondence of Rev. Manasseh Cutler, LL.D.* Ed. William Parker Cutler and Julia Perkins Cutler, 2 vols. Cincinnati: Robert Clarke, 1888.

Danet, Brenda, and Tamar Katriel. "No Two Alike: Play and Aesthetics in Collecting." *Interpreting Objects and Collections.* Ed. Susan M. Pearce. London: Routledge, 1994, 220–239.

Darlington, William. *Memorials of John Bartram and Humphry Marshall.* Ed. Joseph Ewan. New York: Hafner, 1967 (¹1849).

Darwin, Charles. *The Autobiography of Charles Darwin.* Ed. Nora Barlow. New York: Norton, 1969 (¹1958).

———. *Charles Darwin's Beagle Diary.* Ed. Richard Darwin Keynes. New York: Cambridge University Press, 1988.

———. *Charles Darwin's Notebooks, 1836–1844: Geology, Transmutation of Species, Metaphysical Enquiries*. Ed. Paul H. Barrett et al. Cambridge: Cambridge University Press, 1987.

———. *The Correspondence of Charles Darwin*. Ed. Frederick H. Burkhardt and Sydney Smith. 10 vols. Cambridge: Cambridge University Press, 1986–.

———. *The Descent of Man, and Selection in Relation to Sex*. Rev. 2nd ed. New York: Appleton, 1927 ([1]1888).

———. *Journal of Researches into the Natural History and Geology of the Countries Visited during the Voyage of H.M.S. Beagle round the World*. New ed. New York: Appleton, 1902 ([1]1845).

———. *The Life and Letters of Charles Darwin, Including an Autobiographical Chapter*. Ed, Francis Darwin. 3 vols London: John Murray, 1888.

———. *The Origin of Species*. Ed. Gillian Beer. Oxford: Oxford University Press, 1996.

Desmond, Adrian, and James Moore. *Darwin*. New York: Norton, 1991.

Dillwyn, Lewis Weston. *Hortus Collisonianus: An Account of the Plants Cultivated by the Late Peter Collinson*. Swansea: W. C. Murray and D. Rees, 1843.

Dudley, Paul. "An Account of the Rattlesnake. By the Honourable *Paul Dudley*, Esq., F.R.S." *Philosophical Transactions of the Royal Society* 32 (1723): 279–295.

Durant, Mary, and Michael Harwood. *On the Road with John James Audubon*. New York: Dodd, Mead, 1980.

Edwards, Jonathan. *Images or Shadows of Divine Things*. Ed. Perry Miller. New Haven: Yale University Press, 1948.

Edwards, William H. *A Voyage up the River Amazon, Including a Residence at Pará*. London: Murray, 1861 ([1]1847).

Eisner, John, and Roger Cardinal, eds. *The Cultures of Collecting*. Cambridge, MA: Harvard University Press, 1994.

Emerson, Ralph Waldo. *The Journals and Miscellaneous Notebooks of Ralph Waldo Emerson*. Ed. William H. Gilman et al. 16 vols. Cambridge, MA: Harvard University Press, 1960–1982.

Ewan, Joseph, and Nesta Ewan. "John Lyon, Nurseryman and Plant Hunter, and His Journal, 1799–1814." *Transactions of the American Philosophical Society* 53.2 (1963): 1–69.

Fagin, N. Bryllion. *William Bartram: Interpreter of the American Landscape*. Baltimore: Johns Hopkins University Press, 1933.

Farber, Paul L. *The Emergence of Ornithology as a Scientific Discipline, 1760–1850*. Dordrecht: D. Reidel, 1982.

———. "The Transformation of Natural History in the Nineteenth Century." *Journal of the History of Biology* 15.1 (Spring 1982): 145–152.

Fiedler, Leslie A. *Freaks: Myths and Images of the Secret Self*. New York: Anchor, 1993 ([1]1978).

Findlen, Paula. *Possessing Nature: Museums, Collecting, and Scientific Culture in Modem Italy*. Berkeley: University of California Press, 1994.

Ford, Alice. *John James Audubon: A Biography*. New York: Abbeville Press, 1988 ([1]1964).

Fothergill, John. *Chain of Friendship: Selected Letters of Dr. John Fothergill of London, 1735–1780*. Ed. Betsy C. Corner and Christopher C. Booth. Cambridge, MA: Belknap, 1971.

Franklin, Benjamin. *Writings*. Ed. J. A. Leo Lemay. New York: Library of America, 1987.

Godman, John D. *American Natural History*. 2 vols. Philadelphia: Carey & Lea, 1826–1828.

Goetzmann, William H. *New Lands, New Men: America and the Second Great Age of Discovery.* New York: Penguin, 1987 (¹1986).

Gopnik, Adam. "Audubon's Passion." *New Yorker,* 25 February 1991: 96–104.

Gould, Stephen Jay. *The Flamingo's Smile: Reflections in Natural History.* New York: Norton, 1985.

———. *The Mismeasure of Man.* New York: Norton, 1981.

———. *Hen's Teeth and Horse's Toes: Further Reflections in Natural History.* New York: Norton, 1983.

Greene, John C. *The Death of Adam: Evolution and Its Impact on Western Thought.* Ames: Iowa State University Press, 1996 (¹1959).

Gross, Alan. "Science and Culture." *American Literary History* 7.1 (Spring 1995): 169–186.

Haines, George W. *Plays, Players, and Playgoers! Being Reminiscences of P. T. Barnum and His Museums. Also, a Graphic Description of the Great Roman Hippodrome and Lives of Celebrated Players.* New York: Bruce, Haines, 1874.

Hall, Captain. "An Account of Same Experiments on the Effects of the Poison of the Rattle-Snake. By Captain *Hall.* Communicated by Sir *Hans Sloane,* Bar. Med Reg &c." *Philosophical Transactions* 34 (1727): 309–315.

Hamilton, Alexander. *Hamilton's Itinerarium.* Ed. Albert Bushnell Hart. St. Louis, MO: Privately printed, 1907.

Hanley, Wayne. *Natural History in America: From Mark Catesby to Rachel Carson.* New York: Quadrangle, 1977.

Haraway, Donna. *Primate Visions: Gender, Race, and Nature in the World of Modern Science.* New York: Routledge, 1989

Harbison, Robert. *Eccentric Spaces.* Boston: Nonpareil, 1988 (¹1977).

Harlan, Richard. *American Herpetology; or, Genera of North American Reptilia, with a Synopsis of the Species.* Philadelphia: n.p., 1827.

Harper, Francis, and Arthur N. Leeds. "A Supplementary Chapter on *Franklinia alatamaha.*" *Bartonia: Proceedings of the Philadelphia Botanical Club* 19 (1937): 1–13.

Harré, Rom. "Some Narrative Conventions of Scientific Discourse." *Narrative in Culture: The Uses of Storytelling in the Sciences, Philosophy, and Literature.* Ed. Cristopher Nash. London: Routledge, 1994. 81–101.

Harris, Neil. *Humbug: The Art of P. T. Barnum.* Boston: Little, Brown, 1973.

Herrick, Francis Hobart. *Audubon the Naturalist: A History of His Life and Times.* 2nd ed., 2 vols. New York: Dover, 1968 (¹1938).

Hicks, Philip Marshall. "The Development of the Natural History Essay in American Literature." Diss., University of Pennsylvania, 1924.

Holbrook, John Edwards. *North American Herpetology; or, A Description of the Reptiles Inhabiting the United States.* 2nd ed. 5 vols. Philadelphia: J. Dobson, 1842–1843 (¹1836–1840).

Holmes, Oliver Wendell. *Elsie Venner: A Romance of Destiny.* 2 vols. Boston: Houghton Mifflin, 1880 (¹1861).

———. *The Works of Oliver Wendell Holmes.* Standard Library Edition. 13 vols. Boston: Houghton Mifflin, 1892.

Hüllen, Werner. "Reality, the Museum, and the Catalogue: A Semiotic Interpretation of Early German Texts of Museology." *Semiotica* 80.3/4 (1990): 265–275.

Humboldt, Alexander von. *Personal Narrative of Travels to the Equinoctial Regions of America, during the Years 1799–1804*. Trans, and ed. Thomasina Ross. 3 vols. London: Henry G. Bohn, 1852–1853.

Illustrated and Descriptive History of the Animals Contained in Barnum & Van Amburgh's Museum and Menagerie Combination. New York: S. Booth, 1866.

Ingensiep, Hans Werner. "Der Mensch im Spiegel der Tier-und Pflanzenseele: Zur Anthropomorphologie der Naturwahrnehmung im 18. Jahrhundert." *Der ganze Mensch: Anthropologie und Literatur im 18. Jahrhundert*. Ed. Hans-Jürgen Schings. Stuttgart: J. B Metzler, 1994. 54–79.

Inman, Arthur. *From a Darkened Room: The Inman Diary*. Ed. Daniel Aaron. Cambridge, MA: Harvard University Press, 1996.

Isaac, Gwyniera. "Louis Agassiz's Photographs in Brazil: Separate Creations." *History of Photography* 21.1 (Spring 1997): 3–12.

James, Henry. *Autobiography*. Ed. Frederick W. Dupee. New York: Criterion, 1956.

James, William. *The Correspondence of William James*. Ed. Ignas K. Skrupskelis, Elizabeth Berkeley, et al. 5 vols. Charlottesville: University Press of Virginia, 1992–.

———. *Essays, Comments, and Reviews: The Works of William James*. Gen. ed. Frederick H. Burkhardt. Cambridge, MA: Harvard University Press, 1987.

———. *The Principles of Psychology*. Cambridge, MA: Harvard University Press, 1983 ('1890).

———. *Writings, 1878–1899*. Ed. Gerald E. Myers. New York: Library of America, 1992.

———. *Writings, 1902–1910*. Ed. Bruce Kuklick. New York: Library of America, 1987.

Jardine, Nicholas, James A. Secord, and Emma C. Spary, eds. *Cultures of Natural History*. Cambridge: Cambridge University Press, 1996.

Jefferson, Thomas. *Writings*. Ed. Merrill D. Peterson. New York: Library of America, 1984.

Jenkins, Charles F. "The Historical Background of Franklin's Tree." *Pennsylvania Magazine of History and Biography* 57 (1933): 193–208.

Jewett, Sarah Orne. *Novels and Stories*. Ed. Michael Davitt Bell. New York: Library of America, 1994.

Jordan, Winthrop. *White over Black: American Attitudes toward the Negro, 1550–1812*. Chapel Hill: University of North Carolina Press, 1968.

Jordanova, Ludmilla, ed. *Languages of Nature: Critical Essays on Science and Literature*. London: Free Association, 1986.

Kalm, Peter, *Peter Kalm's Travels in North America: The English Version of 1770*. Ed. Adolph B. Benson. 2 vols. New York: Dover, 1964 ('1937).

Karp, Ivan, and Steven D. Lavine, eds. *The Poetics and Politics of Museum Display*. Washington, D.C.: Smithsonian Institution Press, 1991.

Kenseth, Joy, ed. *The Age of the Marvelous*. Hanover, NH: Hood Museum of Art, Dartmouth College, 1991. 81–101.

Kipperman, Mark. "The Rhetorical Case against a Theory of Literature and Science." *Philosophy and Literature* 10 (1986): 76–83.

Klauber, Laurence M. *Rattlesnakes: Their Habits, Life Histories, and Influence on Mankind*. 2nd ed. 2 vols. Berkeley: University of California Press, 1972 ('1956).

Kolodny, Annette. *The Lay of the Land: Metaphor as Experience and History in American Life and Letters*. Chapel Hill: University of North Carolina Press, 1975.

Kunhardt, Philip B., Jr., Philip B. Kunhardt III, and Peter W. Kunhardt. *P. T. Barnum: America's Greatest Showman.* New York: Knopf, 1995.

Lacépède, Bernard Germain Ètienne, comte de. *Histoire naturelle des quadrupèdes ovipares et des serpents, par M. Le Comte de la Cepede, Garde du Cabinet du Roi; des Academies & Societés Royales de Dijon, etc.* 2 vols. Paris: Hôtel du Thou, 1788–1789.

La Fontaine, Jean de. *Fables et oeuvres diverses de J. La Fontaine.* Ed. C. A. Walckenaer. Paris: Firmin Didot Frères, 1865.

Latour, Bruno. *Science in Action.* Cambridge, MA: Harvard University Press, 1987.

Leighton, Ann. *American Gardens in the Eighteenth Century: "For Use or for Delight."* Boston: Houghton Mifflin, 1976.

Lenoir, Timothy, and Cheryl Lynn Ross. "The Naturalized History Museum." *The Disunity of Science: Boundaries, Contexts, and Power.* Ed. Peter Galison and David J. Stump. Stanford: Stanford University Press, 1996. 370–397, 511–516.

Levine, George. "Darwin and the Problem of Authority." *Raritan* 3 (1984): 30–61.

Lévi-Strauss, Claude. *Tristes Tropiques: An Anthropological Study of Primitive Societies in Brazil.* Trans. John Russell. New York: Atheneum, 1967 (¹1955).

Lindfors, Bernth. "P. T. Barnum and Africa." *Studies in Popular Culture* 7 (1984): 18–25. Lindsey, Alton A., ed. *The Bicentennial of John James Audubon.* Bloomington: Indiana University Press, 1985.

Linnaeus. *A General System of Nature, through the Three Grand Kingdoms of Animals, Vegetables, and Minerals, Systematically Divided into their Several Classes, Orders, Genera, Species, and Varieties, with their Habitations, Manners, Economy, Structure, and Peculiarities.* Trans. William Turton, M.D. 7 vols. London: Lachington, Allen, 1806.

———. *Systema naturae per regna tria naturae, secundum classes, ordines, genera, species, cum characteribus, differentiis, synonymis, locis.* 10th ed., 2 vols. Holmiae: Impensis Direct. L. Salvii, 1758–1759.

Locke, David. *Science as Writing.* New Haven: Yale University Press, 1992.

Looby, Christopher. "The Constitution of Nature: Taxonomy as Politics in Jefferson, Peale, and Bartram." *Early American Literature* 22 (1987): 252–273.

Lovejoy, Arthur O. *The Great Chain of Being: A Study of the History of an Idea.* Cambridge, MA: Harvard University Press, 1936.

Low, Susanne M. *An Index and Guide to Audubon's* Birds of America*: A Study of the Double-Elephant Folio of John James Audubon's* Birds of America*, as Engraved by William H. Lizars and Robert Havell.* New York: Abbeville Press, 1988.

Lurie, Edward. *Louis Agassiz: A Life in Science.* Baltimore: Johns Hopkins University Press, 1988 (¹1960).

———. *Nature and the American Mind.* New York: Science History Publications, 1974.

Lyon, Thomas J., ed. *This Incomperable Lande: A Book of American Nature Writing.* New York: Penguin, 1991 (¹1989).

Marshall, Humphry. *Arbustum Americanum: The American Grove; or, An Alphabetical Catalogue of Forest Trees and Shrubs, Natives of the American United States, Arranged According to the Linnean System.* Philadelphia: Joseph Cruikshank, 1785.

Masterson, James R. "Colonial Rattlesnake Lore, 1714." *Zoologica: New York Zoological Society* 23.9 (1938): 213–216.

Mather, Cotton. "An Extract of Several Letters from *Cotton Mather,* D.D., to *John Woodward,* M.D. and *Richard Waller,* Esq., S.R. Secr." *Philosophical Transactions* 29 (1714–1717): 62–71.

Matthiessen, Peter. *Wildlife in America.* New York: Penguin, 1987.

Meisel, Max. *A Bibliography of American Natural History: The Pioneer Century, 1769–1865.* 3 vols. New York: Hafner, 1967 ('1924–1929).

Melville, Herman. *Redburn, White-Jacket, Moby-Dick.* Ed. G. Thomas Tanselle. New York: Library of America, 1983.

Merwin, W. S, *Travels.* New York: Knopf, 1994.

Michaux, Francois André. *The North American Sylva; or, A Description of the Forest Trees of the United States, Canada, and Nova Scotia, considered particularly with respect to their use in the Arts, and their introduction into Commerce; to which is added a description of the most useful of the European Forest Trees.* Trans. Augustus L. Hillhouse. 3 vols. Paris: C. d'Hautel, 1817–1819.

Miller, Lillian B., et al., eds. *The Selected Papers of Charles Willson Peale and His Family.* 4 vols. New Haven: Yale University Press, 1983–1996.

Miller, Lillian B., and David C. Ward, eds. *New Perspectives on Charles Willson Peale: A 250th Anniversary Celebration.* Pittsburgh: University of Pittsburgh Press, 1991.

Miller, Perry. *Errand into the Wilderness.* Cambridge, MA: Belknap, 1984 ('1956).

Moore, L.Hugh. "The Aesthetic Theory of William Bartram." *Essays in Arts and Sciences* 12.1 (March 1983): 17–35.

Morton, Samuel George. *Hybridity in Animals and Plants Considered in Reference to the Question of the Unity of the Human Species.* New Haven: B. L. Hamlen, 1847.

Muschamp, Edward A. *Audacious Audubon: The Story of a Great Pioneer, Artist, Naturalist, and Man.* New York: Brentano's, 1929.

Nelson, E. Charles. *Aphrodite's Mousetrap: A Biography of Venus's Flytrap.* Aberystwyth: Boethius, 1990.

Newman, John. *Fascination; or. Philosophy of Charming: Illustrating the Principles of Life in Connection with Spirit and Matter.* New York: Fowlers & Wells, 1847.

Nott, Josiah Clark, and George R. Gliddon. *Types of Mankind; or, Ethnological Researches Based upon the Ancient Monuments, Paintings, Sculptures, and Crania of Races, and upon Their Natural, Geographical, Philological, and Biblical History.* Philadelphia. Lippincott, Grambo, 1854.

Orosz, Joel J. *Curators and Culture: The Museum Movement in America, 1740–1870.* Tuscaloosa: University of Alabama Press, 1990.

Paton, Lucy Allen. *Elizabeth Cary Agassiz: A Biography.* Boston: Houghton Mifflin, 1919.

Peale, Charles Willson. *Discourse Introductory to a Course of Lectures on the Science of Nature with Original Music Composed for, and Sung on, the Occasion. Delivered in the Hall of the University of Pennsylvania, November 8, 1800.* Philadelphia: Zachariah Poulson, 1800.

———. *Introduction to a Course of Lectures on Natural History.* Philadelphia: Francis and Robert Bailey, 1800.

Peale, Charles Willson, and A.M.F.J. Palisot de Beauvois. *Scientific and Descriptive Catalogue of Peale's Museum.* Philadelphia: Smith, 1796.

Pearce, Susan M., ed. *Interpreting Objects and Collections.* London: Routledge, 1994.

Peck, Robert McCracken. "William Bartram and His Travels." *Contributions to the History of*

North American Natural History. Ed. Alwyne Wheeler. London: Society for the Bibliography of Natural History, 1983. 35–45.

Perry, Ralph Barton. *The Thought and Character of William James, as Revealed in Unpublished Correspondence and Notes, together with His Published Writings.* 2 vols, Boston: Little, Brown, 1935.

Pike, Zebulon Montgomery. *The Journals of Zebulon Montgomery Pike, with Letters and Related Documents.* Ed. Donald Jackson. 2 vols. Norman: University of Oklahoma Press, 1966.

Plinius Secundus. *The Historie of the World. Commonly called. The Naturall Historie of C. Plinius Secundus.* Trans. Philemon Holland. 2 vols. London: Adam Islip, 1601.

Plummer, Gayther L. "*Franklima alatamaha* Bartram ex Marshall: The Lost *Gordonia* (Theaceae)." *Proceedings of the American Philosophical Society* 121.6 (December 1977): 475–482.

Poe, Edgar Allan. *Poetry and Tales.* Ed. Patrick F. Quinn. New York: Library of America, 1984.

Porter, Charlotte M. "The Drawings of William Bartram (1739–1823), American Naturalist." *Archives of Natural History* 16.3 (1989): 289–303.

———. *The Eagle's Nest: Natural History and American Ideas, 1812–1842.* Tuscaloosa: University of Alabama Press, 1986.

Pratt, Mary Louise. *Imperial Eyes: Travel Writing and Transculturation.* London: Routledge, 1992.

Pulteney, Richard. *A General View of the Writings of Linnaeus.* Ed. George Maton. London: J. Mawman, 1805 ('1781).

Purcell, Rosamond Wolff, and Stephen Jay Gould. *Finders, Keepers: Eight Collectors.* New York: Norton, 1992.

Pursh, Frederick. *Flora Americae Septentrionalis: or, A Systematic Arrangement and Description of the Plants of North America.* 2 vols. London: Printed for White, Cochrane, 1814.

Rafinesque, Constantine Samuel. *A Life of Travels and Researches in North America and South Europe; or, Outlines of the Life, Travels, and Researches of C. S. Rafinesque.* Philadelphia: Printed for the author, 1836.

____, ed. *Atlantic Journal and Friend of Knowledge: A Cyclopedic Journal and Review of Universal Science and Knowledge.* 1832.

Regis, Pamela. *Describing Early America: Bartram, Jefferson, Crèvecoeur, and the Rhetoric of Natural History.* De Kalb: Southern Illinois University Press, 1992.

Richards, Robert J. "The Personal Equation in Science: William James's Psychological and Moral Uses of Darwinian Theory." *Harvard Library Bulletin* 30.4 (1982): 387–425.

Richardson, Edgar P., Brooke Hindle, and Lillian B. Miller. *Charles Willson Peale and His World.* New York: Harry N. Abrams, 1983.

Ritvo, Harriet. *The Animal Estate: The English and Other Creatures in the Victorian Age.* Cambridge, MA: Harvard University Press, 1987.

Rousseau, Jean-Jacques. *The Confessions of J. J. Rousseau: With the Reveries of the Solitary Walker, Translated from the French.* 2 vols. London: J. Bew, 1783.

Sacco, Ellen. "Racial Theory, Museum Practice: The Colored World of Charles Willson Peale." *Museum Anthropology* 20.2 (1997): 26–32.

Sargent, Charles Sprague. *Manual of the Trees of North America.* 2nd ed. 2 vols. New York: Dover, 1965 ('1922).

Saxon, A. H. *Barnumiana; A Select, Annotated Bibliography of Works by or Relating to P. T. Barnum.* Fairfield, CT: Jumbo's Press, 1995.

———. *P. T. Barnum: The Legend and the Man.* New York: Columbia University Press, 1989.

Schofield, Robert E. "The Science Education of an Enlightened Entrepreneur: Charles Willson Peale and His Philadelphia Museum, 1784–1827." *American Studies* 30.2 (Fall 1989): 21–40.

Scholnick, Robert J., ed. *American Literature and Science.* Lexington: University of Kentucky Press, 1992.

Seelye, John. "Beauty Bare: William Bartram and His Triangulated Wilderness." *Prospects* 6 (1981): 37–54.

Sellers, Charles Coleman. *Charles Willson Peale.* New York: Charles Scribner's Sons, 1969.

———. *Charles Willson Peale with Patron and Populace: A Supplement to* Portraits and Miniatures by Charles Willson Peale *with a Survey of His Work in Other Genres.* Transactions of the American Philosophical Society ns 59.3 (1969).

———. *Mr. Peale's Museum: Charles Willson Peale and the First Popular Museum of Natural Science and Art.* New York: Knopf, 1980.

———. *Portraits and Miniatures by Charles Willson Peale.* Transactions of the American Philosophical Society ns 42.1 (1952).

Semonin, Paul. "'Nature's Nation': Natural History as Nationalism in the New Republic." *Northwest Review* 30.2 (1992): 6–41.

Short, Thomas. *Medicina Britannica; or, A Treatise on Such Physical Plants, as Are Generally to be found in the Fields or Gardens of Great-Britain: Containing a Particular Account of their Nature, Virtues, and Uses. With a Preface by Mr. John Bartram, Botanist of Pennsylvania, and his Notes throughout the Work. . . .* 3rd ed. Philadelphia: B. Franklin, 1751.

Shteir, Ann. *Cultivating Women, Cultivating Science: Flora's Daughters and Botany in England, 1760 to 1860.* Baltimore: Johns Hopkins University Press, 1996.

Silvers, Robert B., ed. *Hidden Histories of Science.* New York: New York Review of Books, 1995.

Skidmore, Thomas E. *Black into White: Race and Nationality in Brazilian Thought.* New York: Oxford University Press, 1974.

Slaughter, Thomas P. *The Natures of John and William Bartram.* New York: Knopf, 1996.

Sloan, Phillip. "The Gaze of Natural History," *Inventing Human Science: Eighteenth-Century Domains.* Ed. Christopher Fox, Roy Porter, and Robert Wokler. Berkeley: University of California Press, 1995. 112–151.

Smallwood, William M., and Mabel C. Smallwood. *Natural History and the American Mind.* New York: Columbia University Press, 1941.

Smith, Carleton Sprague. "William James in Brazil." *Four Papers Presented in the Institute for Brazilian Studies, Vanderbilt University.* Nashville: Vanderbilt University Press, 1951. 97–138.

Smith, John. *Travels and Works of Captain John Smith.* Ed. Edward Arber. New ed. by A. G. Bradley. 2 vols, Edinburgh: John Grant, 1910.

Sollors, Werner. *Neither Black nor White yet Both: Thematic Exploration of Interracial Literature.* Oxford: Oxford University Press, 1997.

Spongberg, Stephen A. *A Reunion of Trees: The Discovery of Exotic Plants and Their Introduction into North American and European Landscapes.* Cambridge, MA: Harvard University Press, 1990.

Stearns, Raymond Phineas. *Science in the British Colonics of America.* Urbana: University of Illinois Press, 1970.

Stein, Roger B. "Charles Willson Peale's Expressive Design: The Artist in His Museum." *Prospects 6* (1981): 139–185.

Stepan, Nancy. "Tropical Nature as a Way of Writing." *Mundialización de la ciencia y cultura national.* Ed. A. Lafuente et al. Madrid: DoceCalles, 1993, 495–504.

Stewart, Frank. *A Natural History of Nature Writing.* Washington, D.C.: Island, 1995.

Stewart, Susan. *On Longing: Narratives of the Miniature, the Gigantic, the Souvenir, the Collection.* Durham: Duke University Press, 1993 (¹1984).

Stone, Roger D. *Dreams of Amazonia.* New York: Penguin, 1986 (¹1985).

Stork, William. *A Description of East Florida with a Journal, kept by John Bartram of Philadelphia, Botanist to His Majesty for the Floridas; Upon a Journey from St. Augustine up the River St. John's, as far as the Lakes. With Explanatory Botanical Notes. Illustrated with an accurate Map of East-Florida, and two plans; one of St. Augustine, and the other of the Bay of Espiritu Santo.* 3rd ed. London: Sold by W. Nicoll . . . and T. Jefferies, 1769.

Streshinsky, Shirley. *Audubon: Life and Art in the American Wilderness.* New York: Villard, 1993.

Swift, Jonathan. *Gulliver's Travels.* Ed. Christopher Fox. Boston: Bedford Books, 1995.

Tharp, Louise Hall. *Adventurous Alliance: The Story of the Agassiz Family of Boston.* Boston: Little, Brown, 1959.

Thoreau, Henry David. *The Natural History Essays.* Ed. Robert Sattelmeyer. Salt Lake City: Peregrine Smith, 1980.

[Thorpe, T. B.] "The Rattlesnake and Its Congeners." *Harper's New Monthly Magazine* 10 (1854–1855): 470–483.

Trachtenberg, Alan. *Reading American Photographs: Images as History, Mathew Brady to Walker Evans.* New York: Noonday, 1989.

Twain, Mark. *Mark Twain's Travels with Mr. Brown, Being Heretofore Uncollected Sketches Written by Mark Twain for the San Francisco Alta California in 1866 and 1867.* Ed. Franklin Walker and G. Ezra Dane. New York: Knopf, 1940.

Tyson, Edward. *Orang-Outang, sive Homo Sylvestris; or, The Anatomy of a Pygmie Compared with that of a Monkey, an Ape, and a Man. To which is added, A Philological Essay Concerning the Pygmies, the Cynocephali, the Satyrs, and Sphinges of the Ancients.* London: Printed for Thomas Bennet . . . and Daniel Brown, 1699.

———. "Vipera Caudisona Americana; or, The Anatomy of a Rattle-Snake dissected at the Repository of the Royal Society in January 1683, by *Edward Tyson,* M.D. Coll. Med. Lond. etc." *Philosophical Transactions of the Royal Society* 13 (1683): 25–44.

Wallace, Alfred Russel. *My Life: A Record of Events and Opinions.* 2 vols. New York; Dodd, Mead, 1905.

———. *A Narrative of Travels on the Amazon and Rio Negro, with an Account of the Native Tribes, and Observations on the Climate, Geology, and Natural History of the Amazon Valley.* 5th ed. London: Ward, Lock & Bow- den, 1895 (¹1853).

Wallis, Brian. "Black Bodies, White Science: Louis Agassiz's Slave Daguerreotypes." *American Art* 9.2 (1995): 39–61.

Walls, Laura Dassow. "Textbooks and Texts from the Brooks: Inventing Scientific Authority in America." *American Quarterly* 49 (1997): 1–25.

Warren, Lavinia. *The Autobiography of Mrs. Tom Thumb [Some of My Life Experiences] by Countess M. Lavinia Magri.* Ed. A. H. Saxon. Hamden, CT: Archon, 1979.

Watt, Ian. *The Rise of the Novel: Studies in Defoe, Richardson, and Fielding.* London: Chatto & Windus, 1957.

Welker, Robert Henry. *Birds and Men: American Birds in Science, Art, Literature, and Conservation, 1800–1900.* Cambridge, MA: Belknap Press, 1955.

Welty, Eudora. *The Collected Stories of Eudora Welty.* New York: Harcourt, Brace, Jovanovich, 1980.

Werner, R. M. *Barnum.* Garden City, N.Y.: Garden City Publications, 1923.

Weschler, Lawrence. *Mr. Wilson's Cabinet of Wonder.* New York: Pantheon, 1995.

Wherry, Edgar T. "The History of the Franklin Tree, *Franklinia alatamaha.*" *Journal of the Washington Academy of Sciences* 18.6 (19 March 1928): 172–176.

White, Gilbert. *The Natural History and Antiquities of Selborne, in the County of Southhampton.* Ed. Richard Mabey. Harmondsworth: Penguin, 1977.

Whitman, Walt. *Complete Poetry and Collected Prose.* Ed. Justin Kaplan. New York: Library of America, 1982.

Wilson, Alexander. *American Ornithology; or, The Natural History of the Birds of the United States. Illustrated with Plates, Engraved from Original Drawings Taken from Nature.* 9 vols. Philadelphia: Bradford & Inskeep, 1808–1814.

Wilson, David Scofield. *In the Presence of Nature.* Amherst: University of Massachusetts Press, 1978.

Wilson, Edmund. *Patriotic Gore: Studies in the Literature of the Civil War.* New York: Norton, 1994 ('1964).

Wonders, Karen. *Habitat Dioramas: Illusions of Wilderness in Museums of Natural History.* Uppsala: Almqvist & Wiksell, 1993.

Wyatt, Thomas. *A Synopsis of Natural History: Embracing the Natural History of Animals, with Human and General Animal Physiology, Botany, Vegetable Physiology and Geology, Translated from the Latest French Edition of C. Lemmonier, Professor of Natural History in the Royal College of Charlemagne, with Additions from the Works of Cuvier, Dumaril, Lacepede, etc. and Arranged as a Text Book for Schools.* Philadelphia: Thomas Wardle, 1839.

INDEX

Note: Page numbers in italics refer to illustrations. Where the designations are different from modern usage, Audubon's birds as well as the plants mentioned by John and William Bartram and Peter Collinson are listed under their (capitalized) original names. Current common names are given in lower case.

ABOUT THE AUTHOR

CHRISTOPH IRMSCHER is a lecturer on English and American Literature and Language at Harvard University, where he also teaches in the Program in History and Literature. He is the author of a book about role-playing in modernist poetry and has written numerous articles on American literature and literary anthropology, published in such journals as *Raritan, Soundings, The Wallace Stevens Journal,* and *Canadian Literature.* Currently he is editing Audubon's writings and drawings for the Library of America. He lives in Somerville, Massachusetts.

ABOUT THE ARTIST

ROSAMOND PURCELL is a leading American photographer who has earned international acclaim. Her work has appeared in numerous collections, including *Finders, Keepers* (her collaboration with Stephen Jay Gould) and *Bookworm.* She is the subject of the documentary *An Art that Nature Makes.*